Studies in Computational Intelligence

Volume 987

The series "Studies in Computational Intelligence" (SCI) publishes new developments and advances in the various areas of computational intelligence—quickly and with a high quality. The intent is to cover the theory, applications, and design methods of computational intelligence, as embedded in the fields of engineering, computer science, physics and life sciences, as well as the methodologies behind them. The series contains monographs, lecture notes and edited volumes in computational intelligence spanning the areas of neural networks, connectionist systems, genetic algorithms, evolutionary computation, artificial intelligence, cellular automata, self-organizing systems, soft computing, fuzzy systems, and hybrid intelligent systems. Of particular value to both the contributors and the readership are the short publication timeframe and the world-wide distribution, which enable both wide and rapid dissemination of research output.

Indexed by SCOPUS, DBLP, WTI Frankfurt eG, zbMATH, SCImago.

All books published in the series are submitted for consideration in Web of Science.

More information about this series at http://www.springer.com/series/7092

Damien Trentesaux · Theodor Borangiu ·
Paulo Leitão · Jose-Fernando Jimenez ·
Jairo R. Montoya-Torres
Editors

Service Oriented, Holonic and Multi-Agent Manufacturing Systems for Industry of the Future

Proceedings of SOHOMA LATIN AMERICA 2021

Springer

Editors
Damien Trentesaux
UPHF, LAMIH UMR CNRS 8201
Polytechnic University Hauts de France
Valenciennes cedex 9, France

Theodor Borangiu
Faculty of Automatic Control
and Computer Science
University Politehnica of Bucharest
Bucharest, Romania

Paulo Leitão
Polytechnic Institute of Bragança
Research Centre in Digitalization
and Intelligent Robotics (CeDRI)
Bragança, Portugal

Jose-Fernando Jimenez
Faculty of Engineering, Department
of Industrial Engineering
Pontificia Universidad Javeriana
Bogota, Colombia

Jairo R. Montoya-Torres
Facultad de Ingeniería
Universidad de La Sabana
Chia, Colombia

ISSN 1860-949X ISSN 1860-9503 (electronic)
Studies in Computational Intelligence
ISBN 978-3-030-80905-8 ISBN 978-3-030-80906-5 (eBook)
https://doi.org/10.1007/978-3-030-80906-5

This Springer imprint is published by the registered company Springer Nature Switzerland AG
The registered company address is: Gewerbestrasse 11, 6330 Cham, Switzerland

Foreword

You have in hands the proceedings of the very first Latin America edition of the workshop SOHOMA organized by the university of Javeriana with the support of the University of La Sabana, Bogota, Colombia. SOHOMA is the name of the series of annual workshops on Service-Oriented, Holonic and Multi-Agent Manufacturing Systems for Industry of the Future organized by well-known research groups of European universities.

The European series of SOHOMA workshops has been launched now more than 11 years ago at ENSAM of Paris, and since 2011, each edition took place yearly in different EU countries: France, Romania, UK, Portugal, Italy and Spain. The proceedings are systematically published by Springer in their series "Studies in Computational Intelligence".

The SOHOMA workshop series aims to:

- Federate worldwide researches led in the field of Industry of the Future, with a special attention paid to the construction of intelligent control architectures integrating technological innovations and considering human and environmental aspects,
- Propose to the community state-of-the-art contributions, which enable the use of SOHOMA papers for young researchers to discover challenging issues and applications in the field of Industry of the Future,
- Federate a community of researchers interested in this field and foster the construction of joint activities, such as research projects and special issues in highly renowned journals.

The format of SOHOMA, a human-size, friendly, open-minded workshop, favouring disruptive papers and strong scientific innovations, has been identified as a key factor of differentiation and of great interest.

The birth of SOHOMA Latin America (SOHOMA LA) came from the idea to create spin-off editions of SOHOMA Europe outside Europe since SOHOMA topics are widely studied everywhere in the world, but because of time, money and environmental footprints, it is hard to make people outside Europe submit and attend.

And so, SOHOMA Latin America was born! This is not SOHOMA that travels around the world, once a year, this is the Latin America edition of SOHOMA that will take place from now on only in Latin America.

Why Latin America? Like often in this situation, it is because of a friendship between several researchers of the European and the south of the American continents, based on long-lasting historical and fruitful relationships. The pandemic made it complicated, but a strong professional behaviour enabled the organization of this first edition of the SOHOMA LA workshop.

As an indicator of its success, this edition contains a large majority of papers from several Latin America countries: Venezuela, Chile, Colombia to name a few. Latin America offers opportunities to develop specific models and applications of the Industry and the society of the Future. Farming 4.0, healthcare 4.0, supply chain 4.0, humanitarian logistics and sustainable development are of great interest for Latin America countries. The design of control architectures based on various approaches such as holonic, multi-agent, cyber-human-physical system and service orientation, that integrate innovative technologies such as cloud computing, cloud manufacturing, industrial Internet of Things, digital twins, artificial intelligence and machine learning technics, enables sound developments of these applications.

From this first SOHOMA LA edition, several pillars can be identified as future research and development works, each of them gaining from being addressed from the previously introduced architectures and technologies:

- The distribution of intelligence not only in production and logistics systems, but everywhere where an interaction with future industrial systems exists, at the societal or environmental levels (e.g. energy-aware production systems, low footprint logistics, local skill-based manufacturing systems, zero-defect production, smart manufacturing control),
- The design of more human-compatible artificial industrial systems, considering not only the potential benefits and the limits of the human integration, but also welfare and ethical aspects,
- The development of service orientation facilitating the development of local economies (e.g. development of the concept of manufacturing as a neighbour, of an integrated swarm of additive manufacturing plants and of cloud networked services using the Manufacturing-as-a-Service model, etc.).

I hope you will enjoy your reading. SOHOMA LA authors will act to contribute to the development of the factory of the future.

Carquefou, France
May 2021

Olivier Cardin

Preface

This volume gathers the peer-reviewed papers presented at the first edition of the International Workshop "Service-Oriented, Holonic and Multi-Agent Manufacturing Systems for Industry of the Future - SOHOMA Latin America 2021" organized during January 27–28, 2021, by the Industrial Engineering Department of Pontificia Universidad Javeriana in Bogota, Colombia, in collaboration with the Universidad de la Sabana, Colombia and the Colombian Computing. The organizers received scientific support from Polytechnic University Hauts-de-France, France, University Politehnica of Bucharest, Romania and Polytechnic Institute of Bragança, Portugal.

The main objective of SOHOMA Workshops is to foster innovation in smart and sustainable manufacturing and logistics systems and in this context to promote concepts, methods and solutions for the digital transformation of manufacturing through service orientation in holonic and agent-based control with distributed intelligence.

Following the workshop's technical program, this book is structured in six parts each one grouping a number of chapters describing research in the lines of the digital transformation in manufacturing, supply chain and logistics, and human integration in the industry of the future: *Part 1*: Smart production control for Industry 4.0, *Part 2*: Digital twins and artificial intelligence applied in industry and services, *Part 3*: Digitalization of design and planning activities, *Part 4*: People and industry in the era of Society 5.0, *Part 5*: Logistics and supply chains and *Part 6*: Optimization in manufacturing and services.

These six thematic sections include papers presenting methodologies, implementing solutions, applications, case studies and scientific literature reviews of service-oriented Information, Control and Communication Technologies (IC^2T) that are integrated in digital process, resource and product models and in frameworks of the type: holonic manufacturing and logistics system, industrial Internet of Things, cloud manufacturing, cyber-physical production systems for smart process and service supervision, control and management.

The SOHOMA Latin America workshop maintains the traditional focus on holonic manufacturing control implemented with multi-agent systems and cloud services for the Industry and society of the Future and its frameworks: cyber-physical production systems and industrial Internet of Things. The scientific event will draw on research developed in the last years in the scientific community SOHOMA as well as from other sources. The main research lines are:

- IT modelling techniques (UML, object orientation, agents,…) for automation systems,
- Holonic architectures and multi-agent frameworks for smart, safe and sustainable industrial systems,
- Intelligent products, orders and systems in industry and services,
- Service orientation of control and management in the manufacturing value chain,
- Issues of agility, flexibility, safety, resilience and reconfigurability in industrial systems with new service-oriented IC^2T: cloud, Web, SOA,
- Cyber-physical production systems and industrial IoT for Industry 4.0.

The application space for these new developments is very broad. SOHOMA Latin America 2021 addresses all aspects of industrial and service systems management, such as:

- Control and management of production, supply chain, logistics, transportation, after-sales services,
- Industrial and after-sale services, maintenance and repair operations,
- Management and control of energy production and distribution.

In the workshop's perspective, the digital transformation of manufacturing and service sectors is placed in the general framework of the "Industry and society of the Future" by including the actual vision and initiatives of developing control and management systems for production processes, logistics activities and services based on a wide-ranging, Internet-scale platform linking scalably and interoperably technology providers, manufacturing system designers, plant operators, supply chains and service providers. This vision enables the emergence of a sustainable Internet economy and society for industrial production and service delivery—strongly connected to the reality, robust at disturbances, agile relative to markets and customer-oriented.

The implementing space presented at SOHOMA Latin America 2021 for these application classes shows the interaction between the physical and informational worlds with integration of humans through agentification, big data processing, knowledge extraction and analytics, as well as virtualization of products, processes and resources managed as Web and cloud services.

A brief description of the book chapters follows.

Part 1 reports recent advances and ongoing research in *Smart production control for Industry 4.0.* In the Industry 4.0 vision, production systems become smart, being supervised and controlled through hybrid architectures with distributed intelligence. This type of architecture uses the holonic manufacturing paradigm by decoupling the decision layer from the control one; the first layer uses intelligent agents that reconfigure optimally in real-time scheduling of operations and their allocation on resources at batch level; also, resource health is monitored continuously. Global decisions are taken on the centralized MES layer based on real-time machine learning algorithms that predict usage costs influenced by resource performances and quality of performed services, monitor and classify resource states to predict anomalies and prevent failures. The distributed control layer keeps reality awareness during production by using digital twins replicated for all resources. In this perspective, a first paper of this section presents a methodology for the introduction of a predictive model in a flexible, self-organized manufacturing system, allowing the system to make improved decisions. In the context of smart production systems, another paper introduces a hierarchical design pattern for Universal Robot controllers to a decentralized architecture supervised by a central supervisor. A new research work proposes a framework for an intelligent automatic storage and retrieval system for which a hybrid control architecture optimizes operations during normal functioning and activates reactive behaviours in case of perturbations. A research wok is reported in this first part of the book that presents a holonic manufacturing system (HMS) based on holonic production units (HPUs) that control a continuous, large-scale water supply system. The cognitive model of the HPU uses the digital twin concept; it is embedded in monitoring tasks, detects unexpected events, evaluates different courses of action and changes process parameters according to technical and safety norms.

Part 2 includes papers devoted to the theme of *Digital Twins and Artificial Intelligence applied in industry and services*. The digital twin (DT) is a digital representation of a physical asset, process or software that merges with technologies like the Internet of Things (IoT), artificial intelligence (AI), machine learning (ML) and operations research in order to promote efficiency in the real world while testing and tracking states, behaviours and activities or projecting control systems in the digital world. Papers in this second part of the book recognize that at present the DT concept evolved to highly advanced modelling, simulation and support to control used in various fields: design, simulation, monitoring, control and maintenance, integrating additional features like: hardware/software-in-the-loop, model-driven scenarios testing, data-driven detection of events, significant deviations and anomalies. Concerning applications of artificial intelligence (AI), a study included in second part explores the use of language models to artificially generate maintenance descriptions and reduce the class imbalance problem when classifying between dominant and recessive disturbances in an industrial dataset. A second paper uses a deep learning approach based on synthetic data to train computer vision and motion planning algorithms used for collaborative robots operating in an automatic packaging system. A survey based on a documentary analysis concerning the typologies of packing assisted by augmented reality (AR) marker-less tracking algorithms and interfaces is included in this book part. Another paper defines the airline workforce scheduling problem as a dynamic scheduling problem that includes uncertainties and unexpected environmental changes. An agent-based simulation model that implements the employee scheduling problem including features such as personnel skills, preferences and employer constraints is presented. Intelligent scheduling decisions are taken by an embedded genetic algorithm that allows the airline to be more competitive in the market and improve its customer service.

Part 3 is dedicated to the *Digitalization of design and planning activities*. In the Industry 4.0 vision, companies regularly scan their surroundings for opportunities and threats, demonstrate appreciation for their own capabilities and weave innovation into the very fabric of their existence through dynamic planning. To enable a business to anticipate on the opportunities and threats, flexibility in adjusting plans is important. As a result, business and technology innovation are indissolubly linked, and the demand for technology-enabled production planning and business planning transformation services is rapidly growing. One of the best production planning strategies is the semi-heterarchical one that on one hand provides cost optimization at global horizon (batch, activity chain, a.o.) and on the other hand features reality awareness and robustness by reconfiguring in real time the operation plans and their allocation on resources in response to technical and/or business changes. Two papers in this third book section report a literature review of production planning approaches developed during the 4th Industrial Revolution. The first paper presents the analytical framework, the research methodology and the results of the systemic review, while the second one provides a summary of the contributions, an discussion of results which show that current production planning approaches do not exploit all the 4.0 tools and technologies. The proposed framework characterizes the approaches in terms of the addressed production planning activities, the planning horizon, the company size, the dimension of agility and the employed means. A survey is also offered in this section about Building Information Management (BIM) interoperability and collaboration between design and construction activities. The reported research aims to improve data reliability and data-based workflows in the Architectural, Engineering and Construction (AEC) industry; the desired transformation of BIM is the deployment of integrated processes and modes as Integrated Project Delivery (IPD). The implementation of a holonic product-driven platform for increased flexibility in production planning is described in another book chapter. A model based on anarchic holonic architecture and embedded intelligence logic provides robust decision-making capacity in a "production lot" in face of disturbances.

Part 4 includes chapters in which recent research in the area of *People and industry in the era of Society 5.0* is reported. The research reported in this section considers a human-centred approach to the design of intelligent manufacturing and service systems, featuring human awareness, while keeping human decision-making in the loop at different levels of automation. The included works focus on: the integration of human factors in system design, the optimization of human resource organization, the improvement of working conditions, the flexible workforce scheduling and the support for humanitarian logistics. A first paper describes an mapping system that uses unmanned aerial vehicle (UAV) in the first stage of damage recognition in post-disaster situation of a specific urban zone. The recognition system is modelled in order to determine the optimal location of the UAV hubs that should be placed before the disaster happens, and the optimal routing of UAVs after the disaster has happened. A second paper develops a local search algorithm for the balanced work shift assignment in a healthcare organization. The staff scheduling is solved with a two-phase reactive heuristic: a constructive one responsible for the ideal allocation and a local search which considers each request and makes shift exchanges in order to achieve a feasible solution; priorities based on KPIs defined by medical staff are used in the local search, rather than a multi-objective approach. A guideline to ensure the ethics of cyber-physical and human systems (CPHS), the so-called human-in-the-loop cyber-physical systems is further proposed. The objective of this paper is to define a guideline to ensure that a CPHS is ethical, considering its whole lifecycle. Two paradigms are suggested: deontology (deciding according to ethical rules "must/must not") and utilitarianism (deciding according to the possible ethical implications, defining possible decisions to make in abnormal situations, potentially breaking deontological rules, e.g., in case of emergency). Another paper in this book section is related to the vision on Society 5.0 that is defined as "A human-centred society that balances economic advancement with the resolution of social problems by a system that highly integrates cyber and physical space". Because food security is considered as a current and future social problem, the authors develop a FarmBot Simulator that supports the development of a control software able to implement different precision agriculture strategies before its deployment into the physical test-bench. A case study was developed to demonstrate that the FarmBot simulator allows the verification of the control software through the implementation of a software-in-the-loop simulation, fulfilling the objective for which it had been designed.

Part 5 groups papers reporting research works in *Logistics and supply chains*. In SOHOMA research, Logistics 4.0 and Supply Chain Management 4.0 or smart supply chain management concern various aspects of end-to-end logistics and supply chain management in the context of Industry 4.0, the Internet of Things, cyber-physical systems, emerging technologies, advanced data analytics and (semi-) autonomous decisions enabled by AI. Papers in this book section are related to the four key drivers of Logistics 4.0: 1) data automation and transparency (end-to-end visibility over SC, logistics control tower, optimization); 2) new production methods (robotized palletizing, vision-based picking, 3D printing); 3) new methods of physical transport (driverless vehicles, autonomous pickers, drones); and 4) digital platforms (shared warehouse and transport capacity, cross-border platforms). Also, the principle of logistics integration is reinforced: creating a unified information field for the SC and providing decision-making processes with quality information. A first paper provides a comprehensive literature review on connectivity through Digital Supply Chain (DSC) management; the academic contributions on connectivity approaches for DSC are analysed and classified according to four main groups: warehousing management, production systems, transportation networks and reverse logistics. A second paper offers an overview of cyber-physical systems applied to the logistics and supply domains. The approach of is developed in three phases: first, a definition of cyber-physical logistics systems (CPLS) it synthetized; then, the paper explores the current literature works on CPLS organized according to the CPS maturity model; third, the authors highlight the challenges and perspectives of cyber-physical logistics systems towards an industrial implementation. An analysis of relevant academic literature published between 2004 and 2020 examining issues related to disruptions in supply chains is included in this part of the book. The paper contributes to the knowledge on business continuity and in particular to the management of disruptions as it relates to supply chain management. Finally, the development of a weighted sum method for multi-objective linear model optimization that takes into account logistics costs, customer service level and CO2 emissions, and the obtained results are reported. This work is related to the inventory routing problem (IRP) that allows supply companies systems to take decisions in two important areas: inventory management and routing.

Part 6 includes contributions to the theory, modelling and implementing solutions for the *Optimization in manufacturing and services*. A traditional theme in SOHOMA research is the design and implementing of smart manufacturing control systems capable of cost optimization at batch level and reality awareness of: resource QoS decline, unexpected events and workspace disturbances. Machine learning (ML) techniques are used: prediction of resource performances based on history and real-time KPI computing from shop floor measurements to optimize batch costs, classification and clustering to detect anomalies in resource behaviour. Optimization refers to global cost functions computed usually from batch execution time and electrical energy consumed. In this smart control model, optimization is based on a semi-heterarchical computational, three-tier scheme that: a) plans the order of product entry; b) schedules operations on products; and c) assigns operations to resources. A first paper of this section describes a mixed-integer linear model that solves the open-shop scheduling problem, with the objective to minimize the maximum completion time of jobs at batch level. A second paper analyses the complexity of the stochastic collaborative joint replenishment problem (S-CJRP); the authors prove that this problem is NP-complete and describe computational experiments. A new methodology based on quantitative and qualitative parameters that optimize the selection of wireless sensor network (WSN) devices is described in this book part. The proposed method performs an analysis of the qualitative parameters of project such as: signal delay, packet traffic congestion, reachability, feasibility, life of communications connexions, throughput, number of packets, system speed, risks, and sensitive points of failure; this analysis is important to prevent failures and high costs for the implementation and operation process. Another paper is devoted to optimizing maintenance policies of computer tomography scanners with stochastic failures. A continuous time Markov chain model is developed to estimate the different states of the equipment; the objective of optimization model is to maximizing the benefit generated by the operating equipment and requiring maintenance. Budget constraints are considered. Two approaches are considered and compared to solve the optimization model: an exhaustive search algorithm evaluates understand the behaviour of the solution surface generated by the objective function, and a meta-heuristic based on gradient-ascent to find near-optimal solutions in a reasonable time. The final chapter of the book describes the implementation of a mixed optimization method in multi-agent system approach that determines farmers' market (FM) locations in urban areas. The proposed method considers the consumer patterns of buying fruits and vegetables from a number of sellers in a certain probability for each household. FM location is a relevant topic because this type of market is a good strategy to address food insecurity in areas with a low density of food supply.

The research works reported in this book suggest that the most natural solutions for the digital transformation of manufacturing and logistics with human integration in the Industry 4.0 and Logistics 4.0 frameworks consist in virtualizing resources, products, processes, services and systems in dual cyber-physical spaces that employ secure networked cloud services. There is need for securing inter-enterprise data networks, such that manufacturing and service data packets are under a centralized logical control reducing the likelihood of data theft and network delay or failure.

The rapid digitalization and integration of plant resources, processes and software control systems caused an explosion in the data points available in large-scale manufacturing and supply chain systems. The degree at which enterprises are able to capture value from processing this data and to extract useful insights from it represents a differentiating factor on short- and medium-term development and optimization of the processes that drive the industrial and service activities.

This consideration reinforces the necessity to bring closer the reality-reflecting part and the decision-making part of industrial control architectures. The research presented in this book assesses reference models and architectures (industrial IoT, cloud manufacturing, production planning, holonic MES, holonic production units) while focusing on their world-of-interest and closing the gap with software controls and information systems. This implies both intensive use of digital twins and of AI techniques such as machine learning for knowledge-based and predictive production planning, product-driven automation and resource health monitoring.

The research works reported in this first SOHOMA Latin America workshop have an important merit: the development of digital control and management systems in industry and services have a profound human dedication and aim at solving societal needs. "Digital twins in holonic production units monitoring water supply systems" in Venezuela, "Work shift assignment of medical staff in healthcare clinics" in Chile, "Location routing for UAV-based recognition system in humanitarian logistics", "Virtual environment for scaled precision agriculture strategies" and "Optimally locating farmer markets in urban areas" in Colombia are few examples of research works serving the partnership of the future—People and Industry in the era of Society 5.0.

All these aspects are discussed in the present book, which we hope you will find useful reading.

Valenciennes cedex 9, France — Damien Trentesaux
Bucharest, Romania — Theodor Borangiu
Bragança, Portugal — Paulo Leitão
Bogota, Colombia — Jose-Fernando Jimenez
Chia, Colombia — Jairo R. Montoya-Torress
May 2021

Contents

Logistics and Supply Chains

Optimization in Manufacturing and Services

Smart Production Control for Industry 4.0

Development of a Predictive Process Monitoring Methodology in a Self-organized Manufacturing System

Laura María López Castro, Sonia Geraldine Martínez, Nestor Eduardo Rodriguez, Luna Violeta Lovera, Hugo Santiago Aguirre, and Jose-Fernando Jimenez(✉)

Department of Industrial Engineering, Pontificia Universidad Javeriana, Bogotá, Colombia
{la.lopez,soniamartinez,rodrigueznestor,lunalovera,saguirre, j-jimenez}@javeriana.edu.co

Abstract. This paper presents a methodology for the introduction of a predictive model in a flexible, self-organized manufacturing system, allowing the system to make improved decisions. Taking into consideration the high efficiency required for product manufacturing, predictive models provide valuable insights for accurate decision-making to optimize ongoing processes for key indicators such as process execution time. Furthermore, these models can be used to generate an appropriate response to possible perturbations in the system. Data analysis tools such as process mining allow developing a methodology that enables predictive process modelling in flexible self-organized manufacturing systems. The simulated system in this study is based on the manufacturing cell AIP-PRIMECA, which is located at Polytechnic University Hauts de France in Valenciennes. The process mining tools Apromore, Nirdizati, and ProM were used for the development of the proposed methodology and its implementation in the simulated system. It is expected that applying the proposed methodology will make the manufacturing system more efficient.

Keywords: Flexible manufacturing system · Self-organized system · Process mining · Predictive process monitoring · Netlogo · Nirdizati · Celonis · Apromore

1 Introduction

A flexible manufacturing system is a fully integrated system with large-scale production capabilities that consists of various hardware and software components that work cooperatively (Tolio, 2009). Such a system also includes interconnected processing workstations intended for the end-to-end creation of products. These workstations may offer a variety of functions including loading or unloading, material transformation and assembly of products, storage, quality testing and data processing. The system can be programmed to produce a batch of a products of a type in a set quantity, followed by automatic switching to another type of products in a different quantity. It should be noted that there is a certain degree of flexibility in the system that allows it to react to both predicted and unforeseen changes during the entire product execution cycle. This type

D. Trentesaux et al. (Eds.): SOHOMA 2021, SCI 987, pp. 3–16, 2021.
https://doi.org/10.1007/978-3-030-80906-5_1

of manufacturing system is widely used in current production environments including: the AIP-PRIMECA Lab. of the Polytechnic University Hauts de France in Valenciennes, France, the manufacturing of NOVALTI aeronautical components in Barakaldo, Spain and the Ford Motors flexible production system, among others.

A self-organizing system is one that has the ability to define the sequence of operation execution for each product, the agents that control the execution of operations, and the path of each product until the sequence is completed. The applicability of process mining in the analysis of self-organized manufacturing systems has been previously verified using algorithms like the so-called alpha – a discovery algorithm which is used to model the actual behaviour of a process based on an event log with a Petri net (Van der Aalst et al., 2016).

Process mining is a discipline that aims to discover, monitor and improve processes through the extraction of knowledge from the record of events stored in information systems (Aguirre Mayorga and Rincón García, 2015). According to published sources like Process Mining (Van der Aalst) and the Manifesto of Processes (IEEE Task Force on Process Mining), process mining is defined as a method of analysing data that was generated during process evolution and related events. Process mining is also considered a discipline that aims to use this generated data to monitor, control and improve processes – whether from a system, business or other entity – through exhaustive analysis of process events which are stored in information systems.

By simulating a self-organized, flexible manufacturing system, a methodology was developed based on process mining tools to perform predictive monitoring that facilitates the system's decision-making and improves the system's performance. Predictive process monitoring improves decision-making by the control of ongoing processes, whereas traditional process monitoring only provides information on process execution time, consumed resources and occurrence of particular events at the end of the execution cycle (Teinemaa et al., 2017). Predictive process monitoring detects possible problems during process execution, which allows timely decision-making for the mitigation of consequences that may take place (Jalonen and Lönnqvist, 2009). The proposed methodology improves both the control performance for self-organized flexible manufacturing systems by treating more efficiently disturbances, and the decision-making process in case of process execution delays.

There are currently available open source prediction tools based on process mining such as Apromore and Nirdizati, which were developed by Tartu University and Queensland University of Technology, Australia (Verenich et al., 2018). Apromore is an advanced web platform for process analysis developed by the business process management community, whereas Nirdizati is a tool that contains several predictive techniques. Some Nirdizati functionalities were included in Apromore, making the use of Apromore more feasible for the methodology proposed in this project.

The objective of this work is to develop a predictive monitoring methodology making use of process mining in order to improve the efficiency of a self-organized manufacturing system. The paper is structured as follows: characterisation of the self-organized manufacturing control system's functioning; design of the predictive monitoring methodology for the self-organized manufacturing control system using process mining; application of the proposed predictive methodology to the simulator of the AIP-PRIMECA manufacturing control system. In its final part, the paper evaluates the impact of the methodology on the efficiency indicators of the manufacturing system.

2 Background

In recent years, various process mining techniques have been applied in numerous industries and organizations around the world. These techniques make it possible to identify failures or improvement opportunities in complex processes that involve a large number of activities, people and information. Some examples of these techniques are described below.

In 2005, a study was carried out with the Dutch Public Works Department with the objective to verify the applicability of process mining techniques in real-world contexts and a new software tool for this type of problem called ProM. For this investigation, information related to the processing of invoices was taken as input information in 14,279 cases, with a total of 147,579 events or activities carried out by 487 employees. The project was developed in view of three perspectives: the process perspective, the organizational perspective, and the case perspective (Van der Aalst et al., 2007).

Other evidence of the applicability of process mining is presented in (Verenich et al., 2018) where the Nirdizati project was first presented. The project aimed to develop a web-based tool for predictive monitoring of business processes. Afterwards, the Nirdizati tool was integrated into a web-based process analysis platform called Apromore. Due to this integration, users can access the event logs stored in the repository – a public folder on the platform – to train a range of predictive models and later use them to visualize various performance indicators of process cases. Predictions generated by the tool can be presented visually on a dashboard or exported for periodic reporting. Based on these predictions, operations managers can identify potential problems promptly and take appropriate corrective action in a timely manner.

Further applications of process mining are described in Jimenez et al. (2018), which discusses a simulation of a flexible manufacturing process based on laboratory facilities. The experiment in this study was based on identifying two conditions, "normal" and "abnormal", where abnormal was associated with a disturbance of one of the cell machines. It is important to mention that the records of each event were stored in a CSV file. To analyze the information obtained in the course of the work, different algorithms were implemented for the rapid verification of the system according to process mining. This enabled decision-making in each of the atypical cases that could arise in the simulated system, providing a basis for the implementation of the algorithm in real-world contexts. The study had a significant impact on the project presented in this paper, since several points, methodologies and analyses from the study were used as a foundation.

In (Trentesaux et al., 2013), a comparative evaluation of programming and control solutions for flexible manufacturing systems was carried out. Different control systems were generated for a given set of input data and were compared to improve overall system performance. A reference control system based on a real production cell was proposed to the industrial control community together with scenarios for benchmarking. This reference system allows evaluation of the results using traditional operational research tools and enables assessment of the robustness of the control system.

In considering each of these works, it is evident that several processes have been discovered, monitored and improved by extracting knowledge from the record of events stored in an information system, which is a key part of process mining. However, unlike these studies, our work offers the ability not only to discover, monitor and improve processes by extracting knowledge from the event log but also to conduct an exhaustive analysis of the extracted data. This analysis can then be used to predict potential events in order to avoid errors long time before they occur.

3 Manufacturing System Used for Methodology Development

The manufacturing system used for the development of the proposed methodology is a flexible manufacturing system located in the AIP-PRIMECA Lab. of Polytechnic University Hauts de France (UPHF) in Valenciennes, France. The system has the essential characteristic of being a self-organized manufacturing system, meaning that it has the ability to dynamically adapt its organization and behaviour. In particular, the system can create and eliminate interactions for a certain process, change its conditions without external intervention, and maintain its internal coherence in the organizational environment (Belisario and Pierreval, 2013). Furthermore, it is a social system made up mainly of autonomous agents, that continuously features behaviours according to the interactions of the processes in progress (Accard, 2018).

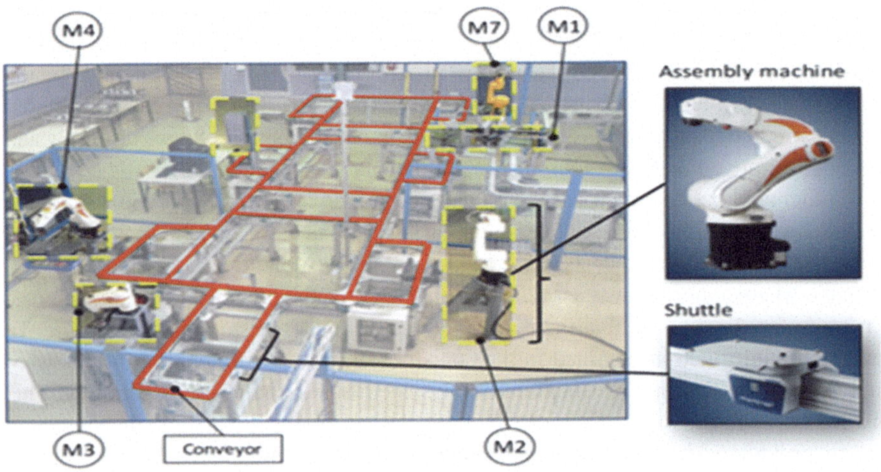

Fig. 1. The flexible manufacturing system cell AIP-PRIMECA located in UPHF, Valenciennes (Jimenez et al., 2018)

A self-organizing system in a flexible manufacturing context, such as the one proposed in (Trentesaux et al., 2013), is capable of determining the sequence of operations to be performed, the agent that performs each operation, and the route of each product throughout the process flow. These actions are generated from the system's own decisions taken in a particular approach. The self-organizing system used in this work was the AIP-PRIMECA manufacturing cell – an assembly system designed to execute a defined set of work processes (Fig. 1). The layout of this manufacturing cell is based on a workstation model where the main components are processing devices (machines, robots), data collecting devices (sensors) and intelligent products (Fig. 2).

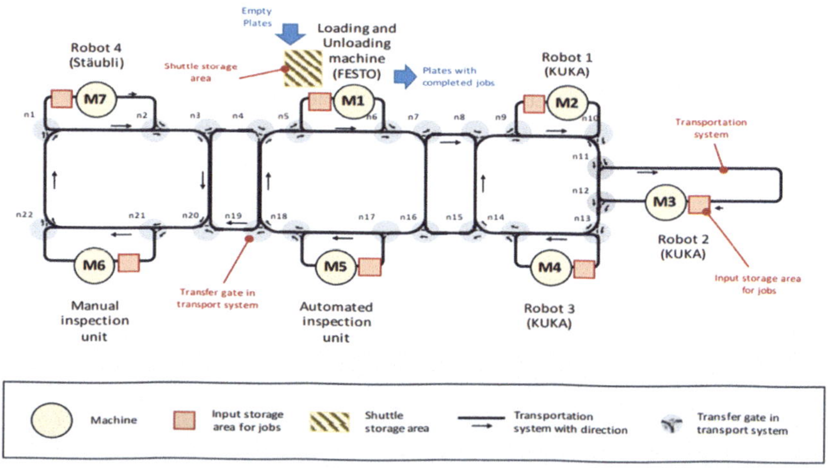

Fig. 2. The layout of the AIP-PRIMECA manufacturing cell

The AIP-PRIMECA system has the ability to make seven types of products – B, E, L, T, A, I and P – where each product has a different assembly sequence and requires the use of different system components. The sequence of operations for each type of product and the machines that must be enabled to execute these operations are presented in Fig. 3. The definition of these products according to the sequence of operations is key element for the analysis of the system behaviour.

a)

Product	Production sequence
B	01-01-01-02-02--05-03-06
E	01-01-01-02-02-04-06
L	01-01-01-05-05-03-03-06
T	01-01-02-04-06
A	01-01-01-02-04-05-03-06
I	01-01-05-03-06
P	01-01-02-04-06

b) Processing times (seconds)

		M5	M2	M3	M4	M6	M7
Axis mounting	01		20	20			20
r_component mounting	02		20	20			20
I_component mounting	03				20		20
L_component mounting	04		20		20		20
Screw mounting	05			20	20		
Check	06	05				05	
Movement	08						

Fig. 3. Product information: a) Production sequence for each product; b) Processing times for each machine M*i*

The production sequence in the manufacturing cell is controlled by a software that generates the sequence of actions for the allocated machines. The execution of each action is predetermined by the code structure created in the NetLogo simulation software (Wilensky, 1999). This software is a programmable modelling environment used to simulate natural and social phenomena, and has sufficient capacity to model complex systems that evolve over time. The modellers contained within the program can instruct hundreds or thousands of independent agents to work. The main advantage of using this software is the ability to observe and analyze the relationship between the micro-level of individual component behaviour and the macro-level patterns that emerge from the interaction of many individuals.

The input data used for system simulation in NetLogo is based on four elements:

1. Agent characteristics – the max. no. of possible elements, specifications of the time delay between each agent, and the speed of products circulating within the system.
2. Component layout (Fig. 4)
3. Products and selection of machines or routes (see Fig. 4); this selection depends on the system's selected operation mode.

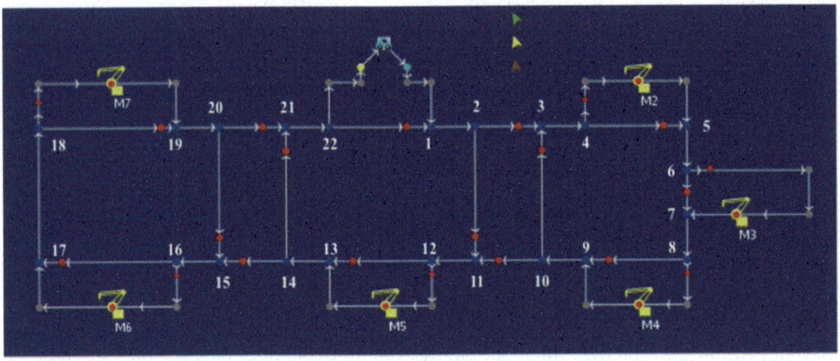

Fig. 4. Layout of the flexible manufacturing cell created with Netlogo software

In this project, four different operation scenarios were analysed: randomly, first available machine (FAM), potential fields, and Control Alloc-Cyrille Pach. After identifying the means by which the scenarios are compared, each phase of the process-based predictive monitoring methodology was performed. It is important to highlight that the main objective of this analysis was to develop a predictive monitoring methodology making use of process mining in order to improve the efficiency of a self-organized manufacturing system.

4 Descriptive Analysis with Process Mining

For each scenario described in the previous section, the same production order was used. Event logs of the different manufacturing system operations were generated. The production of 1400 units (200 items of each of the 7 types of products) was simulated. Each event log of the scenarios was analysed through process mining, using the following main measurement indicators:

- Production rate: $R_p = R_{max}(1 - R_d)$, where R_{max} is the max. volume of produced items and R_d is the number of defect products per time interval
- No. of variants: $nCr = n!r!(n - r)!$, where n, r are respectively the set of all possible decisions in allocation methods, and the no. of decisions per allocation method
- Cycle time: 1/throughput rate = 1/[(units produced or tasks completed)/time]
- Standard deviation of cycle time
- Utilization percentage per machine

Table 1. Comparative table of indicators obtained through process mining in different scenarios

Scenario	Production rate, prod/min	No. variants	Time cycle (sec)	Standard deviation	Machine utilization (%)					
					M2	M3	M4	M5	M6	M7
Pach allocation	1.69	15	182	27.72	96.6	73.31	64.48	10.31	3.80	79.95
First available machine	0.9	6	370	116.15	64.04	56.90	41.01	7.48	0	4.66
Potential fields	1.62	14	184	48.95	92.69	80.81	65.46	12.52	9.98	61.91
Random	0.52	122	658	258.85	28.38	28.06	20.78	2.52	1.83	19.79

According to the comparison of indicators between scenarios, it is possible to characterize the system's behaviour by common factors observed in all scenarios. For example, it was observed that the element that adds time to the process is the transfer of products between machines for the execution of their respective operations – especially the transfers to the system entrance before carrying out operation 1 and the transfer that occurs before carrying out operation 2.

In general terms, the resources with the highest use percentage are M2, M3, M4, and M7 whereas M5 and M6 are less used. This result suggests that the performance of the system could be improved if the activities assigned to each resource are redistributed. Alternatively, each of the resources could be enabled for the execution of all operations. However, in non-flexible manufacturing systems, this solution would be associated with investment costs and additional restrictions for particular cases.

The association between product type and processing time is another important factor to consider. In all the scenarios examined, the products with the shortest processing time were of type P, T and I. This is not surprising since these product types require only five operations, which is the fewest number of operations across all product types. However the number of required operations for a given product type is not the only determinant of the processing time; the order in which operations are executed must also be considered.

Finally, it was determined that the best scenarios for this methodology were the control Alloc-Cyrille Pach and potential fields scenarios based on their production rates and their relative variance from the mean execution time per case. Additionally, these scenarios have a more effective use of resources compared to the random and FAM scenarios based on the level of use of the machines. The application of predictive analysis in the control Alloc-Cyrille Pach and the potential fields scenarios was expected to generate appropriate and relevant information for the improvement of the production process. Conversely, the random and FAM scenarios have poorer performance because of their high variability and/or low production rate.

5 Predictive Analysis with Process Mining

To develop a predictive methodology for each scenario, it was necessary to analyze three samples from each of the scenarios. The first sample consists of the first 28 records of finished products, the second consists of the first 49 records of finished products, and the third includes all records of finished products (Fig. 5). The first two tests were performed with sample sizes that were a multiple of seven since the system produces seven types of products. The purpose of carrying out the analysis with three samples was to evaluate the models obtained in each case and define a specific model for the scenario accordingly. The more samples obtained from the general analysis, the greater the probability of generating an accurate model. It should be noted that the analysis was based on the prediction of the next activity to be carried out on the production line according to the type of product being produced.

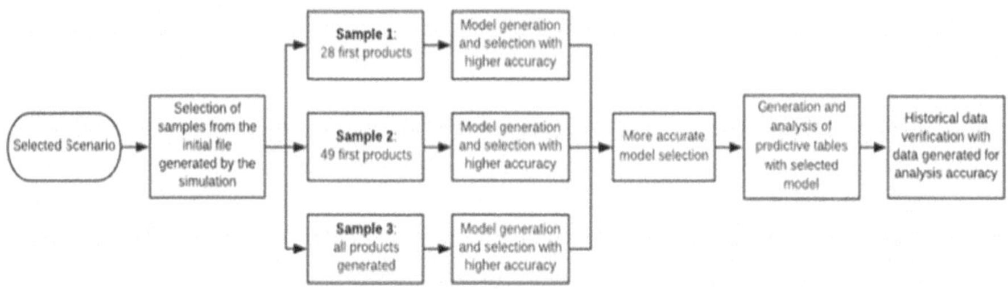

Fig. 5. Process flow for the analysis of each scenario

As shown in Fig. 6, the process begins with the selection of one of the four scenarios defined above. Once selected, historical data from that scenario for the processing of the 1400 products was generated.

After the processing routes for each product in the scenario were determined, the three required samples of 28, 49 and all products were generated. Each of the samples was analysed by different models and the results were verified. The predictive model with the highest accuracy was then selected. The selected model for each scenario was used to generate predictive data that were then compared with the historical data produced by the simulation and the accuracy of the prediction made by the model was evaluated.

5.1 Model Generation and Selection

To choose the model for each scenario, accuracy was used as the selection criterion. A graphic and numerical analysis was performed for each model. In the numerical analysis,

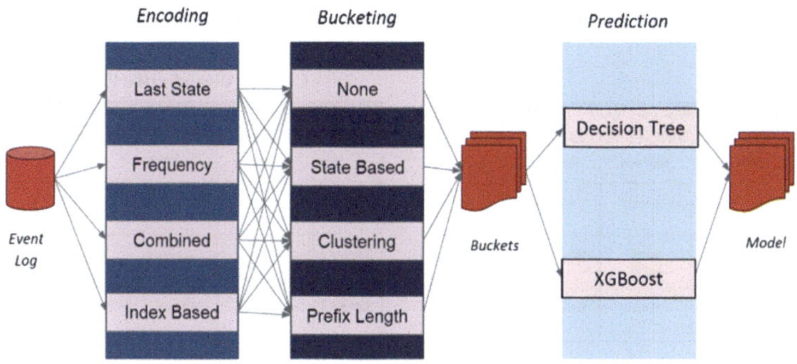

Fig. 6. Training of models with the samples for the generation of the "next activity" prediction for each scenario

the model with an accuracy closest to 1 or 100% was chosen. In the graphic analysis, all the available models were compared, and the model with the graph having the highest similarity to the original data (i.e., the graph whose slope did not significantly vary relative to the original) was selected. To generate the model, a combination of three characteristics were used: encoding, grouping and classifier type. As shown in Fig. 6, the main focus was to predict the next activity by executing all possible combinations for each of the three samples. This process was repeated for each of the scenarios.

The predictive analysis consists of selecting one element from each of the three variables described below.

1. *Encoding*: Transformation of data formulation through the rules or norms of a predetermined code or coding language.
2. *Bucketing*: A data organization method that breaks down the space from which spatial data is drawn into regions called repositories. Conditions for the choice of region boundaries include the number of objects a region contains or the spatial design of the regions with the intention of minimizing overlapping or coverage.
3. *Prediction*: The selection of a method for the predictive analysis of the previously selected components (encoding and bucketing).

Once the models generated and their respective numerical and graphic analysis conducted for each sample and each scenario, the results shown in Table 2 were obtained.

When determining the sample results for each scenario, the model with the highest accuracy was selected as shown in Table 3.

Once the models for each scenario were obtained, the next stage of the analysis according to the methodology in Fig. 6 was to verify that the selected model is coupled to the behaviour of the activities in each scenario. For this, it was necessary to conduct an evaluation simulating the "real-time" behaviour of the process with the selected model.

Table 2. Model training from the three samples for "next activity" generation in each scenario

Randomly

Selection of the model with the highest Accuracy - Randomly				
	Encoding	Bucketing	Prediction	Accuracy
Sample 1	Index Based	Clustering	XGBoost	0,9881
Sample 2	Index Based	Prefix Length	Desicion tree	0,9886
Sample 3	Index Based	Prefix Length	Desicion tree	0,9820

Control Alloc - Cyrille Pach

Selection of the model with the highest Accuracy - Cyrille				
	Encoding	Bucketing	Prediction	Accuracy
Sample 1	Frequency	Clustering	XGBoost	0,869
Sample 2	Index Based	Clustering	XGBoost	0,8677
Sample 3	Frequency	Clustering	XGBoost	0,9046

First Available Machine

Selection of the model with the highest Accuracy - FAM				
	Encoding	Bucketing	Prediction	Accuracy
Sample 1	Index Based	State Based	XGBoost	0,9702
Sample 2	Index Based	Prefix Length	Decision tree	0,9659
Sample 3	Index Based	Prefix Length	XGBoost	0,9658

Pontencial Fields

Selection of the model with the highest Accuracy - PF				
	Encoding	Bucketing	Prediction	Accuracy
Sample 1	Combined	Clustering	XGBoost	0,9226
Sample 2	Index Based	Clustering	XGBoost	0,9205
Sample 3	Combined	None	XGBoost	0,8680

Table 3. Model selected by scenario.

Selected Model with the highest Accuracy **per scenario**				
	Encoding	Bucketing	Prediction	Accuracy
Randomly	Index Based	Prefix Length	Decision tree	0,9886
FAM	Index Based	State based	XGBoost	0,9702
CEP	Combined	Clustering	XGBoost	0,9226
Cyrille	Frequency	Clustering	XGBoost	0,9046

5.2 Generation and Analysis of Results for Each Scenario

Once the model was chosen, a graphic representation of the main metrics or key performance indicators (KPIs) that determine the completion of the activities of a process was created in the form of a "predictive dashboard"; this dashboard is used to generate and present the analyses of the final stage of the predictive methodology (Fig. 7).

For the creation of predictive dashboards for the different scenarios, a step execution process was carried out as shown in Fig. 8, in order to obtain the most accurate results. Although the dashboards are not presented in this paper, they were a useful tool for the analysis of results. For each model, one of the samples was defined as the "base sample" and used for the training of the chosen model depending on the scenario for which the sample was generated. The base sample was then used to generate the prediction models and validated by a "complete sample" test. The models generated predictions from the data as a probability table or a prediction table. These samples were defined as follows:

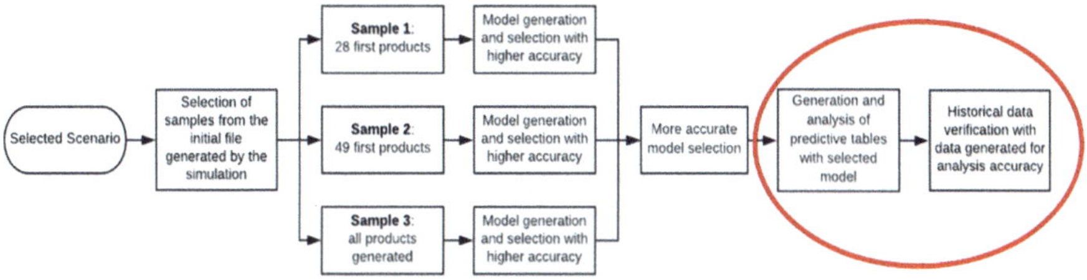

Fig. 7. Selection of the process for analysis

- Base sample = 28 products sample (Sample 1)
- Complete sample = all products sample (Sample 3)

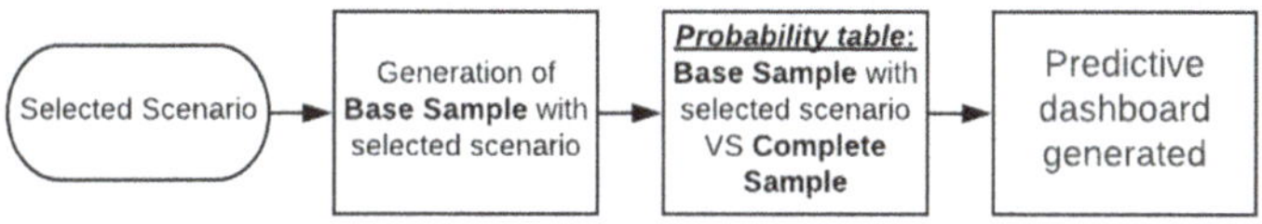

Fig. 8. Dashboard generation process

It should be noted that although the complete sample was used for the validation of the models, the total number of products studied was between 50 and 70. In other words, each model was trained with a sample of 28 products selected from the studied scenario and validated with between 50 and 70 products selected from the complete sample of historical data (generated for 1400 products). For prediction techniques such

Table 4. Summary table of prediction of activities and comparison with real data

Control Alloc- Cyrille Pach		Predicted Activity						
		Movement	Oper. 1	Oper. 2	Oper. 3	Oper. 4	Oper. 5	Oper. 6
Activity performed	Movement	1499	0	2	0	3	94	0
	Oper. 1	0	203	0	0	0	0	0
	Oper. 2	5	0	73	0	0	0	0
	Oper. 3	36	0	0	6	0	14	0
	Oper. 4	8	0	0	0	36	0	0
	Oper. 5	0	0	0	0	0	56	0
	Oper. 6	78	0	0	0	0	0	0

Potencial Fields		Predicted Activity						
		Movement	Oper. 1	Oper. 2	Oper. 3	Oper. 4	Oper. 5	Oper. 6
Activity performed	Movement	1375	0	0	0	0	0	0
	Oper. 1	0	164	0	0	0	0	0
	Oper. 2	0	0	63	0	0	0	0
	Oper. 3	0	0	0	45	0	0	0
	Oper. 4	0	0	0	0	35	0	0
	Oper. 5	0	0	0	0	0	45	0
	Oper. 6	61	0	0	0	0	0	0

First Available Machine		Predicted Activity						
		Movement	Oper. 1	Oper. 2	Oper. 3	Oper. 4	Oper. 5	Oper. 6
Activity performed	Movement	626	0	0	0	0	0	0
	Oper. 1	0	0	0	0	0	0	0
	Oper. 2	0	0	0	0	0	0	0
	Oper. 3	0	0	0	0	0	0	0
	Oper. 4	0	0	0	0	0	0	0
	Oper. 5	0	0	0	0	0	0	0
	Oper. 6	0	0	0	0	0	0	0

Randomly		Predicted Activity						
		Movement	Oper. 1	Oper. 2	Oper. 3	Oper. 4	Oper. 5	Oper. 6
Activity performed	Movement	675	0	0	0	0	0	0
	Oper. 1	26	0	0	0	0	0	0
	Oper. 2	0	0	0	0	0	0	0
	Oper. 3	0	0	0	0	0	0	0
	Oper. 4	0	0	0	0	0	0	0
	Oper. 5	1	0	0	0	0	0	0
	Oper. 6	0	0	0	0	0	0	0

Note that the scenarios for which a predictive model is more accurate are the control Alloc-Cyrille Pach and the potential fields scenarios. This is consistent with what was observed and described in the descriptive analysis presented in Sect. 4, where the low task time variability and the high production rate of these scenarios suggest the potential for the development and application of a prediction model as decision trees or discriminant analysis, the results must typically be validated with data that has not been used for the construction of the model to assess its reliability (Van Der Aalst, 2016). For this reason, it was necessary to also carry out the appropriate model validation.

For each of the scenarios studied, the results of the validation are shown in Table 4, which provides a comparative table of the predicted and actual activities.

Conversely, the FAM and random scenario models do not generate appropriate predictions.

6 Results and Discussion

The scenarios used for the development of the methodology proposed in this work involve decisions regarding the selection of the next machine in order to comply with the production process. These scenarios were simulated, and the resulting event logs used to perform the analyses – both descriptive and predictive – were considered. When conducting these analyses, two scenario classifications were found: scenarios that adapt to the process mining analysis and those that do not adapt. The random and FAM scenarios generated few encouraging results when this methodology was applied, because of their low production rate and high standard deviation. These models have a high degree of variability in their processes; thus, there were almost no predictions generated for these models based on historical data. Conversely, the CEP and Cyrille Pach scenarios have a high production rate and a low standard deviation; thus, these models have a low degree of variability in their processes, being suitable for the application of the proposed methodology.

The application of process mining allowed the identification of process variants and limiting resources. Although these factors are directly linked to the configuration of operations, products and production orders, the value of process mining was verified in a production context. In this case, it was possible to determine that machines M2, M3 and M7 are limiting resources, since all manufactured products must use some of them. Likewise, it was determined that the activity that generates process delays is related to transfers of products between machines, especially the movement of a product from its entry into the system to the machine that will execute the first operation. These results allow improving operational decision-making; for example, one could consider enabling more resources to carry out operation 1 and/or 2 which are those performed by the M2, M3, and M7 machines that increase the task length.

Although the scope of this project was only an interface allowing the visualization of predictions of future activities to be carried out by each machine, the use of information in real-time would allow the development of predictive indicators that are of great value to industrial users. The implementation of this methodology in suitable scenarios would has an impact on flexible manufacturing control system. Since a training model has been developed that provides probabilities and predictions for the next machine to be used in a process, it is possible to generate a forecast of several process elements:

- *Availability of resources per machine*
  Using the methodology proposed in this work, it is possible to predict the most efficient and reliable resources necessary in a manufacturing system. This prediction could be

used to reduce the number of missing parts or down times by assigning optimally the configuration of resources in the self-organized manufacturing system at the start of the manufacturing process.

- *Task time*
  By visualizing the activities or events in particular cases, it is possible to estimate the remaining time required for the completion of a certain product. This type of information allows the improvement of production plans and can be used to generate policies for appropriate use of resources.
- *Production rate*
  From the task time and the number of jobs being performed at a given time, it is possible to generate a predictive indicator for the production rate. This indicator will allow the identification of inefficiencies in the system in advance, allowing the system to make autonomous decisions or provide the operator the information needed to better configure production according to the needs and demand at a given time.
- *Bottlenecks*
  Knowing the activities or operations that each resource is capable to perform as well as the operations for the processes to be realized, it is possible to develop an indicator for intensively used resources. This indicator, with a certain degree of reliability, can be used to identify resources that are limiting a process at a given time. In conjunction with other indicators, this one will allow a better understanding of future process states and will facilitate an improved allocation of operations on available resources to avoid having large amounts of products affected by increased time cycles.

7 Future Work

The simulator and Apromore mining tool will be integrated in future research work. This would allow Apromore to generate predictions in real-time while the simulator generates data; this will ensure making predictions and analysing information in a more efficient and effective way, since the information is kept on a single platform to allow easier management. The output of the forecasts could be presented using the predictive dashboard described in this work.

References

Accard, P.: Criticality: how changes preserve stability in self-organizing systems. Organiz. Stud. SAGE, (2018). https://doi.org/10.1177/0170840618783342

Aguirre Mayorga, H., Rincón García, N.: Minería de procesos: desarrollo, aplicaciones y factores críticos, Cuadernos de Administración **28**(50), 137–157 (2015). https://doi.org/10.11144/Javeriana.cao28-50.mpda

Belisario, L.S, Pierreval, H.: A conceptual framework for analyzing adaptable and reconfigurable manufacturing systems. In: Proc. of 2013 Int. Conference on Industrial Engineering and Systems Management (IESM), pp. 1–7. https://bit.ly/2Qx731O (2013)

Jalonen, H., Lönnqvist, A.: Predictive business – fresh initiative or old wine in a new bottle. Manag. Decis. **47**(10), 1595–1609 (2009). https://doi.org/10.1108/00251740911004709

Jimenez, J., Zambrano, G., Aguirre, S., Trentesaux, D.: Using process-mining for understating the emergence of self-organizing manufacturing systems, IFAC-PapersOnLine **51**(11), 1618–1623 (2018), ScienceDirect, https://www.sciencedirect.com/science/article/pii/S24058963183 1382X

Teinemaa, I., Dumas, M., Rosa, M.L., Maggi, F.M.: Outcome-oriented predictive process monitoring: review and benchmark, ACM Trans. Knowl. Discov. Data **13**(2) (2017) https://doi.org/10.1145/3301300

Tolio, T.: Design of Flexible Production Systems. Springer (2009). https://www.springer.com/gp/book/9783540854135

Trentesaux, D., et al.: Benchmarking flexible job-shop scheduling and control systems. Control Eng. Pract. **21**(9) 1204–1225 (2013). https://www.sciencedirect.com/science/article/abs/pii/S0967066113000889

Van der Aalst, W., et al.: Business process mining: an industrial application. Inform. Syst. **32**(5) 713–732 (2007), Elsevier. https://www.sciencedirect.com/science/article/abs/pii/S03064 37906000305

Van der Aalst, W.: Process Mining: Data Science in Action. Springer (2016). https://www.springer.com/gp/book/9783662498507

Verenich, I., Mõškovski, S., Raboczi, S., Dumas, M., La Rosa, M., Maggi, F.M.: Predictive Process Monitoring in Apromore. In: Mendling, J., Mouratidis, H. (eds.) CAiSE 2018. LNBIP, vol. 317, pp. 244–253. Springer, Cham (2018). https://doi.org/10.1007/978-3-319-92901-9_21

Wilensky, U.: NetLogo. Center for Connected Learning and Computer- Based Modeling, Northwestern University, Evanston, IL. http://ccl.northwestern.edu/netlogo/ (1999)

A GEMMA-Based Decentralized Architecture for Smart Production Systems

Jose Daniel Hernandez[1], David Andres Gutierrez[2], and Giacomo Barbieri[1](✉)

[1] Department of Mechanical Engineering, Universidad de los Andes, Bogota, Colombia
{jd.hernandezr1,g.barbieri}@uniandes.edu.co
[2] Xcelgo A/S, Ry, Denmark
dags@xcelgo.com

Abstract. Within the Industry 4.0 paradigm, production systems have shifted to Smart Production Systems controlled through decentralized architectures. To enable the development of decentralized architectures supervised from a central coordinator, common approaches and vocabularies are desirable for the automation software, since they would facilitate the exchange of information among the different manufacturing entities. To face this challenge, a GEMMA-based decentralized architecture is introduced in this paper and validated by integrating a PLC-controlled system with a Universal Robot for the generation of a decentralized architecture supervised by a central coordinator. Through a lab case study, it is demonstrated how the proposed approach enables the implementation of a decentralized architecture generating a standard interface for the management of the operational modes of production systems.

Keywords: Industry 4.0 · Smart production system · Universal robot · Decentralized architecture · GEMMA

1 Introduction

The Industry 4.0 paradigm is defined as the combination of modern technologies and novel methodological approaches for the resolution of current and (near-)future industrial challenges [1]. In this context, traditional production systems have shifted to smart production systems since reconfigurable-, adaptive-, and evolving- factories are necessary for achieving the required mass-customization and personalization [2, 3].

Smart Production Systems (SPSs) are fully integrated collaborative manufacturing systems capable to respond in real-time to the changing demands and conditions of the factory, the supply network, and the customer needs as effect of their digitalization [4]. To guarantee the required flexibility, SPSs are characterized by decentralized decision-making [5], their control architectures having shifted from centralized to decentralized ones [6].

The centralized control indicates a single 'intelligent' component that controls all the manufacturing entities. Modern SPSs consist of subsystems that are integrated through material, energy and information flows; although a centralized control framework offers

D. Trentesaux et al. (Eds.): SOHOMA 2021, SCI 987, pp. 17–29, 2021.
https://doi.org/10.1007/978-3-030-80906-5_2

the best performance, it is computationally demanding and not fault tolerant [7]. In decentralized architectures, each local controller ignores the state of the other subsystems and does not communicate with other controllers but takes decisions independently or eventually with the support of a central coordinator [8]. Switching from centralized to decentralized control can lead to a more robust production environment, in which deviations from standard operating conditions can be detected [9].

To enable the development of decentralized architectures supervised by a central coordinator, common approaches and terminology are desirable for the automation software because they would facilitate the exchange of information among the different manufacturing entities [10]. Related to this issue, the authors of [11] propose a methodology for the PLC (Programmable Logic Controller) management of operational modes (OMs) of production systems (PSs) using the GEMMA guideline. In the Digital Twin context, the methodology has been demonstrated to generate a standard interface that facilitates the exchange of information between PLCs and digital process models. The methodology consists of: (i) GEMMA-GRAFCET representation specifying the OMs of PSs without the occurrence of emergent behaviours; (ii) Hierarchical Design Pattern (HDP) for the development of hierarchical PLC code from specifications expressed in accordance with the proposed GEMMA-GRAFCET representation. The HDP generates hierarchical PLC code to separate the management of the OMs (common for each PS) from their nested behaviour (specific to each PS). In line with the principle of decentralized architecture, this paper presents two novelties with respect to the previous work:

- Up to now, the HDP was limited to the generation of PLC code. In this paper, we adapt the HDP approach to the generation of control code for Universal Robot (UR)-controllers. Hence, the first novelty reported in this paper is the definition of a *Hierarchical Design Pattern for UR-controllers* from specifications expressed in accordance with a GEMMA-GRAFCET representation.
- By integrating a PLC-controlled system with an UR3[1] robot, we demonstrate that the proposed GEMMA-GRAFCET methodology can be adopted to implement a *Decentralized Architecture Supervised by a Central Coordinator* (DASCC). As shown in Fig. 1, the control code deployed in the local controller of each PS is separated in two hierarchical layers: the 'GEMMA layer' and the 'Nested behaviour'. The GEMMA layer is interfaced with the 'Central Coordinator' and characterized by a standard interface for the management of the OMs, avoiding thus the development of emergent behaviours.

Given the above elements, the paper is structured as follow: the initial GEMMA-GRAFCET methodology is summarized in Sect. 2, while the Hierarchical Design Pattern for UR-controllers is described in Sect. 3. Section 4 applies the methodology to a lab case study for the generation of a Decentralized Architecture Supervised by a Central Coordinator. The results obtained are discussed in Sect. 5, while Sect. 6 presents the conclusions and sets the directions for future work.

[1] https://www.universal-robots.com/it/prodotti/robot-ur3/.

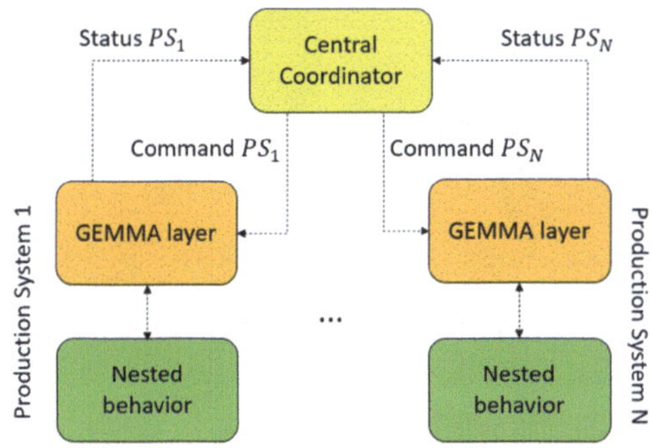

Fig. 1. GEMMA-based decentralized architecture with a central coordinator

2 GEMMA-GRAFCET Methodology

In this Section, the GEMMA-GRAFCET methodology proposed in [11] is resumed since this work extends it. Section 2.1 illustrates the GEMMA-GRAFCET representation for the specification of the OMs of PSs, while Sect. 2.2 describes the HDP for the generation of PLC code from specifications expressed in accordance with the defined GEMMA-GRAFCET representation.

2.1 GEMMA-GRAFCET Representation

GEMMA (Guide d'Étude des Modes de Marche et d'Arrêt) is a checklist which allows to graphically define all the start and stop modes of a system and their evolution [12]. However, GEMMA is partially specified with respect to the syntax and semantics causing misunderstandings and possible emergent behaviours [11]. To semantically specify the GEMMA guideline, we proposed its implementation with the GRAFCET international standard (IEC 60848 [13]).

The GRAFCET model of the GEMMA guideline is shown in Fig. 2. A Step defines a state that does not contain a nested behaviour. A macro-step indicates a state whose outgoing transitions can only fire when the exit step of the nested behaviour is active, while the outgoing transitions of an enclosing step can fire independently from the active step of the nested behaviour. The initial step indicates from which state the system starts to operate. Reset transitions are implemented since in GRAFCET the nested behaviour must be initialized each time the super-ordinate state is entered. When a state has more than one outgoing transition, a number is placed on each transition to indicate its priority. Finally, a negation operator can be placed on the priority number to change the behaviour of the composite state with respect to the one illustrated in the representation – when the 'negated' transition is evaluated. Comparing the GEMMA guideline [14] with our model (Fig. 2), it can be noticed that the obtained representation consists in a one-to-one translation of the GEMMA graphical notation. Due to its implementation with the GRAFCET modelling language, this representation can be used to specify the OMs of PSs without the development of emergent behaviours.

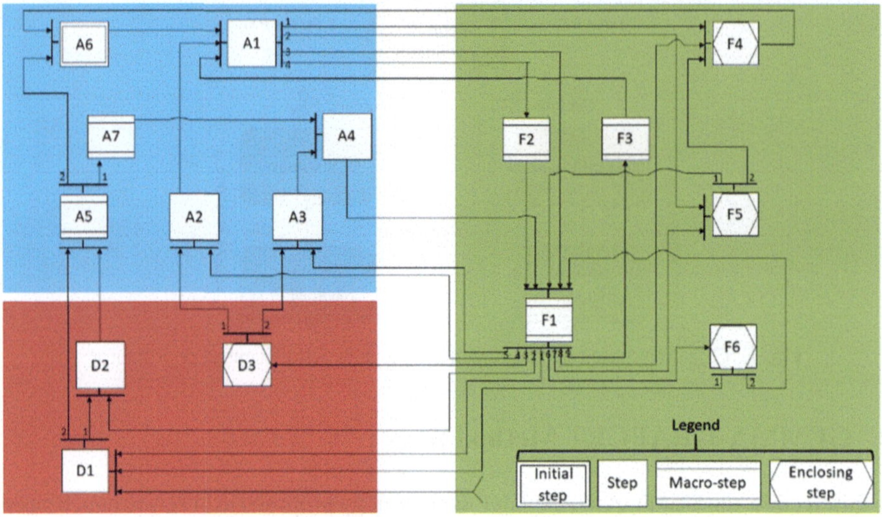

Fig. 2. GRAFCET model of the GEMMA guideline

2.2 Hierarchical Design Pattern for PLC Code

The proposed HDP for PLC code is illustrated in Fig. 3 for the F2 macro-step. Function Blocks (FBs) are adopted for hierarchically structuring the code. One program is generated to implement the behaviour of the GEMMA layer (see Fig. 1) and one FB for each 'GEMMA state' that contains a nested behaviour; i.e. macro-steps and enclosing steps. Only the GEMMA program is scheduled through a periodic task, while the execution of the nested behaviour is invoked from the 'GEMMA active state' by calling the corresponding FB.

The GEMMA layer is converted into Structured Text (ST) PLC code. Since the philosophy of the GEMMA guideline is to have only one state active at a time, a scalar State variable is used as a discriminator of a switch statement; i.e. `CASE..OF` in ST. Transitions are evaluated by means of `IF..THEN` constructs and if a transition is fireable, the new state is assigned. To implement the proposed scheduling strategy, each 'GEMMA state' invokes the FB that contains its nested functionality. Finally, entry actions are performed by using an `Entry` flag and an `IF..THEN` construct that resets the flag once the entry action has been performed.

To manage reset transitions and macro/enclosing steps, two flags are introduced (`Initialization` and `Complete`), respectively declared as input and output variables of each FB. `Initialization` is utilized for implementing the reset transition behaviour. `Initialization` is set on the entry action of each 'GEMMA state' and reset on the continuous action, while the `Complete` flag is used to implement the macro-step behaviour. The outgoing transitions of a macro-step have an additional `AND` condition consisting in the `Complete` flag. This flag is set within the exit step of the nested behaviour and reset during its reset transition. An enclosing step does not have the `Complete` flag condition on the outgoing transitions.

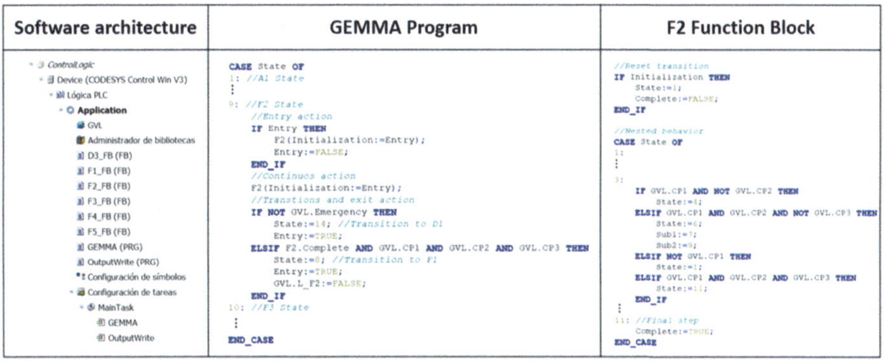

Fig. 3. HDP illustrated for the F2 macro-step

3 Hierarchical Design Pattern for UR Controllers

The HDP for UR-controllers is introduced to generate control code defined according to the GEMMA-GRAFCET representation described in Sect. 2.1. The design pattern is shown in an illustrative example that implements the specifications defined in Fig. 4. These specifications mimic a GEMMA diagram that includes initial step S1, macro-step S2, enclosing step S3, and S4, S5 steps. Finally, an 'or divergence' is implemented as outgoing transition of the initial step S1, in which T2 has higher priority than T1.

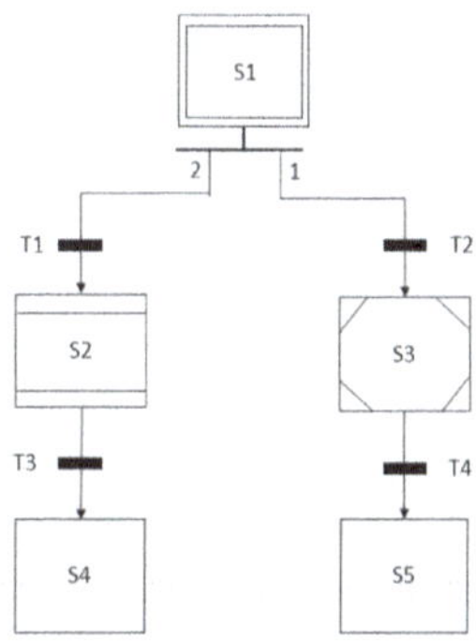

Fig. 4. GEMMA-GRAFCET specifications of the illustrative example

In the proposed HDP, the code is written with the *UR PolyScope*[2] interface which provides better readability than the low-level *UR script*[3] machine code.

Since FB cannot be implemented in UR, programs and sub-programs are used to hierarchically structure the code. We propose to generate one program for implementing the behaviour of the GEMMA layer and one sub-program for each 'GEMMA state'

[2] https://www.universal-robots.com/articles/ur-articles/ur-polyscope-move-with-respect-to-a-custom-featureframe/.

[3] https://www.universal-robots.com/download/.

that contains a nested behaviour; i.e. macro-steps and enclosing steps. Only the `GEMMA` program is executed, while the nested behaviour is invoked from the 'GEMMA active state' by calling the corresponding *sub-program*.

To convert the GEMMA layer into UR-code, a scalar `State` variable is used as a discriminator of a switch structure. Transitions are evaluated – according to their priority – by means of `IF..THEN` constructs and if a transition is fireable, the new state is assigned. Within the continuous behaviour, each 'GEMMA state' invokes the sub-program that contains its nested functionality.

Concerning the *interaction* in between the GEMMA layer and the *nested behaviour*, the code must be able to handle: (i) reset transitions; (ii) macro-steps and enclosing steps. To manage *reset transitions*, we propose the use of an `Initialization` flag declared as global variable (i.e. `installation variable` in UR), since UR does not allow to generate input and output variables for the sub-programs. `Initialization` is set on the exit action of each 'GEMMA state' and reset on the nested behaviour after the initialization operations have been completed. Comparing this pattern with the HDP for PLC code (Sect. 2.2), the declaration of `Initialization` as global variable avoids the necessity to implement an entry behaviour for each 'GEMMA state'.

To differentiate between an enclosing and a macro-step, the `Check Continuously` option is used for evaluating the condition of the `IF..THEN` construct. When this option is active, the condition is verified during each cycle-time of the controller enabling the implementation of an *enclosing step*. When this option is not active, the condition is verified only when the nested behaviour has been completed enabling the implementation of a *macro-step*.

Transitions are defined into a script file named `Transition` that is called in the variable declaration section; i.e. `Before Start` sequence in UR. A function is generated for each transition and each function contains a Boolean expression composed with the variables exchanged with the external environment.

Listing 1 shows the code contained within the `GEMMA` program for the implementation of the GEMMA-GRAFCET representation of the illustrative example. On the *variable declaration* section, `State` and `Initialization` are declared as global variables and the `Transition` script is invoked. `State` is initialized with the value of `1` in accordance with the specifications of Fig. 4, while `Initialization` is `TRUE` to activate the token for the execution of the initialization actions within the nested behaviour. Then, a switch structure is used for modelling the *state evolution*. For enclosing and macro-steps, the sub-program that implements the state nested behaviour is invoked as continuous action of the active state. *Outgoing transitions* are evaluated and the `sync()` function is inserted to generate a cyclic scheduling of the program. When priorities are specified, conditions are evaluated in monotone decreasing order of priority since UR Polyscope does not allow the use of `ELSEIF` operators. Therefore, an `IF..THEN` construct is implemented for each transition. In case that more than one transition is active, the higher priority transition will overwrite the `State` variable. Finally, if the code of `S2` and `S3` is compared, it can be noticed that the outgoing transition of `S3` has the `Check Continuously` option active. This option allows the implementation of an *enclosing-step*, while its absence the implementation of a *macro step*.

Listing 2 shows the code common to each *sub-program*. The `Initialization` flag is used for initializing the behaviour as soon as the state is entered. Then, the nested behaviour is implemented.

Finally, Listing 3 illustrates an example of transition function generated within the `Transition` script. It can be noticed that the external variables are read directly from their memory register position. We decided to separate the transition use (`GEMMA` program) from its guard condition (`Transition` script) to enhance code readability.

Listing 1: `GEMMA` program: UR PolyScope code that implements the GEMMA-GRAFCET specifications of the illustrative example. It can be noticed that `T1` is first verified, since `T1` has lower priority than `T2`.

```
// Before Start Sequence
Set State=1
Set Initialization=True
Script: Transitions.script

//GEMMA behaviour

Switch State
Case 1 //State S1
if T1()
      set State=2
    if T2()
      set State=3
Case 2 //State S2
    Call S2
    if T3()
      Set State=4
Set Initialization=True
Case 3 //State S3
    if not T4() /Check continuously=True
      Call S3
    if T4()
      Set State=5
Set Initialization=True
Case 4 //State S4
Case 5 //State S5
sync()
```

Listing 2: Example of sub-program for the implementation of state nested behaviour.

```
If Initialization
  //Initialization actions
  Set Initialization=False
//Nested behaviour
```

Listing 3: Example of definition of transition T1 within the Transition script.

```
def T1():
  return read_port_bit(#MemoryRegisterPos)
End
```

4 Decentralized Architecture Supervised from a Central Coordinator

A laboratory case study was developed to demonstrate that the proposed methodology enables the implementation of a DASCC architecture. An automatic machine for *filling and encapsulating bottles* was selected (Fig. 5). The machine consists of three stations: (i) transport and feeding; (ii) dosing and filling; (iii) encapsulating. Two stepping slat conveyors allow the simultaneous operation of the three stations. Extension of pneumatic cylinders H and B respectively determines the incremental advance of the input and filling conveyors. The *transport and feeding station* (station 1) is constituted by pneumatic cylinder A that is responsible for the feeding of the bottles from the input to the filling conveyor. The *dosing and filling station* (station 2) is composed by a volumetric dispenser actuated from pneumatic cylinder C, and by an on/off valve (D) used to open and close the liquid supply. The *encapsulating station* (station 3) consists of an UR3 robot that picks the cap and places it on the top of the bottle. Finally, magnetic limit switches are used to signal the position of the pneumatic cylinders and light barrier sensors to indicate the presence of a bottle on each station.

Stations 1 and 2 are controlled through a PLC, while station 3 through an UR-controller. A *DASCC* architecture must be implemented to make the system work by integrating the PLC and the UR-controller through a *Central Coordinator*.

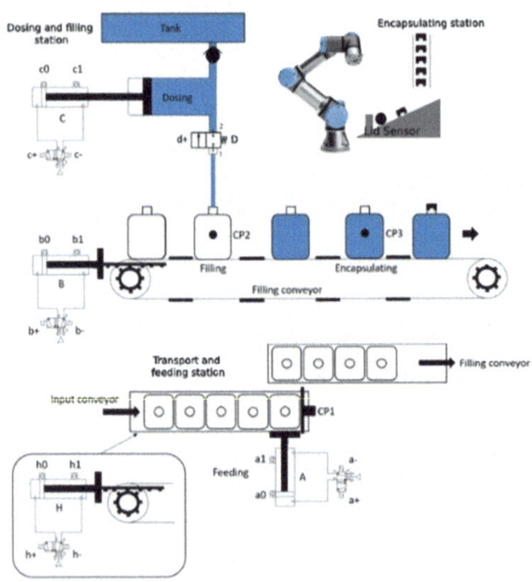

Fig. 5. Schematic representation of the filling and encapsulating machine

The *GEMMA-GRAFCET representation* illustrated in Sect. 2.1 was utilized to specify the OMs of stations 1 and 2, along with the ones of station 3 (Fig. 6). After the specification of the GEMMA layer, the *nested behaviour* of each state was defined. The defined nested behaviour and the GEMMA-GRAFCET representation of station 1 and 2 are made available to the reader[4] but are not illustrated since they do not represent the focus of the work.

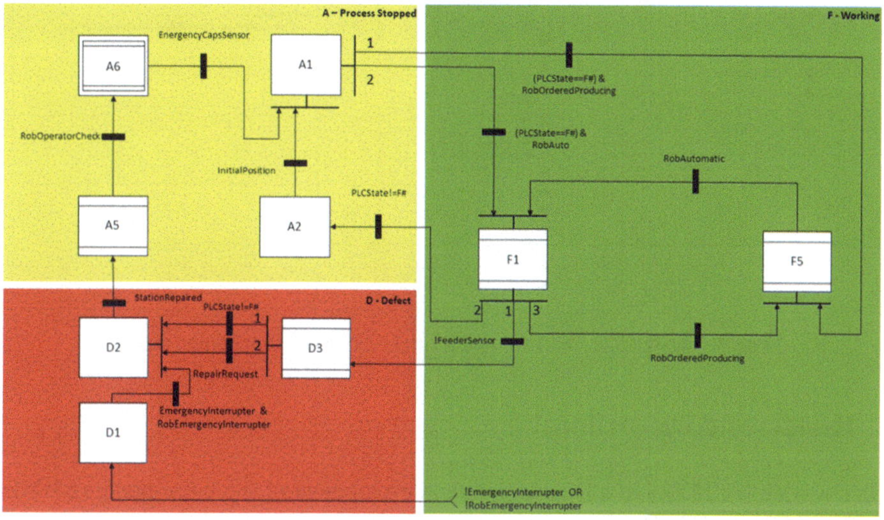

Fig. 6. GEMMA-GRAFCET specifications for the station 3 (UR-controller)

Then, the HDP presented in Sect. 2.2 was applied for the conversion of the specifications of stations 1 and 2 into PLC code written in the CoDeSys[5] programming environment. The *HDP for UR-controller* was utilized to generate the control code of station 3 and its application is next illustrated.

The *software architecture* was developed by generating one program for implementing the behaviour of the `GEMMA` diagram and one sub-program for each macro-step and enclosing step.

The `GEMMA` program is executed, while the sub-programs are invoked from the `GEMMA` program as shown in the left-hand side of Fig. 7 for the F1 state. Next, the *nested behaviour* of the macro-steps and the enclosing steps were generated. The image in the central part of Fig. 7 illustrates the code developed for the F1 state. Finally, the transition functions were defined in the `Transition` script as shown in the right-hand side of Fig. 7.

4 https://www.researchgate.net/publication/344575303_Software_Spec_UR_case_study.

5 https://www.codesys.com/.

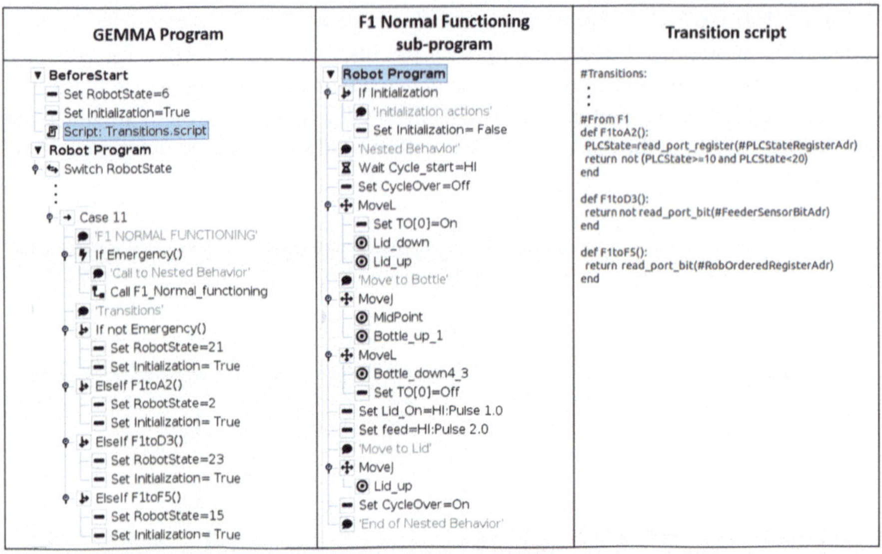

Fig. 7. Design pattern illustrated for the F1 Normal functioning state

5 Results and Discussions

After writing the PLC and UR code and the development of the corresponding HMIs (Human Machine Interfaces), the *DASCC architecture* was built.

A `Machine HMI` was generated and scheduled through a CoDeSys program. The `Machine HMI` acts as the *Central Coordinator* by supervising and coordinating the PLC and UR controllers. We are conscious that in a proper DASCC architecture the code of the manufacturing entities runs on different controllers. However, we decided to implement the Central Coordinator within the PLC, since this choice reduces the number of interconnected software components without compromising the verification of the approach.

Then, we modelled the filling and encapsulating machine in Experior[6] and we connected it with CoDeSys and URSim for the implementation of a *Virtual Commissioning Simulation* [15]. The implemented DASCC architecture is shown in Fig. 8 and videos[7] are made available to the reader showing different use case scenarios.

[6] https://xcelgo.com/.

[7] https://www.youtube.com/watch?v=O0S1rn8C-fQ&list=PLXnpmcz3YbS-0oiNMLt6u5roNr9gsC6el.

The following considerations can be generated by analysing the developed code and the obtained machine behaviour:

- *HDP for UR-controllers*: the UR code presents the following characteristics:
  - *Hierarchical*: use of one program and different sub-programs for separating the nested behaviour of each 'GEMMA state' from the one specified at the GEMMA hierarchical layer
  - *Readable*: separation of the transition utilization from its guard condition
  - *Modular*: use of sub-programs to encapsulate the nested behaviour of each 'GEMMA state' and standard interface in between the `GEMMA` program and the nested behaviour; i.e. `Initialization` flag. The use of global variables is contrary to the principle of modular programming but had to be implemented since UR does not allow to generate input and output variables for the sub-programs
- *DASCC architecture*: a decentralized architecture was implemented by interfacing a PLC-controlled system with an UR3 robot through a Central Coordinator. The developed architecture is more flexible than a centralized one since production can continue even if the robot is being repaired, see the 'Failure Management' use case video. Moreover, the GEMMA-GRAFCET representation generates a common vocabulary for the OMs of PS facilitating the exchange of information between the manufacturing entities and the Central Coordinator.

The only drawback can be found in the HDP for UR controllers. If this latter is compared with the one for PLC [11], it can be noticed that in the UR case it is not possible to customize the macro-step/enclosing step behaviour based on the considered outgoing transition. A strategy for implementing this behaviour is left as future work.

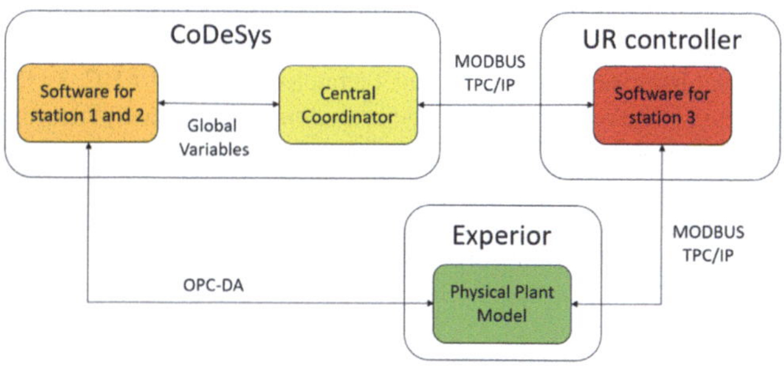

Fig. 8. Implemented DASCC architecture

6 Conclusions and Future Work

This paper introduces a *Hierarchical Design Pattern for UR-controllers* able to generate code from specifications expressed in accordance with a previously proposed GEMMA-GRAFCET representation. The developed code is hierarchical, readable and modular, and allows the development of state-based behaviour in UR-robots without the need of third-party plugins. Through the implementation of a lab case study, it is demonstrated how the presented GEMMA-based approach can be utilized for the generation of a *Decentralized Architecture Supervised from a Central Coordinator*. The approach provides a common vocabulary for the OMs of PSs facilitating the exchange of information in between the manufacturing entities and the Central Coordinator.

In the context of Smart Production Systems, the developed *GEMMA-based decentralized architecture* constitutes a preliminary concept that in the future should be further validated and improved. Some future works identified are:

- *Physical production system*: the GEMMA-based decentralized architecture has been applied to a lab case study through a virtual commissioning simulation. The proposed approach will be implemented in a (laboratory) physical production system to be further validated
- *Customizable state behaviour*: in the current version of the HDP, the behaviour of composite states cannot be personalized based on the considered transition. A solution for this issue will be investigated.

References

1. Lu, Y.: Industry 4.0: a survey on technologies, applications and open research issues. J. Ind. Inf. Integr. **6**, 1–10 (2017)
2. Dotoli, M., Fay, A., Miśkowicz, M., Seatzu, C.: An overview of current technologies and emerging trends in factory automation. Int. J. Prod. Res. **57**(15–16), 5047–5067 (2019)
3. Hu, S.J.: Evolving paradigms of manufacturing: from mass production to mass customization and personalization. Procedia CIRP **7**, 3–8 (2013)
4. Feeney, A.B., Weiss, B.: Smart Manufacturing Operations Planning and Control Program. National Institute of Standards and Technology, Gaithersburg (2015)
5. Lu, Y., Morris, K.C., Frechette, S.: Current standards landscape for smart manufacturing systems. National Institute of Standards and Technology, NISTIR **8107**, 39 (2016)
6. Dilts, D.M., Boyd, N.P., Whorms, H.H.: The evolution of control architectures for automated manufacturing systems. J. Manuf. Syst. **10**(1), 79–93 (1991)
7. Liu, S., Zhang, J., Liu, J., Feng, Y., Rong, G.: Distributed model predictive control with asynchronous controller evaluations. Can. J. Chem. Eng. **91**(10), 1609–1620 (2013)
8. Boccella, A.R., Piera, C., Cerchione, R., Murino, T.: Evaluating centralized and heterarchical control of smart manufacturing systems in the era of Industry 4.0. Appl. Sci. **10**(3), 755 (2020)
9. Radziwon, A., Bilberg, A., Bogers, M., Madsen, E.S.: The smart factory: exploring adaptive and flexible manufacturing solutions. Procedia Eng. **69**, 1184–1190 (2014)
10. Vogel-Heuser, B., Diedrich, C., Pantförder, D., Göhner, P.: Coupling heterogeneous production systems by a multi-agent based cyber-physical production system. In: 2014 12th IEEE International Conference on Industrial Informatics (INDIN), pp. 713–719 (2014)
11. Barbieri, G., Gutierrez, D.A.: A GEMMA-GRAFCET methodology to enable digital twin based on real-time coupling. Procedia Comput. Sci. **180**, 13–23 (2021). in press
12. ADEPA: GEMMA (Guide d'Etude des Modes de Marche et d'Arrêt). Technical report. Agence nationale pour le Développment de la Production Automatisée (1981)
13. International Electrotechnical Commission: GRAFCET specification language for sequential function charts, IEC 60848 (2002)

14. Alvarez, M.L., Sarachaga, I., Burgos, A., Estévez, E., Marcos, M.: A methodological approach to model-driven design and development of automation systems. IEEE Trans. Autom. Sci. Eng. **15**(1), 67–79 (2016)
15. Lee, C.G., Park, S.C.: Survey on the virtual commissioning of manufacturing systems. J. Comput. Des. Eng. **1**(3), 213–222 (2014)

A Hybrid Control Architecture for an Automated Storage and Retrieval System

Jose-Fernando Jimenez[1(✉)], Andrés-Camilo Rincón[1], Daniel-Rolando Rodríguez[1], and Yenny Alexandra Paredes Astudillo[2]

[1] Pontificia Universidad Javeriana, Bogotá, Colombia
j-jimenez@javeriana.edu.co
[2] Polytechnic University Institution Grancolombiano, Bogotá, Colombia

Abstract. An automated retrieval and storage system, or ASRS, is a system that handles automatically the replenishment, picking, and delivery of products of a warehouse system. The current trend of ASRS is to develop less-rigid structures and continuous operations to improve the functionality, efficiency, and reactivity during its operation. However, the dynamic market changes and internal perturbations are characteristics that need to be handled by the ASRS control system to ensure the expected flexibility. The objective of this paper is two-fold. First, it proposes a framework of an ASRS control system, based on a hybrid control architecture that provides efficient operation during normal functioning and reactivity features in case of perturbations. Second, for validation purpose, a platform is constructed that simulates the virtual and physical level of the control system by means of a digital twin that represents the products and the resources composing the ASRS.

Keywords: ASRS · Warehouse system · Hybrid control architecture · Predictive decision model · Reactive decision model · Digital twin emulation

1 Introduction

Recently, competitiveness and new technological advances have changed the global economy, allowing the optimization of industrial operations. New solutions and technological enablers contribute to enhance industrial systems and improve the efficiency of order fulfilment. The digital transformation, or the fourth industrial revolution, permits the migration of production and supply chain processes to flexible, reconfigurable, and intelligent systems. Some technological advances are: Internet of Things (IoT), Big Data analytics, horizontal and vertical enterprise integration, Artificial Intelligence (AI), a.o. In the manufacturing and logistics sector, flexible manufacturing systems, material handling systems, automatic warehouse systems, and automated storage and retrieval systems are examples of control systems in which these advanced technologies are implemented and continuously improved.

Nowadays, automated retrieval and storage systems (ASRS) have replaced a number of independent loading, handling and transportation mechanisms such as roller conveyors, belts, pallet trucks, stackers and forklifts. An ASRS is an automated system that

D. Trentesaux et al. (Eds.): SOHOMA 2021, SCI 987, pp. 30–41, 2021.
https://doi.org/10.1007/978-3-030-80906-5_3

handles mechanically the replenishment, picking and delivery of products within a warehouse system. These systems are computer-controlled mechanisms that move between racks and physical storage spaces to store or retrieve goods from the warehouse facility. In particular these systems improve the capability and speed of the storage/retrieval process, reduce human intervention in the process and minimize the errors in the process (Rodríguez and Martín 2010). In general, the ASRS uses storage locations for a large number of products/boxes, to which a networked crane or AGV moves along racks to execute the corresponding storing or retrieving actions. Figure 1 illustrates some examples of ASRSs.

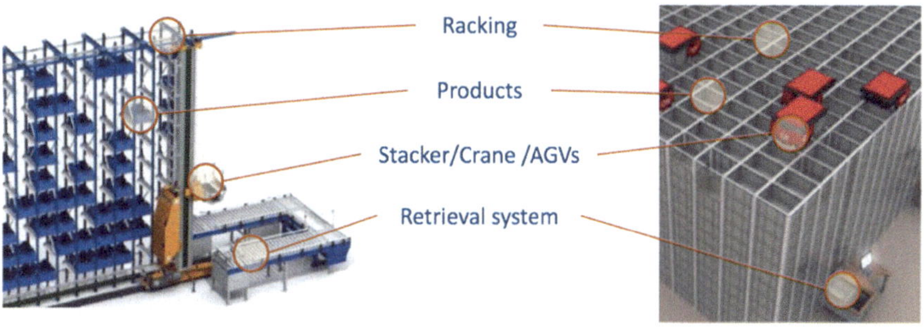

Fig. 1. Examples of ASRS systems with general components

An ASRS deep lane with an AGVs with lifting features is a new type of system that has the potential of featuring flexibility characteristics that are needed for optimization and reactive behaviour (Fig. 1 right). This ASRS is composed of a set of networked, communicating, and collaborative Robots or AGVs that, by moving on a grid on top of deep lane racks, collect and deliver the products stacked in several locations (Laurence and Tobón 2013). According to Jacobsen (2011), the implementation of these systems provides efficient use of space, increased inventory capacity and process throughput, reduced labour cost and high flexibility. However, these ARSRs face the challenge on the overall control of the numerous AGVs that have individual goals of storing or retrieving products. These AGVs are functioning simultaneously within the same grid environment; each AGV must operate collision-free relative to the other AGVs, optimally and initiate reactivity actions if needed. The control system of this type of ASRS is a challenge because it must ensure good communication, optimal operating modes, and strategies to face unexpected situations.

For these reasons, the paper proposes a control system for this type of ASRS using the digital twin concept and multi-agent modelling for optimality and reactive behaviour. The objectives of this paper are two-fold. First, a framework of a distributed control system is proposed, based on a hybrid control architecture that provides efficient operation during normal functioning and reactivity features in case of perturbations. Second, for validation purpose, a platform is designed to simulate the virtual and physical level of the control system, based on a digital twin that represents the products and the resources composing the ASRS. This paper is organized as follows. Section 2 reviews a background of control system approaches for ASRS with AGVs considering predictive and reactive decisional models. Section 3 presents our approach, featuring a hybrid control system for an ASRS based on a distributed architecture and considering a model-driven digital twin. In Sect. 4, the experimentation protocol on a case study is executed in order to validate the solution and potential benefits of the distributed control architecture with digital twin, conducted in agent-based simulation software. Results are presented in Sect. 5. Finally, Sect. 6 rounds up the conclusions and the main considerations to further research.

2 Background

2.1 Hybrid Control Architectures

Recent years have seen a growing trend towards dynamic control of computational and physical systems by using hybrid control architectures (Jimenez *et al.* 2017). These architectures are characterized by a conjoint control of the system: predictive/proactive and reactive techniques for decision making. The predictive/proactive techniques are generally based on operational research and exact methods to address an assigned problem, and generally seek optimality of the entire system. The reactive techniques are generally based on artificial intelligence or heuristics methods and provide a fast response when it is needed. Therefore, the control system starts with an initial configuration based on the predictive/proactive technique, and changes over time in order to provide the functionality and capacity needed (Youssef and ElMaraghy 2007). When a perturbation occurs, the control system tries to maintain its performance based on a reactive technique, and recover to the initially expected performance by executing adequate changes in the control architecture (Terkaj et al.). Such an approach seems to be able to manage the control solution to tackle the complexity and uncertainty of most types of controlled systems.

2.2 Control Systems for ASRS

This section presents a set of approaches dealing with the controls system of automated storage and retrieval systems, specifically of the ASRS deep lane with lifting AGVs. From the literature reviewed, these approaches are based on solving the AGVs routing and redirecting the AGV route at unexpected disruptions. The approaches found are metaheuristics algorithms, multi-agent systems, simulation with Petri net models, and digital twin implementation.

The first set presented are metaheuristics approaches. These approaches focus on the routing of the AGVs as the transportation time and the allocation of tasks are crucial to the overall performance. Tu *et al.* (2017) argue that a genetic algorithm is a viable method to find the route for each AGV, the number of vehicles affected by the area and the demands to be dispatched. Qing *et al.* (2017) proposed an improved Dijkstra algorithm to solve the path-planning problem on an ASRS considering a rectangular environment; the equidistant shortest paths are found by adding running time to the path-planning evaluation. Bertolini *et al.* (2019) propose a simulated annealing (SA) procedure that optimizes performance during the retrieving phase of an automated warehouse by translating the request of customers, described by item code, quality, and weight, into a list of jobs. Liu *et al.* (2018) present a hybrid metaheuristic where a MILP model and a taboo algorithm are used to minimize the total collection time, including transportation and retrieval. These approaches are predictive/proactive techniques to solve the efficiency problem, that use planning software for the execution of the ASRS operations. However, these methods lack reactivity features to support unexpected disruptions that may occur during process control.

The second set refers to hybrid control architecture approaches. These approaches are focused on operation planning and control; they embed the behaviour of the components of the ASRS such as products, AGVs and task orders, offering solutions to react at unexpected perturbations. Saidi-Mehrabad *et al.* (2015) present a hybrid control architecture that uses an ant colony algorithm for efficiency by computing the vehicle routing plan to minimize the execution time of the AGV. Estrada *et al.* (2019) suggest the use of simulated agents with a MILP model to schedule the routes of the AGVs and a priority rule for adapting the routes when two AGVs possibly collide. These approaches use the multi-agent behavior to monitor the execution of the ASRS operations and, besides, include a predictive/proactive technique for planning and programming the routes, and a reactive technique to handle possible disruptions.

The third set presented refers to simulation approaches based on Petri nets. These approaches are focused on piloting the execution of ASRS operations. Ventura and Lee (2008) present simulations to identify the number of active AGVs, analysing the mobility and flexibility characteristics. Xu *et al.* (2019) have a similar approach developing research to identify the maximum number of tasks and restrictions that each AGV can handle without forcing the model and reach good performances. Sun *et al.* (2018) propose a route generation algorithm based on the construction of a Resource Oriented Timed Petri Net model to change dynamically AGVs routes and minimize the AGVs' risk of collision and deadlocks. These approaches use techniques that allow identifying bottleneck situations and exploit the control capability of the ASRS to avoid them.

The last set presented are digital twin (DT) approaches. These approaches include digital twins in the ASRS to increase information gathering for improved real-time task execution, and to detect rapidly unexpected events. Muñoz Alcázar (2019) apply the digital twin concept to synchronize data obtained from the physical world in real-time with decision making in the DT much faster than real time, and then sending the actions to be performed to the ASRS control. Răileanu *et al.* (2020) propose a four-layer DT generic architecture including: (1) data collection and edge processing, (2) data transmission, (3) data aggregation and (4) analysis and decision making for the control of a shop floor product conveyor. Ding (2019) propose a product transfer model in supply chains that use the digital twin concept, and analyse the usage and performances obtained with this system. These approaches rely on the connectivity and information sharing model supported by digital twins, considering that real-time decisions must be obtained from real time warehouse data acquisition and processing.

From these reviewed approaches, it results that they aim at minimizing travel times or response times of the AGV, as most contributions are focused on route planning. Yet, some advances refer to control architectures and develop approaches based on agent-simulation to improve the reactivity of AGVs during execution. However, very little is known about the hybrid or distributed control of ASRSs, and research perspectives may be directed to explore and understand the dynamics of these physical systems. Considering this fact, the present work describes the design of a hybrid control architecture for ASRS using the digital twin concept and evaluates the performance of agent-based simulation software.

3 Hybrid Control Architecture for an ASRS

A framework of hybrid control architecture that controls an ASRS is proposed. The control architecture is designed to provide an efficient operation during normal functioning and to allow reactive actions in case of unexpected perturbations. In particular, the task for the control system is the retrieval of products from the ASRS considering the following actions: receiving a retrieval order with all the products to collect, performing product allocation to the available AGVs, plan the route path according to the minimum distance principle, guiding the AGVs along planned route-path, and changing/adapting in real time the AGV routing plan in order to avoid collisions during execution. This section is divided between the digital twin component and the hybrid control architecture. On one hand, the digital twin component is described in terms of the main characteristics of the atomic components in the control architecture. On the other hand, the hybrid control architecture and the relation with the digital twin are presented.

3.1 Digital Twin Component

The proposed control architecture is based on the paradigm of distributed control systems (Trentesaux, 2009). The main characteristic of this systems is the distribution of the control in different components in order to feature reactivity in case of deviations from the planned program. The control components were modelled as digital twins. A digital twin is a virtual instance of an object, a physical system or software component, that is continually updated with the performance, maintenance and health status data of the physical twin throughout the life cycle of the system (Madni *et al.* 2019).

In this proposal, based on the Pollux architecture (Jimenez *et al.* 2017), three different atomic components are considered within the hybrid control architecture:

- The *global decisional entity* (GDE): represents a coordination component responsible of deciding the operations and tasks of the system with global efficiency.
- The *local decisional entity* (LDE): represents the cell with physical products, and is responsible for connecting the cell and physical objects in a digital twin relationship.
- The *resource decisional entity* (RDE): represents the AGVs.

The GDE does not represent a digital twin but a task coordination activity. In this respect, the LDE and RDE digital twins are illustrated in Fig. 2.

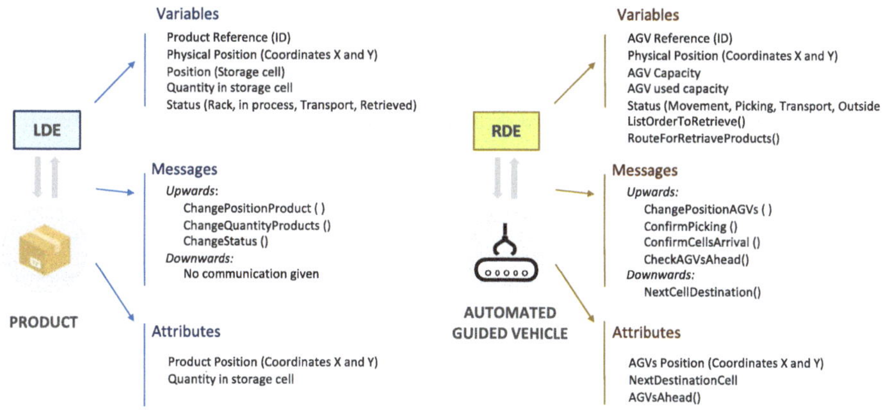

Fig. 2. Digital twin and attributes of a Product and Resource item of an ASRS

3.2 Hybrid Control Architecture

The system is designed as a hybrid control architecture, which is distributed and selects dynamically the optimal operation by help of the coordination components; it uses also reactive decisional techniques to handle abnormal situations. Readers can refer to Jimenez *et al.* (2017) for further information about this type of control.

The proposed control architecture for the ASRS guides the operating of the retrieval system by collecting the products from a picking order list, minimizes the makespan of product retrieving, and directs the AGVS in case of collision risk. The control architecture is divided and located on three different layers (see Fig. 3): the coordination layer, the operation layer, and the physical layer.

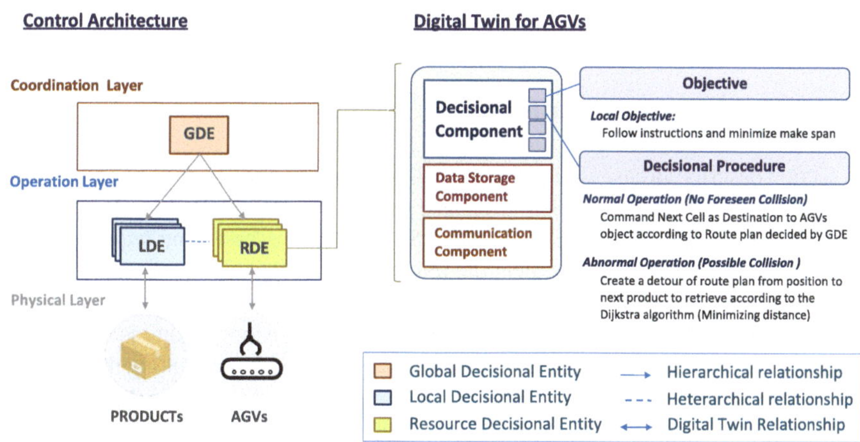

Fig. 3. Control architecture of the proposed ASRS control and detailed description of the AGV digital twin

The main characteristic of the control architecture is the capability of predictive and reactive decision-making. From the predictive perspective, the control architecture contains a tabu search algorithm used to minimize the makespan of retrieval by deciding the allocation of products for each AGV and its route plan for retrieval. From the reactive perspective, the control architecture contains a monitor for detecting perturbations and a Dijkstra algorithm module that searches alternative routes in case a collision is possible. The control architecture is organized as follows:

- **The Coordination layer** hosts the components that hold the global information of the system, the global task, and are responsible for the global efficiency of the system. For the case study presented, this layer is built with a single GDE that has embedded a tabu search algorithm to minimize the makespan of the retrieval process. The solution of this metaheuristic is the set of commands instructed to the operating components for the next level (i.e. product and resource).
- **The Operation layer** hosts the components that control the physical object in a digital twin relationship; it holds only partial information of the system, has a local objective and it is responsible for the completion of partial objectives derived from diving a global objective (e.g., retrieving the assigned product). For this case study, it contains a set of LDE that represents the digital twin of the product to be retrieved and the RDE that represents the digital twin of the AGVs that will collect the assigned products. While the LDEs are passive components that only acknowledge positioning and completion of the task to the system, the RDEs are active components that are piloting the physical AGVs by instructing the motions and actions to be followed by the physical device.
- **The Physical layer** hosts the physical components that follow the orders issued by the digital twin. There exist active, passive, and inactive components such as the AGVs, the products, and the ASRS rack structure. The inactive components are those components in the physical world that do not have images in the digital world, but have indirect interaction with active and passive components in the physical world.

4 Case Study and Validation

This section describes a case study of an ASRS, considering only the retrieval process. In particular, the case study is an ASRS deep lane with AGVs with lifting features as shown in Fig. 1 right. The decision-making in this ASRS consists in assigning required orders to the AGVs from a list of orders, plan the retrieval route from the recollection point of the ASRS racking to the product locations, and executing the recollection minimizing the recollection makespan. On one hand, the predictive decision-making is to optimize the assignment of orders to the AGVs and optimizing the routes for picking the assigned products. On the other hand, the reactive decision-making must control the piloting of the AGVs for avoiding traffic bottlenecks and collisions.

For testing purpose of the proposed architecture, the studied ASRS consists of a grid of 1050 cells in a 30 $\times$ 35 cell layout. In this area, the AGVs can move freely between the cells only along x or y direction and return to the recollection cell once the picking of all products is completed. The task is to retrieve products from an entire delivery order, which is assigned to the AGVs. The start of this process is when the first AGV leaves the entry cell and the completion of the process is when the last AGVs arrives at the entry cell. Figure 4 illustrates the rack, the AGVs, the cells, the products, the entry point and retrieval point in the ASRS.

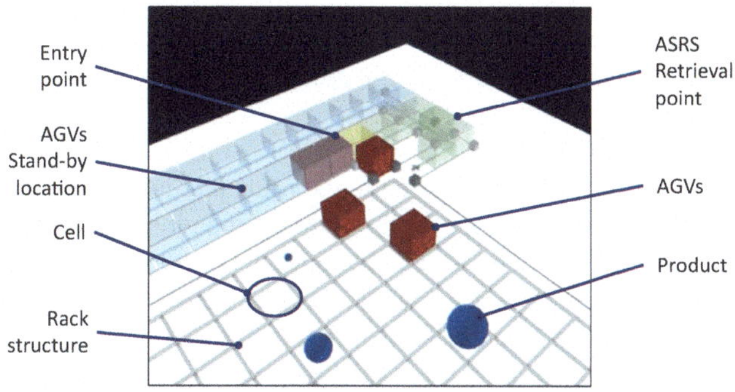

Fig. 4. Entry and exit points in the ASRS, including physical components

This system considers the ASRS's characteristic, which it related to the physical world and the digital world. The control architecture adapted for this case study is shown in Fig. 5.

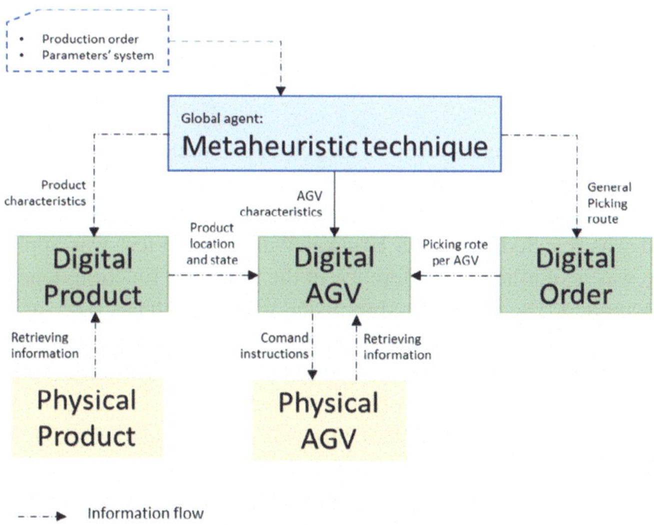

Fig. 5. ASRS control architecture for the case study

The architecture contains a global agent for coordination purpose and the digital twin representation of AGVs, orders and products. The digital twin emulates all the components of the physical environment which allows visualization of the system's performances at a certain time moment. In the case study discussed the simulated system has 5 AGVs, a set of instance orders with several products scenarios and a track of 30 x 35 cells to which the AGVs travel from the entry point to the retrieval point, collecting the assigned products.

The global agent gets the routing solution for the AGVs computed by a taboo algorithm and transfers it to a NetLogo model; the instructions to the AGVs and products are communicated through each twin.

The experiment consists of processing a production order in two main scenarios: predictive and reactive:

- Case A: the order is completed according to the route planned from the requirements known before. This solution, computed to minimize the global retrieval makespan, is the current routing as long as perturbations and unexpected events do not occur. In this case, the time to complete de order by the metaheuristic is compared with the time obtained by the NetLogo model to identify the difference between them (the time computed by metaheuristic does not consider pick up, dispatch and other times.
- Case B: unforeseen events occur, and the system must have a reactive response. To test this case, the following scenario has been set up: AGV #1 has been routed by the predictive model provided by the global agent; an empty AGV #2 blocks a certain position; if the route of AGV #1 includes this point, the control must change the route. The failures of AGV #2 occur in the first third of the total travel time and in the second third. After collecting the data about the completion times for the normal and disturbed scenarios into the NetLogo model, they were compared (see Fig. 6).

5 Results

Considering that the global agent receives the solution obtained with the taboo algorithm, some heuristics were developed to give the first solution that will be improved by the metaheuristics. In an experiment where each heuristic was considered as a factor, the normality test was rejected, so that a Kruskal-Wallis (KW) test was applied to check whether there was a significative difference in the objective function depending on the two heuristics applied. The result was significant, with a P-value < 0.05; using a boxplot it was shown that Clarke and Wright algorithm (CW) achieved the best performance.

The routing plan obtained by the metaheuristic was applied to the NetLogo model. This allowed identifying whether the objective function changes function of the number of orders; an ANOVA test revealed that this was true (P-value < 0.05). Taking into account and considering the possibility to have different numbers orders, a regression was used to determine the value of the completion (Y) for each case. Equation 1 shows the expression defining Y; an order of 30 products and x AGVs were considered.

$$Y = 6679.6 - 386.8x \tag{1}$$

The differences between the result of the objective function obtained with the metaheuristic and the same routing plan applied in NetLogo model for each case (orders from 30 to 120 products shown in Fig. 7) demonstrate that the NetLogo model times are

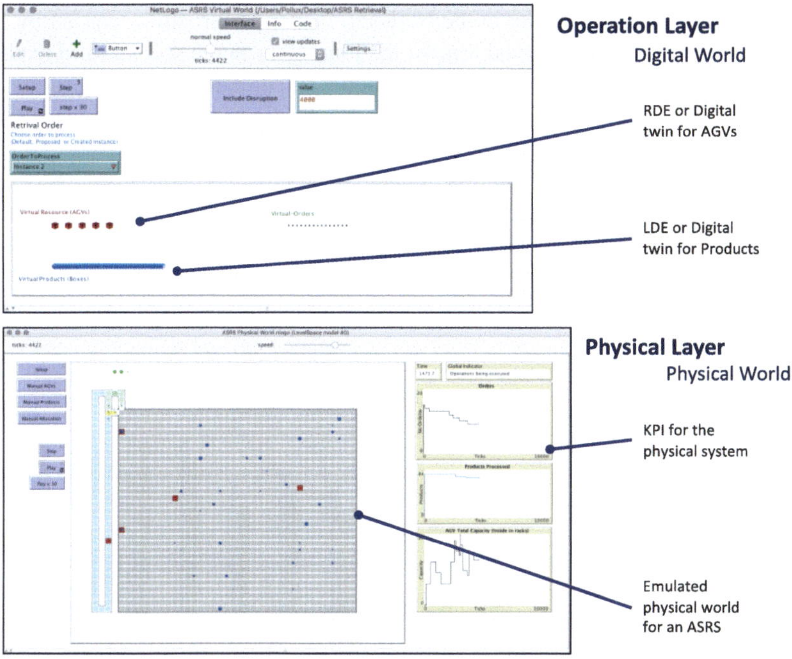

Fig. 6. Emulation of the physical ASRS and the corresponding controlling digital twin, created on the NETLOGO agent-based simulation software.

greater compared to the metaheuristic times because the latter do not include the time intervals for AGV's departure, load and download.

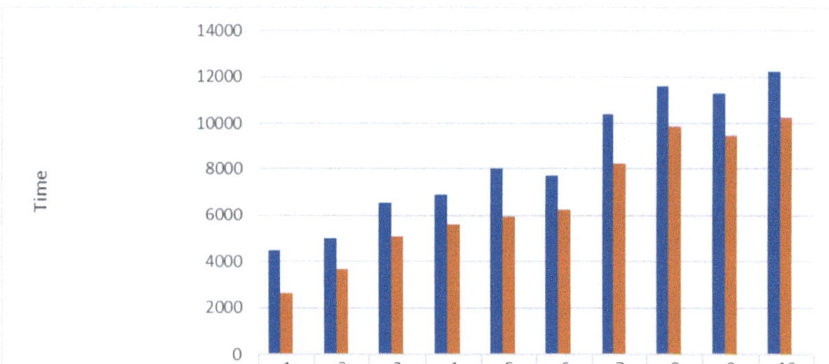

	1	2	3	4	5	6	7	8	9	10
NetLogo model	4505	5021	6541	6919	8018	7714	10375	11576	11312	12250
Metaheuristic	2647	3693	5089	5608	5940	6263	8238	9872	9480	10272
% respect to metaheuristic	70%	36%	29%	23%	35%	23%	26%	17%	19%	19%

Fig. 7. Comparison between NetLogo model and Metaheuristic results

On the other hand, including in the NetLogo model perturbations caused by AGV failures in the first third of the total time of the journey or in the second third leads to a bigger time difference between the NetLogo model and metaheuristic, the average time increase being about 10% when the failure occurs in the first part of the journey. Figure 8 shows how a failure in the first third of the journey for each case in the NetLogo model causes increases in the completion time related to the model without disturbances.

It is also evident that the implementation of the digital twin in the ASRS control architecture is a good alternative to solve scenarios with disturbances for which a predictive model would not provide solutions.

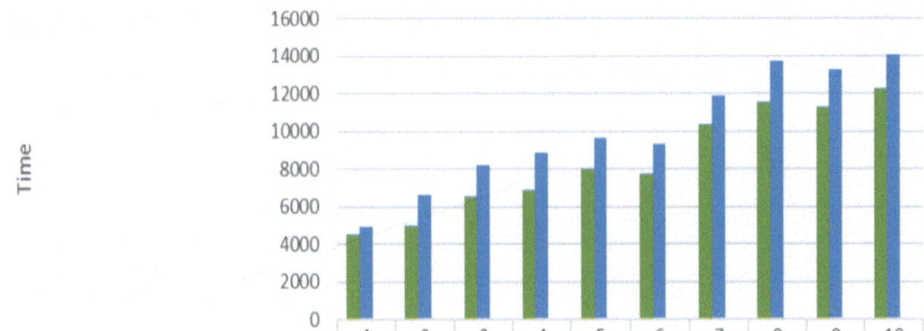

	1	2	3	4	5	6	7	8	9	10
NetLogo model without distrubance	4505	5021	6541	6919	8018	7714	10375	11576	11312	12250
NetLogo model with distrubance	4910	6601	8203	8837	9637	9291	11898	13725	13269	14027
% respect to model without distrubance	9%	31%	25%	28%	20%	20%	15%	19%	17%	15%

Fig. 8. Comparison between NetLogo model with and without disturbances

6 Conclusions

This paper proposed a framework of an ASRS control system, established on an architecture that allows operating in normal circumstances (predictive model) and under unexpected circumstances (reactive model) to complete the retrieval order. Additionally, to validate the proposal and using simulation based on agents, a software system was built, emulating the physical and virtual worlds. The digital twin is embedded in controlling and coordinating the physical world - a type of ASRS with AGVs.

References

Bertolini, M., Mezzogori, D., Zammori, F.: Comparison of new metaheuristics, for the solution of an integrated jobs-maintenance scheduling problem. Expert Syst. Appl. **122**, 118–136 (2019)

Ding, D., Han, Q.L., Wang, Z., Ge, X.: A survey on model-based distributed control and filtering for industrial cyber-physical systems. IEEE Trans. Industr. Inf. **15**(5), 2483–2499 (2019)

Estrada, S., Ramirez, J.P., Uribe, F.: Modelo predictivo-reactivo para el control de un sistema de recolección de productos en bodega (AS/RS) (Unpublished undergraduate thesis). Pontificia Universidad Javeriana. Bogota, Colombia (2019)

Jacobsen, J.: Warehouse automate with AGVs. Beverage Ind. **102**(8), 64–66 (2011)

Jimenez, J.F., Bekrar, A., Zambrano-Rey, G., Trentesaux, D., Leitão, P.: Pollux: a dynamic hybrid control architecture for flexible job shop systems. Int. J. Prod. Res. **55**(15), 4229–4247 (2017)

Laurence, M., Tobon, J.: Desarrollo de una plataforma para la actualización automática de inventarios del sistema de almacenamiento automático AS/RS del centro tecnológico de automatización industrial CTAI, Pontificia Universidat Javeriana (2013). https://repository.javeriana.edu.co/handle/10554/6328

Liu, J., Zhong, X., Peng, X., Tian, J., Zou, C. Design and implementation of new AGV system based on two-layer controller structure. In: Proceedings 37th Chinese Control Conference (CCC), pp. 5340–5346. IEEE Digital Library (2018). https://doi.org/10.23919/ChiCC.2018.8483515

Madni, A.M., Madni, C.C., Lucero, S.D.: Leveraging digital twin technology in model-based systems engineering. Systems **7**(1), 7 (2019). https://doi.org/10.3390/systems7010007

Muñoz Alcázar, J.: Aplicacion del concepto de gemelo digital a un scada Industrial, Ph.D. thesis, Universitat Politècnica de València (2019). http://hdl.handle.net/10251/126015

Qing, G., Zheng, Z., Yue, X.: Path-planning of automated guided vehicle based on improved Dijkstra algorithm. In: 2017 29th Chinese Control and Decision Conference (CCDC), pp. 7138–7143. IEEE, May 2017

Răileanu, S., Borangiu, T., Ivănescu, N., Morariu, O., Anton, F.: Integrating the digital twin of a shop floor conveyor in the manufacturing control system. In: Borangiu, T., Trentesaux, D., Leitão, P., Giret Boggino, A., Botti, V. (eds.) SOHOMA 2019. SCI, vol. 853, pp. 134–145. Springer, Cham (2020). https://doi.org/10.1007/978-3-030-27477-1_10

Rodríguez, R., Martín, J.: Simulación de un sistema AS/RS. Treballs Docents Curs **1**, 55–58 (2010). ISSN:1889-4771

Saidi-Mehrabad, M., Dehnavi-Arani, S., Evazabadian, F., Mahmoodian, V.: An ant colony algorithm (ACA) for solving the new integrated model of job shop scheduling and conflict-free routing of AGVs. Comput. Ind. Eng. **86**, 2–13 (2015)

Sun, S., Gu, C., Wan, Q., Huang, H., Jia, X.: CROTPN based collision-free and deadlock-free path planning of AGVs in logistic center. In: Proceedings 15th International Conference on Control, Automation, Robotics and Vision, ICARCV 2018 (2018)

Terkaj, W., Tolio, T., Valente, A.: Focused flexibility in production systems. In: ElMaraghy, H. (ed.) Changeable and Reconfigurable Manufacturing Systems, pp. 47–66, Springer, London (2009). https://doi.org/10.1007/978-1-84882-067-8_3

Trentesaux, D.: Distributed control of production systems. Eng. Appl. Artif. Intell. **22**(7), 971–978 (2009)

Tu, J., Qian, X., Lou, P.: Application research on AGV case: automated electricity meter verification shop floor. Ind. Robot. **44**(4), 491–500 (2017). https://doi.org/10.1108/IR-11-2016-0285

Ventura, J., Lee, C.: A study of the tandem loop with multiple vehicles configuration for automated guided vehicle systems. J. Manuf. Syst. **20**(3), 153–165 (2008)

Xu, W., Guo, S., Li, X., Guo, C., Wu, R., Peng, Z.: A Dynamic scheduling method for logistics tasks oriented to intelligent manufacturing workshop. Math. Probl. Eng. **2019**, 1–18 (2019)

Youssef, A.M., ElMaraghy, H.A.: Optimal configuration selection for reconfigurable manufacturing systems. Int. J. Flex. Manuf. Syst. **19**(2), 67–106 (2007)

Digital Twin in Water Supply Systems to Industry 4.0: The Holonic Production Unit

Juan Cardillo Albarrán[1], Edgar Chacón Ramírez[1(✉)], Luis Alberto Cruz Salazar[2,3], and Yenny Alexandra Paredes Astudillo[4]

[1] Universidad de Los Andes, Mérida, Venezuela
echacon@ula.ve
[2] Technical University of Munich, Munich, Germany
[3] Universidad Antonio Nariño, Bogotá, Colombia
[4] Politécnico Grancolombiano, Bogotá, Colombia

Abstract. Industry 4.0 (I4.0) and Digital Twin (DT) bring together new disruptive technologies, increasing manufacturing productivity. Indeed, the control of production processes is fast becoming a key driver for smart manufacturing operations based on I4.0 and DT. In this connection, intelligent control such as the Holonic Manufacturing Systems (HMS) generates distributed or semi-heterarchical architectures to improve both global efficiency and manufacturing operations' reactiveness. Still, previous studies and HMS applications often have not dealt with continuous production processes, such as water treatment applications, because of the complexity of continuous production (a single fault can degrade extensively and can even cause a breakdown of production). This work describes a HMS architecture applied to continuous systems, based on Holonic Production Units (HPU). This unit's cognitive model allows building a DT of the unit employing a hybrid dynamic system. This HMS detects events within the environment through a DT, evaluating various courses of action, and changing the parameters aligned to a mission. The DT was created by a simulated model of a water supply system, considering three scenarios: normal condition and two disrupted scenarios (the unexpected increase of demand and water quality degradation). The experiments apply agent-based modelling software to simulate the communication and decision-making features of the HPU. The results suggest that the construction of a holarchy with heterogeneous holons is potentially able to fulfil I4.0 requirements by DT of a WSS.

Keywords: Digital Twin · Holonic Manufacturing Systems · Holonic Production Unit · Industry 4.0 · Water Supply System

1 Introduction

According to the annual report of the World Economic Forum on Global Risks in 2018 [14], the crisis around water is one of the ten main social risks in the

D. Trentesaux et al. (Eds.): SOHOMA 2021, SCI 987, pp. 42–54, 2021.
https://doi.org/10.1007/978-3-030-80906-5_4

world, being a challenge for the hydrological industry to meet its objectives efficiently, maintaining good coordination between the links that make up the process and ensuring a minimum of resource losses. Even with the priority of preserving profitable operation, the World Bank estimates that 25% to 30% of water production is lost due to failures. Water Supply Systems (WSS) aim to provide drinking water to a population from sources of raw water. The "operations area" in the WSS is associated with the water collection, purification (make it potable), distribution, wastewater recovery, which is a product model that contains nine links: five for WSS (potable water) and four for Wastewater System [13]. WSS permanently faces two problems; the first one is the coordination of the process belonging to the WSS value chain in order to provide the best possible service to the users and the second one is to be able to minimize the losses in the system. Although there are adverse seasonal conditions that decrease (low water) or increase (rains, storms) the quantity and quality of raw water, WSS must be prepared to face these variations that, to a greater or lesser degree, influence the quality of service. Climate change contributes to an additional level of uncertainty for the WSS and develops unexpected situations that have an image of the whole process, like the *Digital Twin (DT)* concept. DT is a digital replica of a physical entity and assets as holons (machines, products, processes, etc.) [3], which allows evaluating the overall behaviour and characteristics of its components and methods.

Considering the complexity of a global model representing the whole WSS, this work proposes a Holonic Manufacturing Systems or HMS (a composition of holons), associated with hybrid systems and a preliminary validation by a simple DT. The global model of the WSS is given by the composition of behavioural models for each unit. These processes are shown in Fig. 1, in the value chain for WSS.

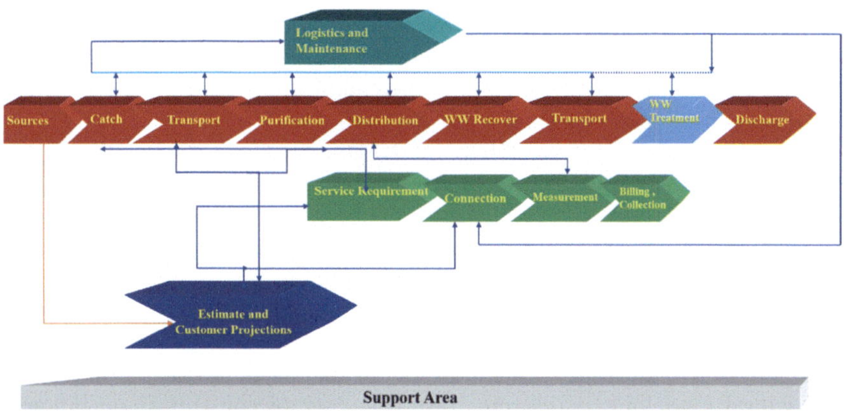

Fig. 1. Value Chain for WSS and its associated processes

The paper is organized as follows: a brief introduction highlighting the importance of WSS and its different problems associated with management, configuration (optimization), and the way to solve it with a DT. In Sect. 2, the structure of smart units, according to the proposal of Industry 4.0 (I4.0), and the modelling schema, which compounds the architecture called Holonic Production Unit or **HPU** (see preliminary works in [4,5]) will be described. Section 3, describes a use case on a WSS configuration to illustrate the modelling technique, and scenarios in Sect. 4 simulate the DT of the HPU. Finally, the conclusions will be given in Sect. 5.

2 Building Models of the Components of Smart Units

In I4.0 one of the most important requirements is the use of the Cyber-Physical Production Systems, which are autonomous systems and can be implemented using the holonic concept [3,5,7]. This work uses the HPU concept to describe the complete behaviour of each unit and this behaviour is constrained by the behaviour of the other units by the exchange of products. The composition of those behaviours gives the global behaviour similar to a Cyber-Physical Production System. The HPU has the following layers: physical (equipment and processes), regulation, supervision, and interaction & logical control. The upper level is a discrete image of the physical process and its regulation and the supervision layers. At this layer the composition allows having the global behaviour of the process [4].

In continuous production processes, the production line is fixed and the configuration of the units is established by the condition of the resources exchanging products: state of the unit and quantity-quality of the product flow. In this case there is only one holarchy that changes dynamically according to the condition of resources, process and products. Here the *cognitive model* (see Fig. 2), within each unit for each skill, is obtained from: the form of processing a product (continuous dynamics), the unit's skills (continuous-discrete-events dynamics), and the form of interaction with other units (dynamics of events), whose holarchy defines a cognitive model given for a hybrid system.

Hybrid Dynamical Systems (HDS) [2,9–11] appear as a tool that facilitates the modelling and control of physical production processes through the sequencing of modes of operation. Each operation mode is represented as a discrete state, and the global dynamics is represented as a jump sequence of operation modes.

The behaviour of a production unit is described as a hybrid system, which has its own monitoring mechanism. Each discrete state of the production unit has its own control law that allows it to stay in that state or reach a target state. In behavioral modelling, a discrete state has a set of constraints on inflows, a set of constraints on outflows, and a control law to achieve or maintain the goal. Each discrete state is an abstraction of the continuous behaviour of the controlled system, and it is necessary to have the event detection mechanisms that guarantee that the continuous dynamics is correctly mapped. Authors of

[11] stated that in the case of interconnected systems, the compound dynamics can be followed through abstractions if the requirements defined above are met. For a mode of operation of a subsystem the output ow must comply with a set of constraints that the receiving system of the ow knows in advance. The capacity to guarantee the restrictions implies that the system changes modes of operation and the receiver must also change modes of operation. The discretization of each subsystem is associated with its outputs. Each subsystem is discretized separately.

An extension of the definition from [10] for a controlled general hybrid dynamic system shown below is used to describe the HPU model:

$$H = (Q,\ X,\ U,\ \Gamma,\ Y,\ f,\ g_c,\ g_f,\ h_e,\ h_i,\ Init,\ Dom,\ E,\ G,\ R), where:$$

$Q \in Z^k$ is a set of *discrete states* and $X \in \Re^n$, is a set of variables for the *continuous state* ;
$U = U_c \cup U_p \in \Re^{m+l}$, is the set of *inputs* with, U_c *control inputs*, and U_p *physical inputs*;
$\Gamma \in \Re^p, \{\Gamma\}, i = 1, \cdots, p$ is a set of *parameters,* associated with the coefficients in both the models and the controllers;
$Y = Y_{in} \cup Y_e \in \Re^{s+r}$ are the *process outputs* where Y_e are *physical* and Y_{ind} are *indicators*;
$f(\cdot,\cdot) : Q \times X \times U \mapsto \Re^n$ is the *vector field*, where $g_c(\cdot)$ is the *control function* and $g_f(\cdot)$ is the *transformation function*;
$h_e(\cdot) : X \times U \mapsto Y_e$ is the *function of the process outputs*; $h_in(\cdot) : X \times U \mapsto Y_{in}$ is the *function of the management outputs*;
$Init \subseteq Q \times X$ is the set of *initial states*;
$Dom(\cdot) \ : \ Q \mapsto 2^X$ is the *domain*;
$E \subseteq Q \times Q$ is the set of *edges*;
$G(\cdot)$ is a *guard condition* and $R(\cdot,\cdot)$ is a *reset map.*

The cognitive model HPU, is comprised of the physical process model, the resource model and the product model. The hybrid model of the HPU, $H_{HPU} = (Q_{HPU},\ X_{HPU},\ U_{HPU},\ \Gamma_{HPU},\ Y_{HPU},\ f_{HPU},\ g_{c_{HPU}},\ g_{f_{HPU}},\ h_{e_{HPU}},\ h_{i_{HPU}},\ Init_{HPU}, Dom_{HPU}, E_{HPU}, G_{HPU}, R_{HPU})$ is obtained through the composition of the abstractions of the behaviour of the hybrid model, like the one defined by Tazaki and Imura (2008) [11]. This is a hybrid model of physical process (H_s), resources (H_r), products (H_p) and a unit management model given as discrete event system ($\sum_{HPU}$) which is a behavioural image of the process, including the start and end condition of a production goal ($Cond - Goal_{HPU}$), that is $H_{HPU}(Comp(\ H_s,\ H_r,\ H_p),\ \sum_{HPU},\ Cond - Goal_{HPU})$. The structure of the cognitive model (hybrid dynamic) in HPU is given in the Fig. 2. The cognitive model is stored on top of the holon and represents the DT of the physical process. The coordination model for a unit is achieved by composing the abstractions of the management model of each unit as in [11].

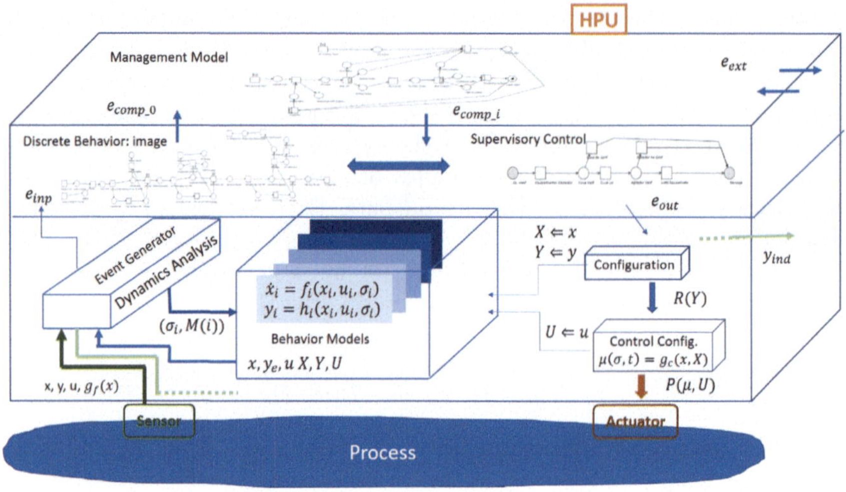

Fig. 2. Continuous and formal models: parametric, non-parametric, heuristic, stochastic.

3 Use Case: The Water Supply Systems and Its Components

The values used in this paper are taken from the WSS of Barquisimeto Edo. Lara Venezuela that supplies drinking water for human consumption (approx.900,000 inhabitants) and its region's industrial sector. The main source of supply is the Dos Cerritos reservoir, and a water treatment plant located in Quibor. Each link in the value chain for the WSS shown in Fig. 1 is an HPU according to the characteristics given in Sect. 2.

This work shows how, from a given physical model for the behaviour of the unit, a linguistic description of the operating conditions and flows of information required for interactions in the unit, the model holarchy is built as a hybrid system that describes the unit's knowledge model. In detail, this will show the conformation of the knowledge model (hybrid model) the impounding reservoir unit. A simulation of the interaction among units, created in the NetLogo tool [1,6], is displayed at the end of Sect. 4.1.

The HPU architecture here starts including the intelligent beings for the resource (RH), mission (MH), and supervision (SH) holons and adds then intelligent decision-makers (agents). Overall, agent-based software -NetLogo- creates a platform on which decision-makers are modelling their environment. HMS and multi-agent systems regularly are considered comparable theories by experts. Multi-agent systems, are a frequently encountered way to apply HMS; however, DT's services should be preferred over an intelligent agent's services by developers of HMS [12].

At the top level, the HMS differentiates intelligent beings (or DT) from agents. DT receives the dependability from realism (grouping various replicas or models), which they mirror, and holonic approach beings that reflect what exists in the world without imposing artificial limitations to this reality, i.e., farther away from classical modelling.

3.1 WSS Components

Impounding Reservoir

The Impounding Reservoir (IR) has an area of 100 ha, with an average capacity of 120 million m^3 and a maximum capacity of 169.8 million m^3. For simplicity, it is assumed a trapezoidal geometry, with a depth of 38 m, the suction head at 640 $masl$. Five capacities, with a height above sea level ($masl$) and storage capacity in millions of cubic meters (MM m^3) are described: (i) Bottom (sludge) at 634 $masl$ with 2 MM m^3, (ii) Critical at 644 $masl$ with 54.5 MM m^3, (iii) Minimum at 654 $masl$ with 71.74 MM m^3, (iv) Normal at 667 $masl$ with 127.11 MM m^3, (v) Maximum at 672 $masl$ with 169.8 MM m^3. Properties and conditions for the operation of the reservoir are shown in Fig. 3. The characteristics and operating conditions of the raw water entering the reservoir are the maximum ow rate of $6.3\,m^3/s$, and minimums of up to $2\,m^3/s$, the ecological ow (ambient return) is $1.18\ l/s$. The permitted water quality conditions are: turbidity < 50 Nephelometric Turbidity Unit or NTU (turbidimeter, max 1000 NTU), apparent colour < 100 Platinum-Cobalt Unit or PCU (spectrophotometer, max 500 PCU), pH (6.5–9) (ph-meter, range 0–14), determined by laboratory. For this case, the type of substrate, anions and cations present, as well as the dissolved oxygen are not included.

Table 1 shows the operating conditions for the development of the behaviour models (guard) for the management of the IR.

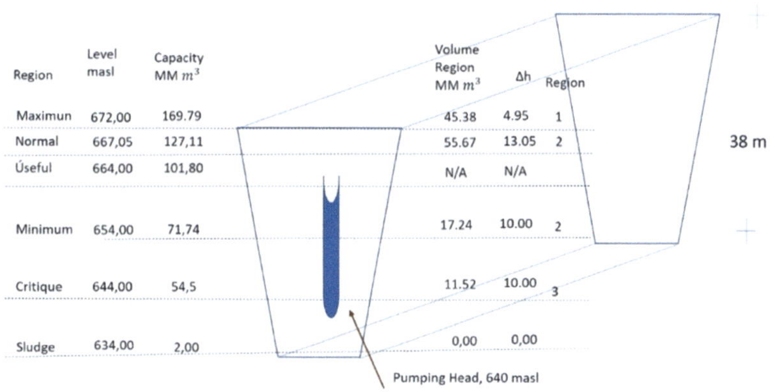

Fig. 3. Schematic for the reservoir

With previous knowledge of both quality and ow of the raw water, at the entrance of the water treatment plant, the Jar Test is carried out. This test consists of emulating (applying) the clotting, flocculation and sedimentation process to several samples of rough water (four usually) and provides the optimal ranges of the coagulant dose according to the pH of raw water in additional the operational conditions in each unit mentioned. Table 1 shows the operating conditions for the development of the behaviour models for the management of the IR.

Table 1. IR operational conditions

Var – Condition	Normal	Degraded	Failure
Q_i $flow(m^3/s)$	$4 < Q < 6.3$	$2 < Q < 4$	$Q < 2$
Level $masl$	$664 < C < 672$	$654 < C < 664$	$C > 672$ or $C < 654$
Turbidity	$T < 15$	$15 < T < 49$	$T > 50$
pH	$6.5 < pH < 8.5$	$5 < pH < 6.5$ $8.5 < pH < 9$	$pH < 5$ $pH > 9$

The reservoir behaviour model is obtained by the composition of three discrete events models described in Petri Nets: the first one corresponds to the behaviour of the entry gate, the second one is the reservoir water level and the last one is the pumping system. The entry gate model is established according to the following conditions: the reservoir level and the quality of the input raw water (see Table 1). For the reservoir level model, only three states are considered: high, medium and low regions in Fig. 3, and the pumping model is established according to the following conditions: inflow and reservoir level of the same figure. The composition of these models defines the reservoir behaviour, as can be seen in the composite model in Petri Nets in the figure. This model has four modes of operation, each of which has an associated differential equation as described in Sect. 2: filling and pumping $dh_i/dt = 1/\rho A(h_i)(q_{mi} - q_{m0})$, filling and no pumping $dh_i/dt = 1/\rho A(h_i)q_{mi}$, no filling and pumping $dh_i/dt = 1/\rho A(-q_{m0})$ and, no filling and no pumping $dh_i/dt = 0$) corresponding to the evolution given in Fig. 4

The model for the management of interactions between units, associated with the management model in Petri Nets establishes the condition of the unit (its state); this is an operating condition based on quantity and quality of stored raw water, quantity and quality of raw water reaching the reservoir and quantity of water towards purification. This model is presented in the right side of Fig. 4 and from now on, only the conditions for the operation of each of the units will be given that allow the construction of their dynamics at discrete events. For each unit a procedure equivalent to the one shown here must be followed to establish the knowledge model.

Transport and Water Purification Plant

Transport refers to the flow of the pumping system that is fixed at 4.5 m^3/s. Thus, depending on the conditions of the IR and the consumption requirements, in the planning activity, the number of pumpings by day and the pumping time (Tp) according to purification capacity are determined.

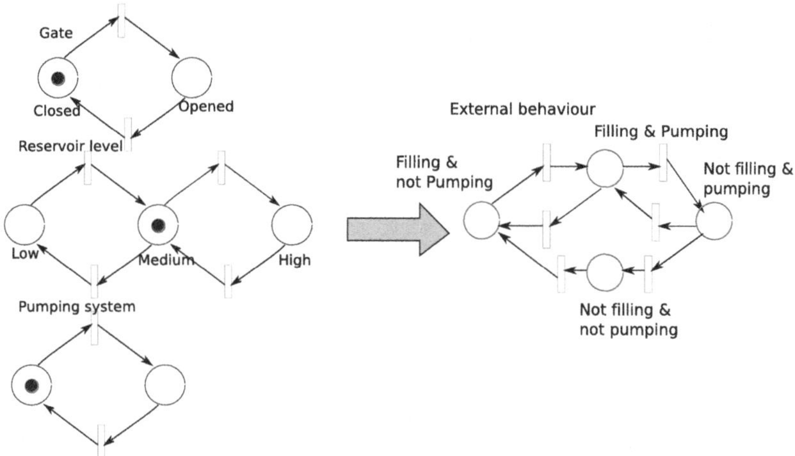

Fig. 4. Petri Nets for the Impond reservoir components and external behaviour

The water purification plant consists of transforming raw water, with turbidity and pH parameters given by laboratory test (see Table 1), into potable water with the following parameters: turbidity < 2, apparent colour < 15, and $6.5 < \text{pH} < 8.5$. The purification process contains six 6 phases (units): coagulation, flocculation, sedimentation, filtration, disinfection, and storage (service reservoir). Each unit performs tests to determine the quality of the water at its exit, as a part of an entire HMS. The next section presents the application of the HMS by the HPU concept for continuous production process. The structure of this section is two-folded. Firstly, it describes the water treatment plant used in this case study, and the application of the HPU concept for this scenario is presented. Secondly, it describes the agent-simulation of the water treatment process considering a set of scenarios for validating the HMS proposed.

Human Integration to WSS

Human integration into intelligent systems is crucial because, despite the automation level, human intervention gives it flexibility. From Fig. 5, a person role may be different depending on its location into the WSS. For this reason and taking as reference the integration and activities of human-in-the-mesh proposed by [8] that consider the A position, the human oversees the system and develops general planning. In the B position, the human analyses information changes and controls the systems. The C position is related to intervention and maintenance operations when it is necessary to adjust or repair a failure of a physical resource. In the last, D position the human can be considered a passive observer, as a user.

In this study, the human is in the A position because his intervention is decisive in understanding the WSS, proposing indicators, and getting knowledge of the system, especially in unexpected situations. Humans are creators of physical systems, managers, and users; therefore, some of their abilities are required such

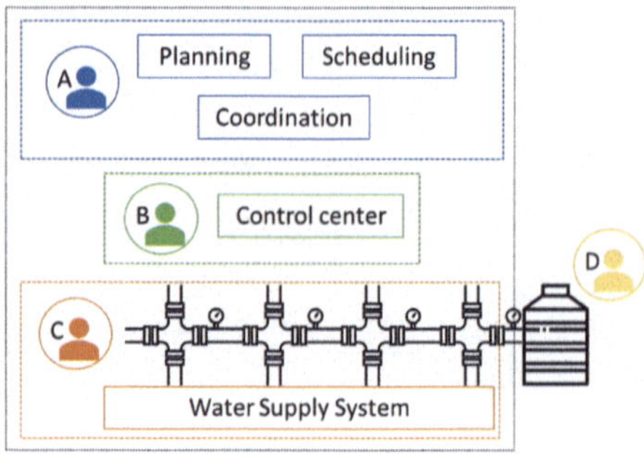

Fig. 5. Four types of human integrations for WSS (A to D)

as perception, cognition, learning, analysis, and decision-making [15]. Here the authors follow the human role as "master", where his creative and intellectual work purposing and driving the system is decisive for the WSS and the DT's conception. They are the creators of cyber systems, so the smartness of cyber systems, no matter how powerful, comes from humans [15].

4 DT of the HPU Simulation for Implementation

For validation of the proposed approach, the HPU concept and the test cases were implemented in a proof-of-concept software called NetLogo [1]. NetLogo is agent-based modelling made to explore and analyze the emergent behaviour of natural, engineering, and social phenomena [6]. Three testing scenarios are considered in the agent model to validate the results. These testing scenarios are conducted in order to demonstrate the vertical integration of the planning and execution process, the reactiveness given in the control task, and the pertinence of the HMS approach proposed for the continuous production processes.

4.1 HMS Architecture and the Simulation Scenarios

The case study presented concerns this paper is the purification of raw water in a water treatment plant. The purification of water, considered a continuous production process, is a complex process where the raw water passes throughout the water plant to be purified to improve quality (i.e., turbidity). However, this process is a complex one; it also has characteristics of a batch production process. In this case study, the water treatment process is composed of a reservoir, four treatment tanks, and a service reservoir. Figure 6 illustrates the simulation of the water treatment plant, the storage capacity of the reservoirs and tanks, the flow rates, and the minimum quality of water to pass to the next component. The tanks are connected through a pipeline that pumps the water between the different elements. The pumping of water starts in the reservoir, passes through the coagulation, flocculation, sedimentation, and filtration tanks of the water

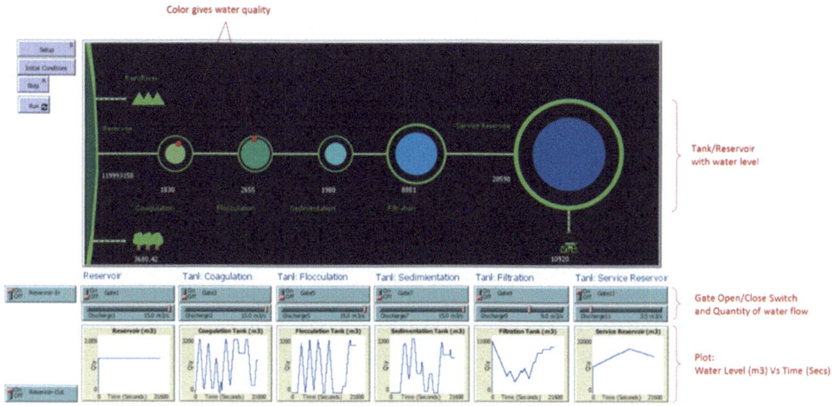

Fig. 6. NetLogo interface for the Reservoir HPU representation

treatment plant. It finishes in the service reservoir for distribution to human and industrial consumption.

The proposed HMS architecture is divided into three levels: The plant level, the image level, and the enterprise level. While the plant and image levels are responsible for the execution of the water treatment process, the enterprise level is responsible for the planning process. The plant-level contains the physical assets of the water treatment plant. In this paper, the physical assets are the physical function of the reservoir, the four treatment tanks, and the service reservoir. The image level contains the HPUs that, similarly to the DT concept, controls the function of the physical assets. The enterprise-level is oriented to evaluate the global performance of the execution, information transfer to the enterprise system, as well as acceptance/rejection of assigned objectives. For the components of the HMS architecture, six HPUs were created for each tank of the case study. As a general description, each HPU is responsible for monitoring events and triggering actions of the corresponding physical asset. On one side, the HPU monitors the water level of the tank/reservoir (measured in m^3), the flow in/out of the tank/reservoir (measured in m^3/s), and the quality of water (measured in NTU). On the other side, the holon supervisor of the HPU could trigger the tank/reservoir gates to the next physical asset, the quantity of chemicals for the purification process (i.e. flocculation, and chlorination) and the timing of each process in the tank, as shown in the next scenarios.

Validation of the HPU in a Normal Condition (Scenario 1)
The first testing scenario is related to the regular operation of the water treatment plant. This scenario is explored to test the proposed approach without any perturbation. It aims to understand the dynamics and normal functioning of the water treatment process. Furthermore, this scenario works as a control scenario for the other testing scenarios, and it will allow testing the reactiveness in case of a perturbation.

Disrupted Condition of the HPU-Water Demand (Scenario 2)
The second testing scenario concerns water demand where a perturbation is detected during the water treatment process. Specifically, the total demand of the water treatment plant increases more than usual surpassing the forecasting and corresponding threshold. This event is considered a perturbation for the normal conditions and requires that the control executed by the proposed HMS architecture reacts to the new situation. Therefore, this scenario aims to devise the functioning of the HMS architecture when it recovers from a disrupted scenario.

Disrupted Condition of the HPU-Quality of Water (Scenario 3)
The third testing scenario is related to the quality of water when a perturbation is detected during the water treatment process. In this case, the perturbation detected is related to the quality of water. As mentioned before, raw water from reservoirs contains a large number of particles that are cleaned in the water treatment process. This haziness of the fluid is measured in turbidity and represents a key test of the water quality. Normally, it is expected that the turbidity is between a low and a high accepted limit. However, depending on the rain, river ow, and reservoir environment, this turbidity could exceed the expected measure. Then, the processing time in the water treatment plant may increase accordingly. The perturbation in this scenario is that the turbidity of the raw water exceeds the turbidity expected. This scenario aims to explore the functioning of the HMS architecture considering that the processing time in the water treatment process (i.e. coagulation, flocculation, etc.) lasts longer than usual. Figure 7 shows an example of possible simulations using NetLogo of the scenarios above mentioned.

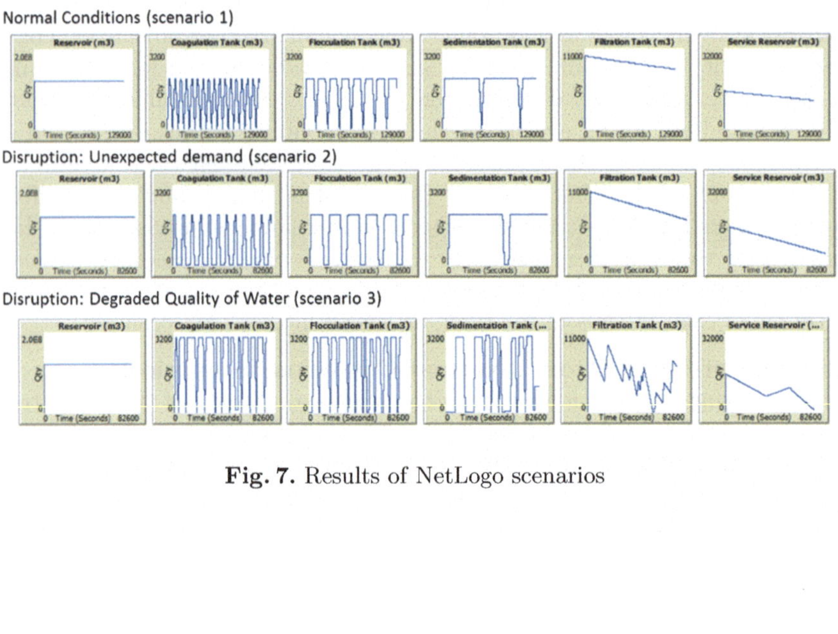

Fig. 7. Results of NetLogo scenarios

5 Conclusions and Future Work

Following the preliminary design in [4], the expanded HMS architecture for continuous processes is given. The HPU is based on the HMS concept and contains three fundamental holons (RH, MH, and SH) [5]. An additional holon, similar to the evolution of ADACOR in [3], allows the supervision and control tasks for each HPU. In this way, the HPUs can resolve unpredictable demands and implement internal fault tolerance behaviour. The mechanisms of high-level cooperation and the establishment of global objectives are achieved through elements that model the state of physical processes and allow maintaining a coherent operation between units. Cooperation between units is achieved by establishing viable global configurations and selecting the optimal one. Each unit is modelled separately, and the composition of models establishes possible operation configurations. This configuration guarantees the continuous operation of the system, even if this is not the best one. A simple DT validates this approach by developing a model based on agent-simulation implemented in NetLogo. The formation of individual behaviour models allows determining global behaviours through simulation. Future work will focus on integrating human-in-the-loop as the holon entity of the WSS.

References

1. Barbosa, J., Leitão, P.: Simulation of multi-agent manufacturing systems using agent-based modelling platforms. In: 2011 9th IEEE International Conference on Industrial Informatics, pp. 477–482. IEEE (2011)
2. Branicky, M.S., Borkar, V.S., Mitter, S.K.: A unified framework for hybrid control: model and optimal control theory. IEEE Trans. Autom. Control **43**, 31–45 (1998)
3. Cardin, O., Derigent, W., Trentesaux, D.: Evolution of holonic control architectures towards industry 4.0: a short overview. IFAC-PapersOnLine **51**(11), 1243–1248 (2018)
4. Chacón Ramírez, E., Albarrán, J.C., Cruz Salazar, L.A.: The control of water distribution systems as a holonic system. In: Borangiu, T., Trentesaux, D., Leitão, P., Giret Boggino, A., Botti, V. (eds.) SOHOMA 2019. SCI, vol. 853, pp. 352–365. Springer, Cham (2020). https://doi.org/10.1007/978-3-030-27477-1_27
5. Cruz Salazar, L.A., Rojas Alvarado, O., Carvajal, J.H., Chacón, E.: Cyber-physical system for industrial control automation based on the holonic approach and the IEC 61499 standard. In: Forum on Specification & Design Languages (2018)
6. Da Silva, R.M., Junqueira, F., Santos Filho, D.J., Miyagi, P.E.: Control architecture and design method of reconfigurable manufacturing systems. Control. Eng. Pract. **49**, 87–100 (2016)
7. Derigent, W., Cardin, O., Trentesaux, D.: Industry 4.0: contributions of holonic manufacturing control architectures and future challenges. J. Intell. Manuf. 1–22 (2020). https://doi.org/10.1007/s10845-020-01532-x
8. Fantini, P., et al.: Exploring the integration of the human as a flexibility factor in CPS enabled manufacturing environments: methodology and results. In: IECON 2016–42nd Annual Conference of the IEEE Industrial Electronics Society, pp. 5711–5716. IEEE, October 2016

9. Henzinger, T.A.: The theory of hybrid automata. In: Inan, M.K., Kurshan, R.P. (eds.) Verification of Digital and Hybrid Systems, pp. 265–292. Springer, Heidelberg (2000). https://doi.org/10.1007/978-3-642-59615-5_13
10. Lygeros, J., Sastry, S., Tomlin, C.: Hybrid Systems: Foundations, Advanced Topics and Applications. Springer, Heidelberg (2012)
11. Tazaki, Y., Imura, J.: Bisimilar finite abstractions of interconnected systems. In: Egerstedt, M., Mishra, B. (eds.) HSCC 2008. LNCS, vol. 4981, pp. 514–527. Springer, Heidelberg (2008). https://doi.org/10.1007/978-3-540-78929-1_37
12. Valckenaers, P.: Perspective on holonic manufacturing systems: PROSA becomes ARTI. Comput. Ind. **120**, 103226 (2020)
13. Water Supplies Department. Core businesses: Drinking water quality (2017). Accessed Jan 2019
14. World Economic Forum: Harnessing the fourth industrial revolution for water. Technical report, World Economic Forum (2018)
15. Zhou, J., Zhou, Y., Wang, B., Zang, J.: Human–cyber–physical systems (HCPSs) in the context of new-generation intelligent manufacturing. Engineering **5**(4), 624–636 (2019)

Digital Twins and Artificial Intelligence Applied in Industry and Services

Artificial Data Generation with Language Models for Imbalanced Classification in Maintenance

Juan Pablo Usuga-Cadavid[1,3](✉), Bernard Grabot[2], Samir Lamouri[1], and Arnaud Fortin[3]

[1] LAMIH UMR CNRS 8201, Arts et Métiers – Institute of Technology, Paris, France
{juan_pablo.usuga_cadavid,samir.lamouri}@ensam.eu
[2] LGP, INP/ENIT, Tarbes, France
bernard.grabot@enit.fr
[3] iFAKT France SAS, Toulouse, France
a.fortin@ifakt.com

Abstract. Harnessing data that comes from maintenance logs may help improving production planning and control in manufacturing companies. However, maintenance logs can contain highly unstructured text data, presenting imbalanced distributions. This hinders the training of Machine Learning (ML) models, as they tend to poorly perform when identifying the underrepresented classes. Thus, this study uses a recent language model called GPT-2 to generate artificial maintenance reports. These artificial samples are employed to mitigate the class imbalance when training a Deep Learning (DL) architecture named CamemBERT. To carry out the experiments, an industrial dataset is used to train eleven DL models with different approaches to tackle class imbalance. Findings suggest that mixing random oversampling with artificial samples improves the performance of classifiers when trained on imbalanced datasets. Finally, results imply that using nucleus sampling when generating artificial text sequences with language models ameliorates the quality of produced data.

Keywords: Natural language processing · Language model · Maintenance · Deep learning · Class imbalance · Artificial data · Industry 4.0

1 Introduction

Valuing data coming from maintenance logs may provide several advantages to performing better production planning and control. For instance, by adapting a production schedule to unexpected disturbances, the engaged delivery dates can still be respected [1]. Machine Learning (ML) has been extensively used in production planning and control research to improve manufacturing systems in the framework of Industry 4.0 [2]. In fact, ML offers a way to harness data from diverse sources such as information systems, equipment sensors, products, customers, etc. to support decision making [2, 3].

Despite the potential advantages provided by ML, the quality of the learning process strongly depends on the dataset employed. In applications such as fraud detection, disease

D. Trentesaux et al. (Eds.): SOHOMA 2021, SCI 987, pp. 57–68, 2021.
https://doi.org/10.1007/978-3-030-80906-5_5

diagnosis or image recognition, data distributions may be strongly skewed towards one of the classes [4]. For example, in the case of a rare disease diagnosis, there will be few examples of patients having a certain disease compared to the number of healthy patients. This naturally induced class imbalance is denominated *intrinsic imbalance*, conversely to *extrinsic imbalance* which occurs when the imbalance is artificially introduced by external factors [4]. Maintenance logs can also present intrinsic imbalance. For example, few issues will lead to a halt in the production process, while the vast majority will not cripple it.

Class imbalance may strongly hurt the performance of ML models, as the learning process tends to be disproportionately influenced by the Overrepresented Class (OC). Thus, the model fails to correctly detect the Underrepresented Class (UC) in most of the cases. This may be unacceptable in some contexts, where not identifying the UC can lead to severe consequences. For instance, not detecting that a production problem will cripple the production line can strongly disrupt the manufacturing process.

Maintenance logs normally contain free-form text data manually provided by technicians. These reports describe the symptoms of events like machine breakdowns and provide guidance to understand the issue. Nevertheless, even if the textual reports encapsulate meaningful information to train ML models, they are highly unstructured: they commonly contain typos, abbreviations, and they may be strongly influenced by jargon. Hence, this research focuses on the use of a recent language model called GPT-2 [5] to generate artificial descriptions of maintenance reports leading to a production halt. The objective will be to use these artificially generated reports to reduce the effect of class imbalance when training a state-of-the-art Deep Learning (DL) model. Such a model will seek to determine whether a maintenance report corresponds to an issue that blocks the production. The task of classifying maintenance reports from their description will be handled as a classification problem in supervised learning. Following the nomenclature used by [6], problems that stop the production process are named dominant disturbances, while others are called recessive disturbances.

The remainder of this article is organized as follows: Sect. 2 provides details about the necessary background and related work. Section 3 presents the employed dataset, tested techniques, and training policies. Section 4 presents the results and discussion. Finally, Sect. 5 concludes this study and provides perspectives on future work.

2 Background and Related Work

2.1 Background

Handling Class Imbalance with Data-level Techniques. According to [4], the techniques that mitigate the effect of class imbalance can be grouped into three categories: data-level, algorithm-level, and hybrid approaches. Data-level techniques modify the training set distribution to reduce the level of imbalance. Algorithm-level methods modify the way ML algorithms perform learning by, for instance, assigning a higher importance to the UCs. Finally, hybrid approaches combine the latter two strategies. This study will focus on the data-level approach, leaving the other two for future research.

Two common techniques employed in the data-level approach are Random Over-Sampling (ROS) and Random Under-Sampling (RUS). ROS randomly resamples the set of UCs with replacement until the training set is nearly balanced. Conversely, RUS randomly removes observations from the set of OCs until achieving balance.

Both ROS and RUS have been extensively compared in the scientific literature. Nevertheless, no sampling method is guaranteed to perform best across all of the domains [4]. In fact, each method has its own advantages and shortcomings: while ROS has proven to better mitigate a class imbalance, it may greatly increase the requirements in terms of computing power and memory usage due to an increase in data. Additionally, it may cause overfitting to the oversampled classes [7]. On the other hand, RUS has outperformed ROS in some scenarios and reduces the training time, but it may discard meaningful information in the training set when excluding observations.

Other techniques such as SMOTE [8] and data augmentation [9] focus on generating artificial samples for the UC instead of resampling from the already existing observations. They have proven to greatly improve the performance of ML algorithms, especially of DL models, which are prone to overfitting.

As stated by [4], most of the research that has been done on DL with a class imbalance has targeted Convolutional Neural Networks (CNNs) and image data for computer vision applications. Thus, this research focuses on the use of data-level approaches to tackle class imbalance in the field of Natural Language Processing (NLP) with DL. More specifically, this is done through the use of recent transformed-based models using attention mechanisms [10], which have greatly improved the state of the art in NLP.

Transformer-Based Architectures in NLP. When working in NLP, choosing how to vectorize text inputs into numeric representations exploitable by ML is important. Since their introduction in 2013 [11], word embeddings obtained through models such as Word2Vec, GloVe or Fasttext have been used extensively. Word embeddings are vector representations of text obtained, for instance, through neural networks. These vectors have improved the state of the art in NLP with respect to older techniques that rely on weighting strategies such as TF-IDF.

Despite the advantages provided by approaches such as Word2Vec, the vectors produced are non-contextualized embeddings. This means that the polysemy of words is ignored. Put differently, a certain word will have the same vector representation no matter its usage, which may be harmful for terms whose meaning depends on context.

To solve this, DL models relying on attention mechanisms like ELMo [12], BERT [13], and GPT-2 [5] have been developed. These architectures are normally called transformed-based models. They are normally pre-trained on several gigabytes of text data to learn meaningful feature representations of language. Also, they produce contextualized word embeddings and use more robust tokenization strategies, such as Byte-Pair encoding [14] which may better handle typos, acronyms, or abbreviations. Thus, this research will focus on these architectures. More specifically, on GPT-2 and a modified version of BERT adapted to French, which is called CamemBERT [15].

GPT-2 is a language model: it can be employed, among other tasks, to probabilistically generate the next words from a given text input. Hence, it will be used to artificially generate maintenance log reports describing problems blocking the production process and reduce the class imbalance. Once these artificial texts are generated, the CamemBERT model will be trained to perform classification. The aim will be to classify whether a description of an issue will lead to a halt in the production.

2.2 Related Work

Mitigating class imbalance problems when training ML and DL models have been an important yet understudied topic in recent research [4, 16]. Most scientific production in the domain has focused on CNNs and image data, leaving significant research gaps regarding other DL architectures and data types [4]. Such is the case for NLP. This section summarizes related work concerning the use of NLP in datasets containing class imbalance. For each study, the imbalance ratio ρ, as used in [4], is provided. This ratio is estimated as presented in Eq. 1. If several datasets were used, the highest imbalance ratio is reported.

$$\rho = \frac{max_i\{|C_i|\}}{min_i\{|C_i|\}} \tag{1}$$

In Eq. 1, C_i is the whole set of observations of class i. Thus, $max_i\{|C_i|\}$ and $min_i\{|C_i|\}$ represent the maximum and minimum class sizes, respectively. For instance, if the largest class has 10000 observations and the smallest has 10 observations, $\rho = 1000$.

In the context of social network security, [17] is focused on the task of recognizing bots on Twitter. As the number of bots is fewer than the number of human accounts, the training set presented a class imbalance of $\rho \approx 4.3$. To tackle this imbalance, a data-level approach was used: a modified generative adversarial neural network [18] was employed to produce artificial observations and train a neural network. The approach outperformed other data-level techniques such as ROS, SMOTE, and ADASYN.

In the field of biomedical research, [19] explored the task of classifying texts containing descriptions about drug-drug pairs, drug-adverse effects pairs and drug-disease pairs, which is a multi-class classification problem. To mitigate the class imbalance ($\rho \approx 26.3$), authors used a data-level approach, i.e. SMOTE, and different corpora from different data sources to train a CNN.

Regarding software development, [20] targeted the bug severity prediction from text reports. The dataset used seven contained levels of bug severity and presented an imbalance of $\rho \approx 45.5$. Authors employed an algorithm-level approach to reduce the disparities between class-sizes and train several FastText [21] classifiers: they developed a hierarchical tree-like architecture to train several binary models. The first model was trained on the largest class versus the other classes altogether. Then, after discarding the largest class, the second model was trained on the second largest class, versus the remaining classes. This process was repeated until only two classes were left. This approach was compared with a standard training, resulting in similar performance.

In the study performed by [22], the aim was to identify the most important factors contributing to the perception of *quality* for a brand. To achieve this, they employed a logistic regression to classify companies between *top brands*. As top brands were less frequent, the imbalance level was $\rho \approx 7.2$, which was corrected through RUS.

Finally, [7] assessed the use of two new loss functions to train deep neural networks in imbalanced datasets: the mean false error and mean squared false error. This algorithmic-level approach was then tested on several datasets of both image and text data containing different levels of imbalance. Regarding NLP, the most severe case concerned document classification for the Newsgroup dataset with an imbalance level of (nearly) $\rho = 20$. Carried experiments compared the performance of neural networks trained with the proposed loss functions to models trained with the mean squared error. Findings suggested that using the new loss functions produced better results.

Despite the recurrent use of data-level approaches, no other study has employed language models to generate artificial text samples and to reduce class imbalance. Furthermore, transformed-based models to perform classification were not used either. To the best of the authors' knowledge, this is the first study using transformed-based models to both generate and classify maintenance logs containing free-form text descriptions.

3 Methods and Materials

3.1 Employed Dataset

The employed dataset comes from the maintenance logs of a company whose industry and name will not be mentioned for confidentiality purposes. Each maintenance log contained the description of the symptoms, the name of the equipment concerned, the importance level of the equipment, and the type of disturbance (recessive or dominant). From these inputs, the equipment name and symptoms are free-text comments, which means that two technicians reporting the same problem on the same machine may not produce the same description. Finally, the importance level was a categorical variable containing three possible values: "essential", "important", and "secondary". The initial dataset contained around 26000 observations. After cleaning the data, 22709 records were kept.

As transformed-based models can handle unstructured text sequences including typos, abbreviations, etc., the choice was to create two new variables (i.e. *Text seed* and *Issue description*) by concatenating the already existing inputs:

1. Text seed: this variable concatenates the equipment name and importance level. It will be used as text seed to generate the artificial samples with the language model.
2. Issue description: this variable concatenates the equipment name, importance level, and symptoms. It will be used to predict the type of disturbance with CamemBERT.

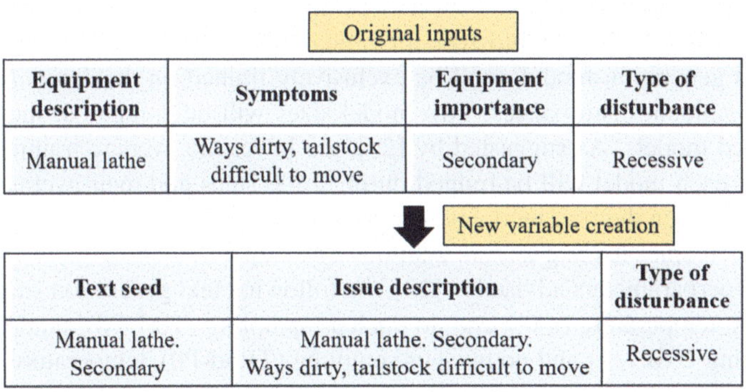

Fig. 1. Creation of the Text seed and Issue description from initial variables

Figure 1 illustrates the variables created through a toy example. Finally, the imbalance level between recessive and dominant disturbances is $\rho \approx 10.6$, being the recessive disturbances largest class.

3.2 Techniques Tested

The proposed method in this study has two main modules: a data generation module and an issue classification module.

The data generation module will use GPT-2, which is a recent language model proposed and pre-trained on around 40GB of text by the authors of [5]. There are four model sizes available: small (124M parameters), medium (355M parameters), large (774M parameters), and extra-large (1558M parameters). For this study, the small and medium architectures will be fine-tuned, compared, and the best one will be selected. The library employed to use GPT-2 is the one proposed by [23].

To generate the artificial maintenance descriptions, two main hyperparameters were explored: the temperature and nucleus sampling. The temperature determines how much randomness will be introduced into the language model choices: the higher the temperature, the higher the randomness. Hence, language models with higher temperatures will tend to create more *creative* text sequences. Nucleus sampling, proposed by [24], helps avoid generating incoherent words by setting a threshold of P. Hence, the cumulative probability distribution is computed for all of the tokens, starting with the most likely ones. After it reaches P, all the of other tokens, which are less likely to be generated, are discarded.

The issue classification module will use CamemBERT, a transformed-based model inspired from the RoBERTa architecture [25]. CamemBERT was pre-trained by the authors of [15] on 138GB of uncompressed text in French employing several GPUs for 17 h. The implementation uses the subclass of CamemBERT, called *CamemBertForSequenceClassification* available in [26], and the code was based on the example proposed by [27]. The following subsection details the training policies of each module.

3.3 Training Policies

First, the initial dataset is split into 75% for training and the remainder for the test. Then, the training set is further split into 10% for validation and 90% for actual training.

The data generation module will be exclusively trained on the *Issue descriptions* leading to dominant disturbances. Two model sizes will be compared: the small and medium sized models. As suggested by [23], the lower the average training loss, the better. Thus, each model will be trained during 2400 steps and their average loss will be compared. When the best model is chosen, it will use the *Text seeds* to generate the artificial text samples leading to dominant disturbances.

Using hyperparameters advised in [23], the following text generation strategies will be employed: temperature of 0.7 and no nucleus sampling (T0.7-P0), random temperature following $U(0.7, 1)$ and no nucleus sampling (TRnd-P0), temperature of 0.7 and nucleus sampling with a threshold of 0.9 (T0.7-P0.9), and random temperature following $U(0.7, 1)$ and nucleus sampling with a threshold of 0.9 (TRnd-P0.9).

The issue classification module will be trained on the training set balanced through the four following strategies: ROS, RUS, artificial data coming from each of the four text generation strategies, and 50% of ROS plus 50% of artificial data. Furthermore, a model trained on the training set with no modifications will be also assessed. The validation set will serve to fine tune the hyperparameters of each model and to select the best one. Then, the best model will be retrained by mixing the training and validation sets and by following the best class balancing strategy. Finally, its performance will be measured with the test set. The eleven models that will be compared are summarized in Fig. 2.

For comparison purposes, the Matthews Correlation Coefficient (MCC) will be used. Recent research has suggested that the F1-score may not be suitable to assess the quality of classifiers in imbalanced datasets [28]. Instead, the MCC is preferred [29]. The MCC ranges from −1 to 1, where 1 represents a perfect classifier.

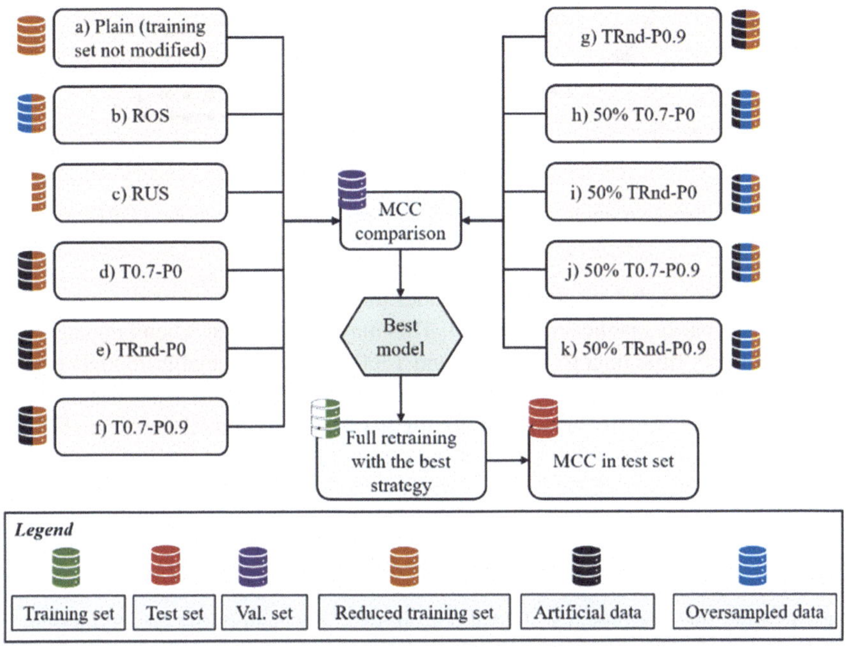

Fig. 2. Training policy for the issue classification module

4 Results

The models were trained using a GPU Tesla P100-PCIE-16GB. GPT-2 was trained using TensorFlow, while the CamemBERT models used PyTorch. As using GPUs introduces randomness, the experiments were run several times: five times for each of the two GPT-2 models and 20 times for each of the CamemBERT models.

4.1 Results for the Data Generator Module with GPT-2

Figure 3 shows the mean of the average training loss across all five runs and the 2400 training steps for the small and medium model.

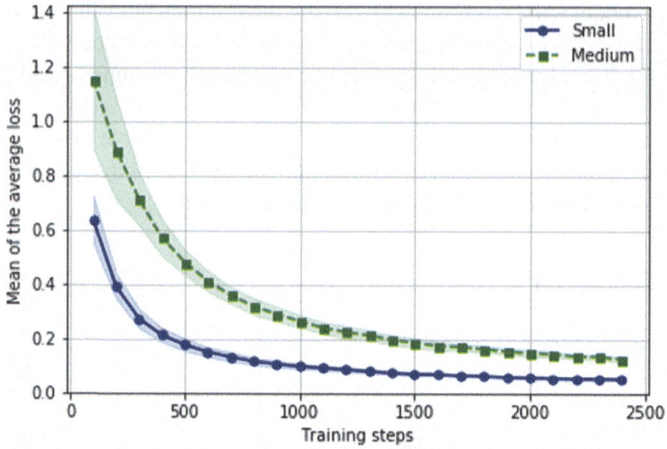

Fig. 3. Mean average training loss for the small (blue) and medium (green) GPT-2 model

Findings suggest that the language model that better fits the text descriptions for dominant disturbances is the small model. This may indicate that when language models have relatively little data to learn (around 1300 examples), smaller models perform better. Thus, the small architecture was used to generate the artificial data for the following experiments. Finally, it is worth noting that the GPT-2 model used from [23] was originally designed for English. Thus, the generated texts dropped all of the uniquely French characters.

4.2 Results for the Issue Classification Module with CamemBERT

Figure 4 shows the box plots for the validation MCC for each of the eleven models.

Table 1 provides further detail on the results, including more common metrics. It presents the average accuracy (Acc.), specificity (S1), sensitivity (S2), and MCC. In this case, the S1 and S2 measure the percentage of recessive and dominant disturbances that were correctly classified, respectively. For each measure, the highest value is highlighted in bold. Models are displayed by decreasing MCC.

Results suggest that using the 50%T0.7-P0.9 (j) approach increases the average MCC. This approach mixed 50% of the resampled observations of the UC with 50% of the artificial examples generated with a stable temperature and nucleus sampling. With respect to the plain model (a), the sensitivity is greatly improved, which means that more dominant disturbances are correctly detected. When compared to ROS (b), results are similar in terms of specificity and sensitivity. However, model j globally improves the results of the classifier, which are observed through a superior MCC.

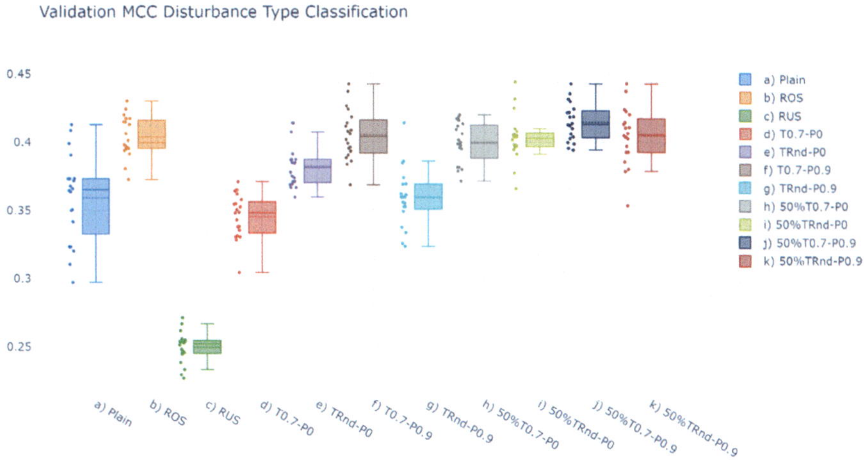

Fig. 4. Validation MCC for the eleven CamemBERT models

Table 1. Average validation accuracy, specificity, sensitivity, and MCC

Model name	Acc.	S1	S2	MCC
j) 50%T0.7-P0.9	0.902	0.930	0.554	**0.415**
f) T0.7-P0.9	0.927	0.972	0.373	0.405
k) 50%TRnd-P0.9	0.895	0.921	0.571	0.405
b) ROS	0.895	0.922	0.565	0.404
i) 50%TRnd-P0	0.901	0.931	0.537	0.403
h) 50%T0.7-P0	0.903	0.934	0.522	0.400
e) TRnd-P0	0.920	0.962	0.394	0.382
g) TRnd-P0.9	0.918	0.963	0.365	0.359
a) Plain	**0.931**	**0.985**	0.251	0.359
d) T0.7-P0	0.919	0.966	0.339	0.346
c) RUS	0.710	0.708	**0.734**	0.250

The results obtained with RUS (c) yielded the best sensitivity, meaning that this is the best model to detect dominant disturbances. Nevertheless, the information loss produced by excluding observations severely penalizes the global performance of the classifier. This also means that the model will fail to detect more recessive disturbances in maintenance, which is not advantageous, either.

The fact that the Plain model (a) achieves the best accuracy and specificity shows why these measures are not well suited to evaluate ML models in imbalanced datasets: the classifier will mainly learn the OC, which will boost its accuracy, even if it has bad performance when detecting the UC.

The findings indicate that nucleus sampling is beneficial to generate meaningful artificial samples. In fact, three out of four models using it achieved good performance, reaching the top 3 MCCs among all of the models. Finally, the fact that employing artificial samples achieved good results even when using a GPT-2 model that was not adapted to French suggests that further improvements could be done with this technique.

The performance of model 50%T0.7-P0.9 (j) is then assessed using the test set. Results are shown in Table 2.

Table 2. Average test accuracy, specificity, sensitivity, and MCC

Model name	Acc.	S1	S2	MCC
j) 50%T0.7-P0.9	0.890	0.920	0.566	0.419

Model j performance in the test set is close to the one presented in the validation set. This suggests that CamemBERT did not overfit the training data and could generalize to the task. This validates the performance of the proposed approach.

5 Conclusion and Future Work

This study explored the use of language models to artificially generate maintenance descriptions and reduce the class imbalance problem when classifying between dominant and recessive disturbances in an industrial dataset. The approach used two state-of-the-art models in NLP: GPT-2 and CamemBERT. GPT-2 was employed to generate the artificial data, while CamemBERT was trained as a classifier to detect whether a maintenance issue would block the production process by analyzing its description. Two versions of GPT-2 were compared: a small and a medium version. The former provided better training performance. Also, the influence of the temperature and nucleus sampling when generating the artificial samples with GPT-2 was assessed. Results suggested that employing nucleus sampling improves the quality of the generated data.

Regarding CamemBERT, the best model was achieved by reducing the class imbalance with a mixture of real and artificial data. Such data was generated by keeping a constant temperature of 0.7 and using a threshold for nucleus sampling equal to 0.9. Test performance validated the results and suggested that there was no apparent overfitting.

Future work will focus on four key aspects: first, the proposed approach is to be compared with algorithmic-level techniques, as increasing the amount of data may not be suitable for applications using massive datasets. In fact, such techniques may further improve the results without increasing the data volume. Secondly, the mix between real and artificial data was arbitrarily set to 50% in this study. This is to be analysed to find relatively good values for this mix. Thirdly, the approach is to be validated using several industrial datasets. Finally, using a version of GPT-2 adapted to French may increase the effectiveness of the approach. This will be further explored in future work.

References

1. Usuga Cadavid, J.P., Lamouri, S., Grabot, B., et al.: Estimation of production inhibition time using data mining to improve production planning and control. In: 2019 International Conference on Industrial Engineering and Systems Management (IESM), Shanghai, China, pp. 1–6 (2019)
2. Usuga Cadavid, J.P., Lamouri, S., Grabot, B., Pellerin, R., Fortin, A.: Machine learning applied in production planning and control: a state-of-the-art in the era of industry 4.0. J. Intell. Manuf. **31**(6), 1531–1558 (2020). https://doi.org/10.1007/s10845-019-01531-7
3. Tao, F., Qi, Q., Liu, A., Kusiak, A.: Data-driven smart manufacturing. J. Manuf. Syst. **48**, 157–169 (2018). https://doi.org/10.1016/j.jmsy.2018.01.006
4. Johnson, J.M., Khoshgoftaar, T.M.: Survey on deep learning with class imbalance. J. Big Data **6**(1), 1–54 (2019). https://doi.org/10.1186/s40537-019-0192-5
5. Radford, A., Wu, J., Child, R., et al.: Language models are unsupervised multitask learners, Corpus ID: 160025533. Semantic Scholar (2018)
6. Wang, C., Jiang, P.: Manifold learning based rescheduling decision mechanism for recessive disturbances in RFID-driven job shops. J. Intell. Manuf. **29**(7), 1485–1500 (2016). https://doi.org/10.1007/s10845-016-1194-1
7. Wang, S., Liu, W., Wu, J., et al.: Training deep neural networks on imbalanced data sets. In: Proceedings of the International Joint Conference on Neural Networks, Vancouver, BC, Canada, pp. 4368–4374. Institute of Electrical and Electronics Engineers Inc. (2016)
8. Chawla, N.V., Bowyer, K.W., Hall, L.O., Kegelmeyer, W.P.: SMOTE: Synthetic minority over-sampling technique. arXiv e-prints (2011). arXiv:1106.1813
9. Shorten, C., Khoshgoftaar, T.M.: A survey on image data augmentation for deep learning. J. Big Data **6**(1), 1–48 (2019). https://doi.org/10.1186/s40537-019-0197-0
10. Vaswani, A., Shazeer, N., Parmar, N., et al.: Attention is all you need. In: Advances in Neural Information Processing Systems, 30. Long Beach, CA, USA, pp. 5998–6008. Curran Associates, Inc. (2017)
11. Mikolov, T., Sutskever, I., Chen, K., et al.: Distributed representations of words and phrases and their compositionality. In: Proceedings of the 26th International Conference on Neural Information Processing Systems. Red Hook, NY, USA, vol. 2, pp. 3111–3119. Curran Associates Inc. (2013)
12. Peters, M.E., Neumann, M., Iyyer, M., et al.: Deep contextualized word representations. arXiv e-prints (2018). arXiv:1802.05365
13. Devlin, J., Chang, M.-W., Lee, K., Toutanova, K.: BERT: Pre-training of deep bidirectional transformers for language understanding. arXiv e-prints (2018). arXiv:1810.04805
14. Sennrich, R., Haddow, B., Birch, A.: Neural machine translation of rare words with subword units. arXiv e-prints (2015). arXiv:1508.07909
15. Martin, L., Muller, B., Ortiz Suárez, P.J., et al.: CamemBERT: a tasty french language model. arXiv e-prints (2019). arXiv:1911.03894
16. Masko, D., Hensman, P.: The impact of imbalanced training data for convolutional neural networks. Corpus ID: 46063904, KTH Royal Institute of Technology, Semantic Scholar (2015)
17. Wu, B., Liu, L., Yang, Y., et al.: Using improved conditional generative adversarial networks to detect social bots on Twitter. IEEE Access **8**, 36664–36680 (2020). https://doi.org/10.1109/ACCESS.2020.2975630
18. Goodfellow, I.J., Pouget-Abadie, J., Mirza, M., et al.: Generative adversarial networks. arXiv e-prints (2014). arXiv:1406.2661

19. Deepika, S.S., Saranya, M., Geetha, T.V.: Cross-corpus training with CNN to classify imbalanced biomedical relation data. In: Métais, E., Meziane, F., Vadera, S., Sugumaran, V., Saraee, M. (eds.) NLDB 2019. LNCS, vol. 11608, pp. 170–181. Springer, Cham (2019). https://doi.org/10.1007/978-3-030-23281-8_14
20. Nnamoko, N., Cabrera-Diego, L.A., Campbell, D., Korkontzelos, Y.: Bug severity prediction using a hierarchical one-vs.-remainder approach. In: Metais E., Meziane F., (eds.) NLDB 2019. LNCS, vol. 11608, pp. 247–260. Springer, Cham, (2019). https://doi.org/10.1007/978-3-030-23281-8_20
21. Bojanowski, P., Grave, E., Joulin, A., Mikolov, T.: Enriching word vectors with subword information. arXiv e-prints (2016). arXiv:1607.04606
22. Kato, T., Tsuda, K.: A management method of the corporate brand image based on customers' perception. Procedia Comput. Sci. **126**, 1368–1377 (2018). https://doi.org/10.1016/j.procs.2018.08.088
23. Woolf, M.: gpt-2-simple (2019). https://github.com/minimaxir/gpt-2-simple. Accessed 30 Apr 2020
24. Holtzman, A., Buys, J., Du, L., et al.: The curious case of neural text degeneration. arXiv e-prints (2019). arXiv:1904.09751
25. Liu, Y., Ott, M., Goyal, N., et al.: RoBERTa: a robustly optimized bert pretraining approach. arXiv e-prints (2019). arXiv:1907.11692
26. Huggingface: Transformers: State-of-the-art Natural Language Processing for TensorFlow 2.0 and PyTorch, (2019). https://github.com/huggingface/transformers. Accessed 1 Feb 2020
27. McCormick, C., Ryan, N.: BERT Fine-Tuning Tutorial with PyTorch, (2019). https://mccormickml.com/2019/07/22/BERT-fine-tuning/. Accessed 1 Feb 2020
28. Hand, D., Christen, P.: A note on using the F-measure for evaluating record linkage algorithms. Stat. Comput. **28**(3), 539–547 (2017). https://doi.org/10.1007/s11222-017-9746-6
29. Delgado, R., Tibau, X.-A.: Why Cohen's Kappa should be avoided as performance measure in classification. PLoS ONE **14**, 1–26 (2019). https://doi.org/10.1371/journal.pone.0222916

Machine Vision for Collaborative Robotics Using Synthetic Data-Driven Learning

Juan Camilo Martínez-Franco(✉) and David Álvarez-Martínez

Universidad de los Andes, Bogotá, Colombia
{jc.martinez10,d.alvarezm}@uniandes.edu.co

Abstract. This paper presents a deep learning approach based on synthetic data for computer vision training and motion planning algorithms to be used in collaborative robotics. The cobot is in this case part of fully automated packing and cargo loading systems that must detect items, estimate their pose in space to grasp them, and create a collision-free pick and place trajectory. Simply recording raw data form sensors is typically insufficient to obtain an object's pose. Specialized machine vision algorithms are needed to process the data, usually based on learning algorithms that depend on carefully annotated and extensive training datasets. However, procuring these datasets may prove expensive and time-consuming. To address this problem, we propose the use of synthetic data to train a neural network that will serve as a machine vision component for an automated packing system. We divide the problem into two steps: detection and pose estimation. Each step is performed with a different convolutional neuronal network configured to complete its task without excessive computing complexity that would be required to perform them simultaneously. We train and test both networks with synthetic data from a virtual scene of the workstation. For the detection problem, we achieved an accuracy of 99.5%. For the pose estimation problem, a mean error for the centre of the mass of 17.78 mm and a mean error for orientation of 21.28° were registered. Testing with real-world data remains pending, as well as the use of other network architectures.

Keywords: Synthetic data · Machine vision · One-class classification · Collaborative robotics · Pose estimation

1 Introduction

Collaborative robots (cobots) present novel opportunities and challenges compared to their industrial counterparts. Sharing the same workspace as humans allows for extreme adaptability to a multitude of tasks with little programming required. However, this versatility comes with control and safe operation challenges. To address them, it has become commonplace to make use of machine learning techniques.

D. Trentesaux et al. (Eds.): SOHOMA 2021, SCI 987, pp. 69–81, 2021.
https://doi.org/10.1007/978-3-030-80906-5_6

These techniques typically rely on annotated datasets with sizes ranging from hundreds to millions of samples. This becomes a problem when the amount of applicable data is reduced. For instance, one could have a dataset consisting of 200 everyday objects photographed from different angles along with their respective position relative to the camera. However, it is possible that only one or two of the objects will be similar to the items that will be handled by a cobot or perhaps none of the objects resemble in a meaningful way the items that will be handled. In this case, a new dataset would have to be introduced, but this process can be time consuming or financially expensive (usually both).

A possible solution for this situation is the generation of synthetic data. The recreation of sample physical workspace in a virtual environment can provide extremely accurate data points. Rotation matrices, position vectors, and camera parameters can be quickly iterated, and the resulting geometry can be rendered to generate a robust, synthetic dataset for computer vision. Physics simulations can be run several times faster than the real world interactions they represent, providing large amounts of data for motion planning strategies.

1.1 Experimental Objective

The goal of this work is to introduce a novel method for the training of a machine vision system that can detect relevant objects in a workspace (captured in the form of RGB-D images) and to estimate the 6DOF pose of these objects. The novel method will consist in the generation of a synthetic dataset of depth images obtained through the rendering pipeline of a popular video game engine. We will explore the use of such a synthetic dataset to train multiple neural networks for specific tasks in a collaborative robotics context.

2 Background

2.1 Collaborative Robots

Collaborative robots are a type of manipulators that are used in hybrid automation tasks, programmable in three different axes of movement. They differ from industrial manipulators in that they share a physical space and operate in tandem with human workers. The degree to which this collaboration occurs involves the use of techniques described in the still-evolving standard ISO/TS 15066 [13]:

- Safety-rated monitored stop (see Fig. 1)
- Hand-guiding operation
- Speed and separation monitoring
- Power and force limiting

Of these techniques, speed and separation monitoring can use a computer and machine vision techniques where image data is processed in such a way that the geometry of a scene is analyzed and separated to identify tools, workers and obstacles.

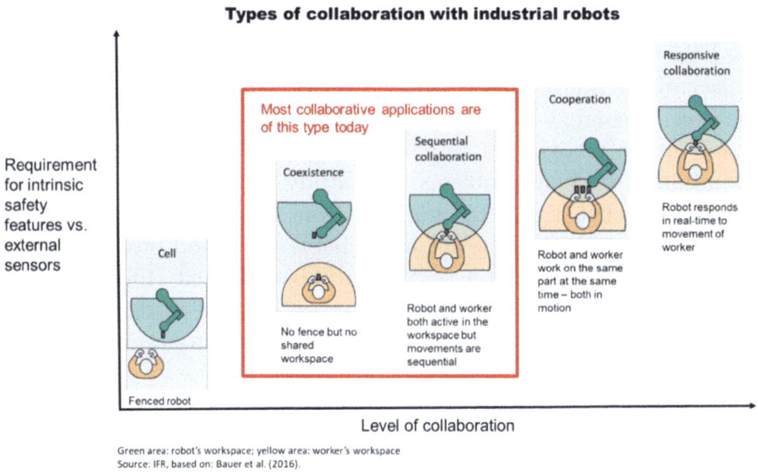

Fig. 1. Cobot saftey features. The higher the level of collaboration, the more complex and crucial it becomes to ensure safety [2]

2.2 Machine Vision

Object detection in computer and machine vision may be modelled as a multi-class classification problem (object is detected vs. object is not detected), or as a one-class classification problem (object is detected vs. null). The former presents an issue that is hard to overcome: extreme similarity between the classes. Images with and without relevant objects are quite similar, given that the objects to be handled will typically occupy only a small area of the total image. This is similar to another computer vision problem: defect detection. The latter has benefited from the one-class classification paradigm [33].

In the field of one-class classification, researchers have developed innovative ways to improve the learning of deep networks. For example, one study considers the creation of a pseudo-negative data on the feature space based on a zero centred Gaussian noise obtaining relevant results for a convolutional neuronal network with a binary cross-entropy loss function [17]. Another approach for deep networks uses two loss functions training a network architecture in parallel [18]. The idea behind training with two loss functions is to produce a network capable of recognizing differentiating features without increasing the intra-variance in the feature space of the positive data class.

Annotation of the pose of objects in an image through deep networks is frequently made with the use of bounding boxes - windows that delimit the object in the image plane. One of the earlier works related to bounding box estimation uses sliding windows and an accumulative strategy of bounding boxes [26], so the model locates the object where the highest number of overlapping bounding boxes occurs. Another study developed the you-only-look-once YOLO algorithm [19], where a grid is overlaid on the image and every cell of the grid has an associated output where it is determined whether in that grid there is an object or not. In a positive case, the output cell contains the information of the bounding box. Finally, another algorithm uses 3D bounding boxes projected in the 2D space of the image [16].

In the specific context of this project, there are approaches to generate large datasets for object recognition [27] but no similar approaches were found for pose estimation in packing problems. One study has developed a deep-learning method to reconstruct a 3D object model from two pictures of the scene [28]. Other researchers studied the relationship between a good pose estimation and the robot task performance in a packing problem context [10]. Finally, we use another approach to estimate the pose of the boxes in the workspace [23]. In this study, the researchers calculate the position of the box from the information of the depth channel.

2.3 Synthetic Data

In a recent lecture on the deep learning state of the art [6], one of the new exciting ideas that were discussed was the training of deep neural networks with synthetic data. Some researchers have proven that training a deep convolutional neuronal network with synthetic images has results comparable to those obtained by training with a real dataset [29]. Even when randomly introducing noise to the dataset, training forced the network to adapt and focus on the essential features, with the model achieving acceptable performance.

3 Methodology

Our methodology has four steps. The first consists of generating synthetic data from a virtual reconstruction of the physical workspace. The second step involves training a classification network to detect whether a box is present on the scene or not. Pose estimation is initiated in the third step, training a location prediction network to obtain the projected 2D coordinates of the vertices of the box. With these coordinates, the fourth step consists of finding a solution of a camera model using the 2D coordinates and a depth map to calculate the final 3D pose.

3.1 Project Workspace

The workspace is composed of a table platform, a UR3 industrial robot, a tripod-mounted Kinect V2 camera, and (optionally) several small wooden cubes with an edge of 5 cm. The lab room where the experimental setup is located measures approximately 28 m^2 (Fig. 2).

Fig. 2. Virtual scene depicting the workspace. The scene includes one UR3 robot.

3.2 Machine Vision

Regarding the camera parameters for the virtual scene, we use the same values as in the available hardware. The Microsoft Kinect V2 has an operative measuring range of 0.5 to 4.5 m. The colour camera has a 1920×1080 resolution, and the depth camera has a 512×424 resolution. The horizontal field of view is 70°, and the vertical field of view is 60°.

3.3 Data Synthesis

Using measurements from the lab room, we built the virtual environment in Blender, a software used for 3D modelling and rendering. In the construction of the scene, we considered all the components of the project that were visible to the camera, so we included a virtual and configurable model of the UR3 robot, see Fig. 2.

Although Blender can produce photorealistic renders, it requires a significant amount of time, in the order of several seconds per frame. Therefore, to generate a dataset large enough for use in training convolutional neural networks, we opted to use the real-time oriented rendering application Unity. Another advantage is that Unity's rasterization engine provides access to the g-buffer, making depth data more readily obtainable than in Blender.

After importing the model in Unity and adjusting the parameters of the virtual camera with the properties of the Kinect V2, we prepared the necessary scripts to record RGB-D pictures with 1920×1080 resolution. A script moves the box randomly while another script captures the photos and saves them to a local folder. When the program execution finishes, a text file is generated with the position of each vertex of the box. It is necessary to clarify that the text file has a two-dimensional position for each vertex. Therefore, for each vertex there is a set of coordinates (y1, y2) saved in a two-dimensional array representing the image plane (Fig. 3).

Fig. 3. Sample of a synthetic RGB image (top) and the corresponding depth channel (right)

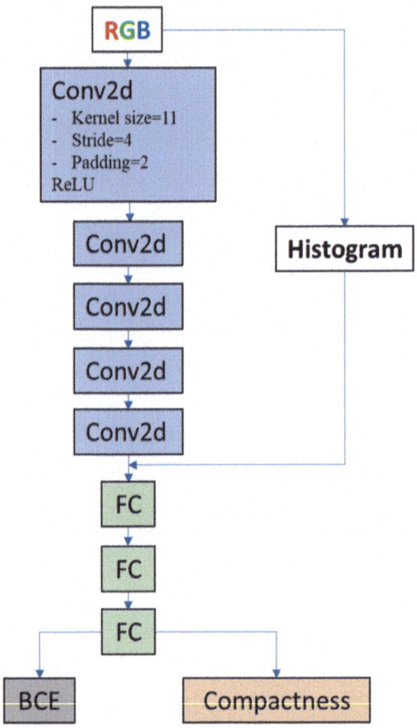

Fig. 4. Architecture of the modified AlexNet. Two loss functions are used to modify the weights of the layers through sequential backpropagation. The color histogram is included in the fully connected layers, resulting in a mix of inferred and handcrafted features

3.4 Training and Testing of the Classification Network

The second step of our methodology is to detect whether there is a box in the scene. This step is essential because, in the overall context, the robot must first know if there is an object to grasp or not. If there is a box, the robot calls the pose prediction network; otherwise, it is not necessary to perform any action. This problem is a one-class detection problem because there is only one class to detect, in this case a box. The labels for model training are assigned as follows: if the image contains a box we assign it the number one; otherwise, we assign a zero. The convolutional neural network we use is a modified AlexNet, Fig. 4. The architecture contains convolutional layers and also fully connected layers where data is combined with the colour histogram of the image. During training, we compute two different loss functions. The first one is the widely used binary cross entropy (BCE) loss, and the second is the compactness loss, as defined in (1).

N being the number of samples, the idea of the two different loss functions is to train the network with the features of a box so it will be able to detect whether there is or not a box in the image. Both loss functions train the network in parallel [18]. We trained the deep network with a dataset of one thousand positive and negative samples (for the binary cross-entropy function) and a dataset of one thousand positive samples (for the compactness function). The behaviour of the loss functions in a training dataset through 350 epochs is shown in Fig. 5.

The work that proposed the compactness loss function also experienced a similar peak as the model began to converge.

3.5 Training and Testing of the Prediction Network

The third step of our methodology is to predict the location of the box. To do this we train an AlexNet to predict the two-dimensional position of each vertex of the box in an RGB image, then we use this information along with the depth channels in a pinhole camera model to calculate the three-dimensional point in the space.

The convolutional neural network that we use for the pose prediction is a conventional AlexNet. The explanatory variable is an RGB image and the response variable is a vector with two coordinates per vertex, the relative position of the pixels is each vertex in the image. Therefore, the result is similar to a 3D bounding box projected in the image plane. The loss function of the prediction network is the L1 function.

In this case, the convolutional neural network must identify the features of each vertex and the associated training process is considerably expensive in computational terms. Therefore, we used a pre-trained AlexNet that was trained with the ImageNet dataset over several epochs, using a transfer learning methodology to make use of the learned features. The pre-trained network is available in the PyTorch package [19].

We performed a fine-tuning process that adjusted the weights of the different layers to our problem [1]. In this process of fine-tuning, we used a set of synthetic data containing two thousand images, with which the model was trained for one

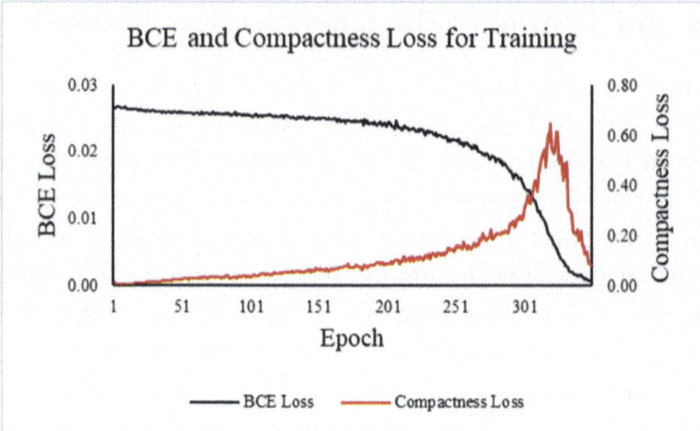

Fig. 5. Progression for BCE and compactness loss. Compactness loss increases initially as the network learns the feature distribution for negative data

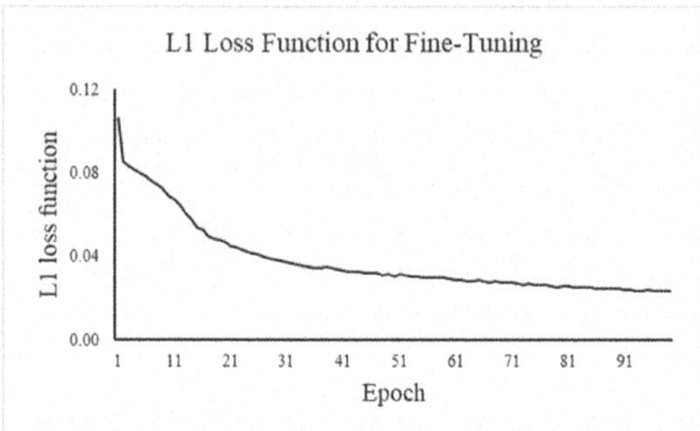

Fig. 6. L1 loss progression for the pose estimation network

hundred epochs. The behaviour of the loss function through epochs is visible in Fig. 6. The prediction of the convolutional neural network is a two-dimensional vector, containing two coordinates for each vertex of the bounding box (see Fig. 7).

$$CompactnessLoss(x, y) = a * b \tag{1}$$

$$L = (l_1, l_2, ..., l_i, ..., l_N) \tag{2}$$

$$l_i = (y_i - x_i)^2 \tag{3}$$

$$a = \frac{1}{N}\sum_{i=1}^{N} l_i \tag{4}$$

$$b = \frac{N^2}{(N-1)^2} \tag{5}$$

$$\begin{pmatrix} y_1 \\ y_2 \end{pmatrix} = \frac{f}{x_3} \begin{pmatrix} x_1 \\ x_2 \end{pmatrix} \tag{6}$$

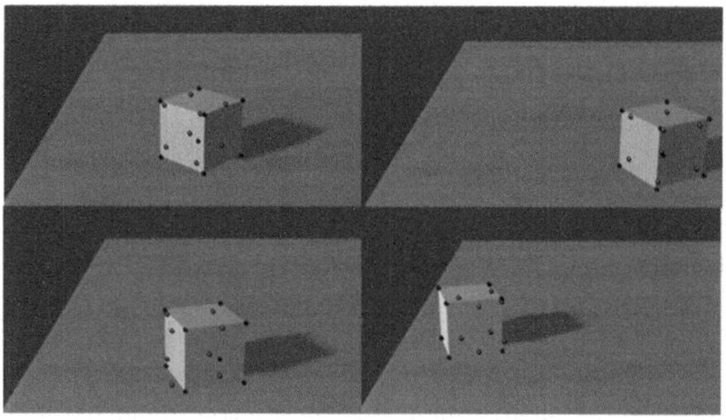

Fig. 7. Ground truth (black points) vs prediction (red points) in test samples (Color figure online)

3.6 Pose Estimation

The prediction network returns the vector with the pixel coordinates of the vertices of the bounding box. With these 2D coordinates, we search in the depth image the pixel values of the vertices. This is equivalent to the depth value x3 that the Kinect V2 records.

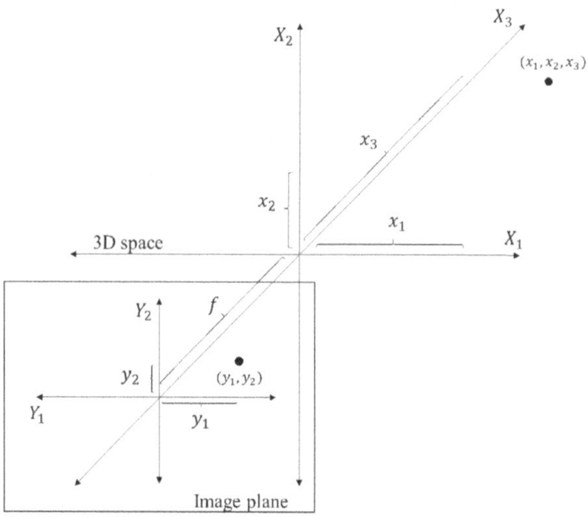

Fig. 8. Geometry of the pinhole camera model. The projection of the 3D point in the image plane is governed by the focal length and the orthogonal distance from the plane to the point

With the projected x and y values and the z coordinate, it is possible to apply the pinhole camera model (Fig. 8) to get the missing coordinates of a three-dimensional (x1, x2, x3) point (2). In the equations above f is the focal length of the camera obtained from the physical properties of the Unity camera and analogous to that of the Kinect, (y1, y2) is a point in the image plane and (x1, x2, x3) is the captured point in space. To recover the orientation of the bounding box, Euler angles are computed using the 3D edges formed by the bounding box vertices. An assumption proposing that the boxes always lie flat on the table simplifies this problem so that only yaw needs to be found.

3.7 Motion Planning

Synthetic data can be exported at different levels of abstraction. We proposed a reinforced learning neural network that creates trajectories around an obstacle given its centre-point and bounding box coordinates as state inputs.

4 Results

The convolutional neural network for the detection problem has an accuracy of 99.5%; this accuracy is achieved with a test sample of two hundred synthetic images. Conversely, the descriptive statistics of the pose estimation error for five hundred test samples are given in Table 1. Although an acceptable performance was achieved for the box detection stage, the behaviour of the BCE Loss indicates that the model is capable of additional learning [8]. Therefore, given the BCE loss function and the accuracy, it is reasonable to believe that the detection problem may be solved with a more simple network architecture than the modified AlexNet model. The algorithm for prediction produced adequate points when compared to other procedures used in bounding box detection works [23]. However, the range of our error is significantly wide because in the predictions of the network some extreme cases (outliers) occur. In these cases, the network predicted a position of a vertex and that predicted point was on the background of the real image; hence the depth value of this point was inconsistent with the depth of the box.

Table 1. Position and orientation error

Error quantity	Mean	Median	Standard deviation
Position (mm)	17.78	16.76	6.37
Orientation (degrees)	21.28	7.01	35.04

5 Conclusions

A machine vision solution based on synthetic data and deep learning for collaborative robots operating in an automatic packaging context was presented. Using synthetic data, we trained a one-class classification CNN that achieves an accuracy of 99.5% in box detection scenarios. The same technique is used to estimate the pose of three-dimensional objects, with a position error of 17.7 mm and an orientation error of 21.28°. The results obtained suggest that synthetic data-driven deep learning is a powerful tool to address the problem of 3D pose estimation in a packing context. Working with synthetic data reduces time and costs and produces very accurate annotations. We generated a database that can be useful to compare performances and results of other studies. This paper also demonstrates that transfer learning is a valuable asset for feature inference. However, it remains to verify how the architecture responds to real-world images captured in different lighting conditions. Although box detection can be easily tested, position and orientation relative to the camera are difficult to measure. Future work will focus on producing annotated real-world data for testing and even training operations to address this problem, as well as on a study that compares synthetic data with real world-data in the training of visual servoing applications.

References

1. Agrawal, P., Girshick, R., Malik, J.: Analyzing the Performance of Multilayer Neural Networks for Object Recognition. University of California, Berkeley (2014)
2. Bauer, W., Bender, M., Braun, M., Rally, P., Scholtz, O.: Lightweight robots in manual assembly – best to start simply! Examining companies' initial experiences with lightweight robots. Frauhofer I40 Study (2016)
3. Deng, J., Dong, W., Socher, R., et al.: ImageNet: a large-scale hierarchical image database. In: IEEE Conference on Computer Vision and Pattern Recognition (2009)
4. Chi, J., Walia, E., Babyn, P., Wang, J., Groot, G., Eramian, M.: Thyroid nodule classification in ultrasound images by fine-tuning deep convolutional neural network. J. Digit. Imaging **30**(4), 477–486 (2017). https://doi.org/10.1007/s10278-017-9997-y
5. Dosovitskiy, A., Fischer, P., Ilg, E., et al.: FlowNet: Learning Optical Flow with Convolutional Networks. University of Freibur and Technical University of Munich (2015)
6. Fridman, L.: Deep Learning State of the Art. MIT (2019)
7. Gaidon, A., Wang, Q., Cabon, Y., Vig, E.: Virtual Worlds as Proxy for Multi-Object Tracking Analysis. Xerox Research Center Europe and Arizona State University. arXiv:1605.06457v1 (2016)
8. Goodfellow, I., Bengio, Y., Courville, A.: Deep Learning (2016). ISBN-13 978-0262035613
9. Han, D., Liu, Q., Fan, W.: A new image classification method using CNN transfer learning and web data augmentation. Expert Syst. Appl. **95**, 43–56 (2018)
10. Hietanen, A., Latokartano, J., Foi, A., et al.: Object Pose Estimation in Robotics Revisited. Tampere University and Aalto University. arXiv: 1906.02783v2 (2019)

11. Hoo-Chang, S., Roth, H., Gao, M., et al.: Deep convolutional neural networks for computer-aided detection: CNN architectures, dataset characteristics and transfer learning. IEEE Trans. Med. Imaging **35**(5), 1285–1298 (2016)
12. Huh, M., Agrawal, P., Efros, A.: What makes ImageNet good for transfer learning? UC Berkeley. arXiv:1608.08614v2 (2016)
13. ISO/TS 15066:2016 - Robots and robotic devices - Collaborative robots. https://pytorch.org/docs/stable/nn.html. Accessed 18 Apr 2020
14. Krizhevsky, A.: One weird trick for parallelizing convolutional neural networks. Google Inc. arXiv:1404.5997v2 (2014)
15. Krizhevsky, A., Sutskever, I., Hinton, G.: ImageNet classification with deep convolutional neural networks. Commun. ACM (2012). https://doi.org/10.1145/3065386
16. Mousavian, A., Anguelov, D., Flynn, J., Kosecka, J.: 3D Bounding Box Estimation Using Deep Learning and Geometry. George Mason University and Zoox Inc. arXiv:1612.00496v2 (2017)
17. Oza, P., Patel, M.: One-Class Convolutional Neural Network. IEEE. arXiv:1901.08688v1 (2019)
18. Perera, P., Patel, M.: Learning Deep Features for One-Class Classification. IEEE. arXiv:1801.05365v2 (2019)
19. Pytorch documentation. https://pytorch.org/docs/stable/nn.html. Accessed 15 Apr 2020
20. Redmon, J., Divvala, S., Girshick, R., Farhadi, A.: You only look once: unified, real-time object detection. In: IEEE Conference on Computer Vision and Pattern Recognition (CVPR) 2016, pp. 779–788 (2015)
21. Reed, S., Akata, Z., Yan, X., Logeswaran, L.: Generative adversarial text to image synthesis. In: Proceedings of 33rd International Conference on Machine Learning, New York, 2016, vol. 48 (2016)
22. Reyes, A., Caicedo, J., Camargo, J.: Fine-tuning Deep Convolutional Networks for Plant Recognition. Laboratory for Advanced Computational Science and Engineering Research, Universidad Antonio Nariño and Fundación Universitaria Konrad Lorenz, Colombia (2015)
23. Rodriguez-Garavito, C.H., Camacho-Munoz, G., Álvarez-Martínez, D., Cardenas, K.V., Rojas, D.M., Grimaldos, A.: 3D object pose estimation for robotic packing applications. In: Figueroa-García, J.C., Villegas, J.G., Orozco-Arroyave, J.R., Maya Duque, P.A. (eds.) WEA 2018. CCIS, vol. 916, pp. 453–463. Springer, Cham (2018). https://doi.org/10.1007/978-3-030-00353-1_40
24. Ros, G., Sellart, L., Materzynska, J., et al.: The SYNTHIA dataset: a large collection of synthetic images for semantic segmentation of urban scenes. In: 2016 IEEE Conference on Computer Vision and Pattern Recognition (CVPR), pp. 3234–3243 (2016)
25. Sangkloy, P., Lu, J., Fang, C., Yu, F., Hays, J.: Scribbler: controlling deep image synthesis with sketch and color. In: 2017 IEEE Conference on Computer Vision and Pattern Recognition, pp. 6836–6845 (2016)
26. Sermanet, P., Eigen, D., Zhang, X., et al.: OverFeat: Integrated Recognition, Localization and Detection using Convolutional Networks. Courant Institute of Mathematical Sciences, New York University. arXiv:1312.6229v4 (2014)
27. Solund, T., Glent, A., Kruger, N., et al.: A Large-Scale 3D Object Recognition dataset. Danish Technological Institute, University of Southern Denmark and Technical University of Denmark (2016)
28. Solund, T., Savarimuthu, T., Glent, A., et al.: Teach it yourself - fast modeling of industrial objects for 6D pose estimation. In: Fourth International Conference on 3D Vision (3DV), 2016, pp. 73–82 (2015)

29. Tremblay, J., Prakash, A., Acuna, D., et al.: Training deep networks with synthetic data: bridging the reality gap by domain randomization. In: CVPR 2018 Workshop on Autonomous Driving, arXiv:1804.06516 (2018)
30. Tsirikoglou, A., Kronander, J., Wrenninge, M., Unger, J.: Procedural Modeling and Physically Based Rendering for Synthetic Data Generation in Automotive Applications. Linkoping University and 7D labs. arXiv:1710.06270v2 (2017)
31. Xiang, Y., Schmidt, T., Narayanan, V., Fox, D.: PoseCNN: A Convolutional Neural Network for 6D Object Pose Estimation in Cluttered Scenes. NVIDIA, University of Washington and Carnegie Mellon University. arXiv:1711.00199v3 (2018)
32. Zhang, H., Xu, T., Li, H., et al.: StackGAN: Text to Photo-realistic Image Synthesis with Stacked Generative Adversarial Networks. ICCV 2017, arXiv: 1612.03242 (2017)
33. Zhang, M., Wu, J., Lin, H., Yuan, P., Song, Y.: The application of one-class classifier based on CNN in image defect detection. Procedia Comput. Sci. **114**, 341–348 (2017)

A Survey on Components of AR Interfaces to Aid Packing Operations

Guillermo Camacho-Muñoz[1(✉)], Humberto Loaiza-Correa[1], Sandra Esperanza Nope[1], and David Álvarez-Martínez[2]

[1] Escuela de Ingeniería Eléctrica y Electrónica, Universidad del Valle, Calle 13 # 100-00, Cali, Colombia
{guillermo.camacho,humberto.loaiza,sandra.nope}@correounivalle.edu.co
[2] Departmento de Ingeniería Industrial, Universidad de Los Andes, Carrera 1 Este No. 19A – 40, Bogotá 11711, Colombia
d.alvarezm@uniandes.edu.co

Abstract. Manual packing operations require a suitable interface to communicate the physical packing sequence (PPS) to the operators involved. The interfaces available in these applications are not much used because of their low effectiveness. Augmented Reality (AR) interfaces can help to improve the communication of the PPS but there is a lack of knowledge about its components and the performance in relation with the packing operation. This paper explores this relation with a method based on a documentary analysis that allows identifying articles and surveys concerning: (1) typologies of packing and its relationship with AR interfaces, (2) components of an AR system, and its relationship with packing operations and (3) tracking algorithms suitable to industrial packing environments. The survey is intended to review techniques in AR and packing operations, that allow the identification of current trends, and the formulation of a taxonomy in marker-less tracking algorithms suitable for packing applications. Authors expect that those formulations serve as guidance for the creation of new solutions in the area.

Keywords: Augmented reality · Packing · Operation · Marker-less tracking

1 Introduction

Packing problems correspond to geometrical assignment problems where small items (also called cargo) must be assigned to a large object (also called container). The purpose of the assignment is to optimize a cost function subject to a set of constraints [4]. This process explores geometric combinations by strategies defined with optimization routines until obtain a loading-pattern, i.e., a formal description of a solution to the assignment problem. Currently, there exist many commercial optimizing-software to assist the computing of loading-patterns [13] and a suitable physical packing sequence (PPS).

D. Trentesaux et al. (Eds.): SOHOMA 2021, SCI 987, pp. 82–94, 2021.
https://doi.org/10.1007/978-3-030-80906-5_7

The traditional interface to communicate the PPS to a human operator has been the so-called paper-based interface, with some complements that include virtual objects in stationary or hand held displays [11]. This solution requires an operator that must continuously switch its attention between the instructions and the item that is been handled. These interfaces have shown low effectiveness [17,19]. As a consequence, packing operators use their own experience and knowledge, instead of following the instructions associated with the computed PPS. This behaviour is associated with sub-utilization of the available space to pack the items, increase in resources (time, labour, vehicles, maintenance, fuel) and risks (product damage, loss of products, re-processing) associated with the packing operation. The indirect consequences include the increase of economic and environmental costs in the logistics operation and the reduction of the related organizations' competitiveness.

An opportunity to improve the interface that communicates the PPS is identified in the Augmented Reality (AR) systems. AR technology increases the visual perception of the user with virtual objects overlaying the real scene. A schematic of this type of interface is presented in Fig. 1. The performance of this interface depends on the complexity and nature of the operation involved [22]. To the best of our knowledge there is no analysis that allows comparing the complexity of the operation with the scope of the AR technology. This article aims to contribute to this analysis by reviewing techniques in AR and packing operations, identifying current trends and formulating a taxonomy of marker-less tracking algorithms suitable for packing applications. This is the main characteristic that differentiates our article from previous reviews in the packing and AR literature. We have selected the survey research approach to build this review. This selection was motivated by the fact that this is an exploratory study to the specific problem of packing aided by AR. The procedure is composed by three main steps: (1) define a set of relevant questions in the areas of (*i*) typologies of C&P problems, (*ii*) components of AR interfaces and (*iii*) marker-less tracking algorithms, (2) find seminal papers in each of the defined questions, based on experience of the authors in the area and search of primary and secondary studies, and (3) analyze the founded papers to resolve the initial questions and identify the current working lines of each one.

The remainder of this paper is ordered as follows. Section 2 presents the structure of a packing system with an AR interface. Section 3 presents related works. The available packing typologies and their relationship with an AR interface are presented in Sect. 4. Components of the AR interface and its relationship with requirements in industrial packing are presented in Sect. 5; Sect. 6 presents a taxonomy of algorithms to resolve the marker-less tracking in AR applications. Lastly, the document presents opportunities and working lines in each of the components of AR systems to aid packing operations in Sect. 7, and conclusions in Sect. 8.

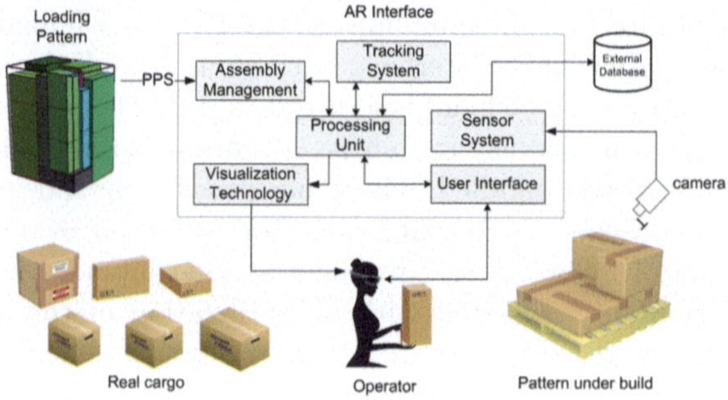

Fig. 1. System with an AR interface to communicate the physical packing sequence (PPS)

2 Overview of a Packing System with AR Interface

A packing system must resolve two operations: (1) compute a loading pattern and (2) assembly the pattern in the container. The first operation assigns the cargo to a container, optimizing a cost function subject to a set of constraints [4]. A solution that satisfies these conditions is called loading-pattern. In the second operation, the cargo is loaded to the container until completing the loading-pattern. The operation requires converting the loading pattern into a physical packing sequence (PPS) the sequence by which each box is placed inside the container in a specific location [27]. The communication of the PPS with AR interface is depicted in Fig. 1. Under this structure, the PPS is an input for the AR interface. The challenge in this system includes multiple areas. From computer vision area, box detection operations are approached in cluttered scenes, which implies the appearance of objects with different spatial properties (positions, orientations and scales), segmentation of objects in scenes with occlusion, estimation of pose with six degrees of freedom (three for position and three for orientation) in dynamic scenes. From the interface design it is necessary to direct efforts to improve the user experience in terms of visualization technology of real and virtual objects, intuitive interaction and management of the assembly process that allows giving precise instructions to the operator at each stage of the packing. The challenge is amplified by considering that the previous operations must be solved in real time, with computing resources that favour user mobility.

3 Related Works

We have identified a low number of publications related with the application of AR interfaces to assist packing process. In [17] the authors assess the advantages of a set of interfaces in the packing operation in terms of psychological and cognitive demands. The outputs of the evaluation were measured in a mental workload scale and a system usability scale. The results allow concluding that AR devices are appropriate to assist the packing, but usability must be improved considerably. The work in [18] evaluates human factors while interacting with three interfaces that allow visualization of pre-computed loading-patterns and assist a human operator in the job of placing heterogeneous cargoes inside a predefined delivery box. The main problem of the paper is to determine the effect of each technology in terms of mental workload, usability, user experience, physical complaints and objective measurements. The results allow to conclude that the technology AR is appropriate and exhibits an acceptable performance in all indicators.

None of the previous works discusses the components of the AR interface (tracking methodologies, sensors used, speed in terms of latency and update rates, accuracy, precision and resolution) or the relationship of AR interfaces with packing typologies. These omissions motivated initiatives to use AR that can enhance the performance as interfaces in packing operations.

4 Typologies of C&P Problems and Its Relationship with AR Systems

Between the available typologies in packing literature [1,10,37], the authors consider that the one in [37] is the most suitable to define the scope of an AR interface in packing operations; in the remainder of this paper, we call it as WHS Taxonomy. This decision is founded on two criteria: (1) cover a wide range of related problems in packing and (2) high acceptance in the academic and industrial fields. WHS Taxonomy proposes a classification in three stages; in this section we will define the scope of an AR interface for each of the stages.

4.1 Stage 1 - Kind of Assignment and Assortment of Small Items

Relative to the kind of assignment the AR interface can address both cases: input minimization and output maximization. This is a consequence of the independence between the optimization criteria and the execution of the PPS. In the assortment of small items [37] defines three possible values: identical, weakly heterogeneous, strongly heterogeneous. Cases of assembling the loading–pattern with identical value have been satisfactorily addressed by automatic packing systems [25,38]; then this variant does not merit an interface based on AR. On the other side, automatic packing systems have shown limitations when dealing with heterogeneity in small items' assortment: (1) a low flexibility to grasp cargoes of different sizes and weights, (2) low performance when arranging objects by the complexity associated to the involved tasks [5] (planning collision-free trajectories and prediction of changes of environment based on forces applied to it). This fact justifies heterogeneity in the assortment of small items for an AR interface in packing operations. The inclusion is suggested without making distinctions of strong or weak levels. This is necessary because some reports allow concluding that weak and strong heterogeneity have been treated in similar proportions (within the specialized literature [37]). Besides there is not a quantitative definition of heterogeneity, which introduces subjectivity when classifying instances of packing problems.

4.2 Stage 2 - Assortment of Large Objects

This stage defines a heterogeneity descriptor between large objects with two values: single large-object or multiple-large objects. AR interfaces can contribute to both values. Additionally, the assumption of rectangular and homogeneous material in the large-objects defined in [37] reduces the complexity of the AR interface at the execution of tracking routines.

4.3 Stage 3 - Dimensionality and Shape of Small Items

AR interfaces can handle any option in dimensionality (1, 2, 2.5, 3) but three-dimensional cases motivated the use of AR in packing operations. Concerning the shape of small items AR interfaces can handle any option, but the most reported shape in packing literature is rectangular [4]. This shape eases registration and poses tracking tasks in the AR system.

Synthesizing, an AR interface in packing can asses any packing problem classified as 3D rectangular and with heterogeneity in the assortment of small items as defined in [37]. We have not identified typologies that allow to classify operations on the assembly of the pattern in the container (i.e. the complexity of movements, the cargo distribution in the working area, or the weight of cargo), except for some related remarks [4,27]. The absence of this kind of typologies makes it challenging to select the most appropriate components at the AR interface. The work in [4] states that a feasible loading pattern cannot always be visualized so that the loading personnel appropriately understand it, and their implementation may be too time–consuming. This kind of pattern could be aided by collaborative AR systems, where multiple users share the space and interact with the same virtual objects. The work in [27] presents a loading efficiency metric that can be used to assess the operator's physical effort to load a complete cargo arrangement when aided by an AR interface.

5 Components of an AR Interface and Its Relationship with Packing Operations

In this section we present the main components of the AR interface adapted to the packing application and classification criteria for each component.

Visualization Technology. This component allows displaying digital information overlaid with the real environment. The AR interface in the packing application has a safety requirement that includes systems that visually perceive the environment in absence of the augmentations and systems that do not restrict the operator's mobility [2]. The first requirement penalizes the video see-through (VST) alternative [30]; the second requirement privileges the wireless displays. Another requirement in packing is the presentation of virtual content in free spaces i.e., the slot in the container where an item must be packed. This can be reached with spatial alternatives but increases the technical requirements [3]. Consequently, the alternative of Optical see-through (OST) is the most suitable for the packing application assistant. Finally, the packing operator requires free hands and perception of depth during all the operation, thus the stereo-HMD is the better alternative. In synthesis, the visualization technology that better satisfies AR interfaces in packing operations is the wireless stereo OST-HMD.

Tracking System. This component localizes a target object in the image sequence and computes a camera pose for this target [33]; the pose includes position and orientation, each with a maximum of three degrees of freedom. The algorithms proposed to resolve the tracking based on optical sensor rely on two approaches: (1) marker-based and (2) marker–less. The first approach uses artificial elements with predefined properties to simplify the localization and pose measurement. This alternative implies fixing a tag in the scene and is currently the most reported in industrial applications for its low computational cost and excellent performance [7]. However, marker-based approaches are sensitive to visual pollution and tracking jitter, owing to the absence of adequate feature points [36]. Therefore, markers are not suitable for packing applications, where it is common to have partial occlusion between cargoes and full occlusion of markers. The second approach resolves the tracking by extracting natural features of images captured from the scene, CAD models of the targets or reference images, and computer vision algorithms [30]. A more in–depth analysis of this approach is presented in Sect. 6.

Sensor System. The primary function of this component is to acquire information about the environment to resolve the tracking problem. The commercial AR devices used in industrial applications include at least two types of sensors: optical and non-optical; usually, the non-optical sensors are embedded into a single PCB called inertial measurement unit (IMU): gyroscope, magnetometer, and accelerometer [7]. The IMU is used in combination with optical sensors to estimate the 6D-pose of an entity. Pros and cons related to the operating principle of these sensors are presented in [16]. In packing operations there is a preference in interfaces that operate with the AR device's built-in sensors, usually IMU and RGB–D camera. This is a consequence of an interest in minimizing complexity in installing, configuring and maintaining the systems [2].

User Interface. The main function of this component is to resolve the interaction between the user and the AR system, enhancing usability and accessibility. This interaction is expected in two-way communication: input, output. Some common inputs are touch surfaces on real or augmented objects, gestures, the direction of gaze, speech recognition, or conventional peripherals based on the hardware of the type: mouse, keyboard, and hand-scanner [23]. In the output group, the main mechanism is the visual augmentation on the scene but this one has been complemented with acoustic cues [14] and force feedback [20].

Processing Unit. This component is responsible for executing basic computations to support diverse tasks, from communication with the external source of data to spatial transformations required to support image processing tasks. [7] reports that in 25% of the industrial implementations, the processing and display are performed on different devices. This is a consequence of the limited resources of

HMD devices for processing, especially in computer vision routines. There are approaches to resolve this limitation based on outsourcing computing as those reported in [6].

Assembly Management. This component is exclusive for AR systems applied in assembly guidance. It is responsible for displaying instructions according to the context of the assembly in such a way that the information displayed and the behaviour of the interface adapt to different contexts. This topic is also known as interactive AR guidance systems for the assembly process [35]. In the packing interface, the functions involved to this module are related to: (1) managing the sequence of instructions, (2) tracking the assembly status, (3) identifying assembly errors introduced by the operator, and (4) requesting updates of the load–pattern and the PPS in case of manual changes inserted by the operator.

6 Algorithms to Resolve the Marker-Less Tracking

In this section, the search was restricted to algorithms handling the tracking of rigid objects and reported in AR applications. The identified algorithms can be grouped in two principal categories [21]: (1) model–based and (2) planar scenes based. The description of these categories and their sub-divisions will be presented in this section.

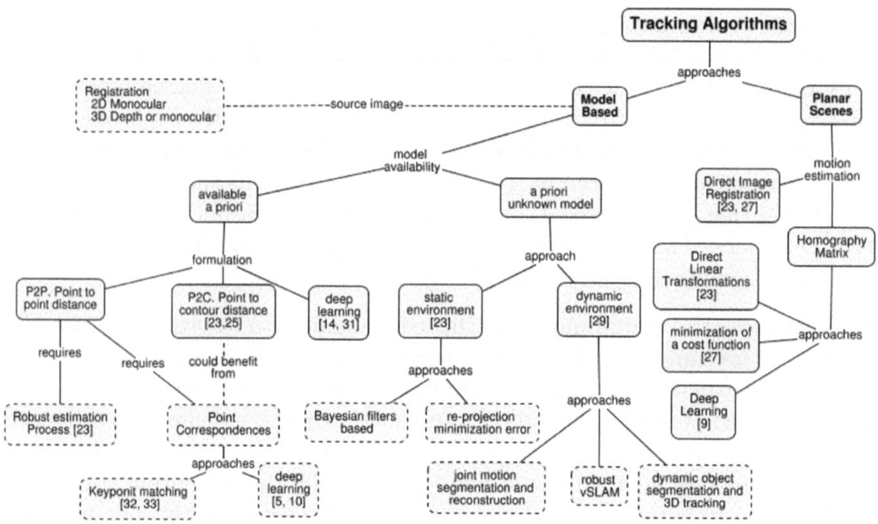

Fig. 2. Tracking Algorithms Taxonomy. Model–based tracking algorithms or 3D–2D estimators and Planar Scenes tracking algorithms or 2D–2D estimators

6.1 Model–Based

This group resolves the tracking problem from two sources: a 3D model of the target object for tracking and the images acquired during the exploration of the environment; these images can be 2D or 3D in case of depth perception; consequently these cases are also known as 3D-2D and 3D-3D estimators, respectively. Figure 2 presents the algorithms available in this group.

Model Available a Priori. There are three frameworks in this branch as depicted in Fig. 2. The point to point distance (P2P) uses a 3-stage pipeline to resolve the tracking: (*i*) compute point to point correspondences (*ii*) robust estimation process and (*iii*) camera pose estimation.

The first stage has been extensively explored in computer vision literature, and currently there are approaches based on key-points extractions [32] and based on deep learning [9]. The key-points are sensible to occlusion but have a low response time, unlike the deep-learning which are more stable in occlusion scenes but exhibit high response time [40]. The second stage deals with spurious data related to perturbations of the type: errors in correspondences, changes in illumination, occlusion phenomena, noise in images. Some reported algorithms are voting techniques, random sample consensus (RANSAC), M–Estimators, and Least Median of Squares (LMedS). The third stage formulates the pose estimation problem from a set of correspondences between image and model. Between the available algorithms, the preferred alternative in AR by accuracy is the PnP based on non–linear minimization and iterative processing [21] for 3D-2D estimations and the ICP algorithm for 3D-3D estimations [28].

The point to contour distance (P2C) uses a 2-stage pipeline: (*i*) projection of the 3D model over the acquired image and (*ii*) contour alignment. The method assumes the knowledge of an initial pose to resolve the first stage. The second stage is resolved by sequence tasks of detecting contour in the projected image, detecting target points in the acquired image, and formulation of a minimization problem between contour and points [21,24]. The technique must fuse the information of edge features with the information given by particular key-points to avoid ambiguities between different edges [15], especially in cluttered scenes.

In the deep learning framework the problem is formulated as a regression in training and testing stages [12,31]. The training stage uses images labeled with the pose of the target object; these images can be acquired by sampling and rendering the 3D model [39]. The test stage uses a tuned learning machine to estimate the pose of the target objects in new images. Its primary disadvantage resides in the high computational load and the low precision of the estimation. Its advantages include the ability to estimate pose in scenes with occlusion of the target object.

Model Not Available a Priori. This case is related to the problem of motion structure in computer vision, where the purpose is to resolve simultaneously the scene structure and the pose estimation of the camera. Between the available solutions, the AR field has shown a preference for algorithms that reside on vision sensors (vision-based Simultaneous Localization And Mapping or vSLAM), single camera, appearance–based re–localization, and real–time response. The available algorithms are classified in two groups, as presented in Fig. 2. The algorithms in the static environment branch assume that the real objects in the scene remain static; otherwise, the acquired 3D models quickly become obsolete, and the algorithm must restart the reconstruction. [21] classifies these algorithms in two approaches: (*i*) based on statistical or Bayesian filters and (*ii*) re-projection minimization error. The first approach performs iterative measurements of features in the scene; these measurements are sequentially integrated and allow updating the probability density associated with the system (camera position and structure of the scene). The second approach formulates a minimization problem over acquired points and re–projected points, until obtaining both the pose estimation and the structure of the scene. The algorithms for dynamic environments are reviewed in [29], where they propose three types of problems: (*i*) robust vSLAM, (*ii*) dynamic object segmentation and 3D tracking, and (*iii*) joint motion segmentation and reconstruction. The first group builds static maps by rejecting dynamic features. The second group focuses on segmentation and tracking dynamic objects ignoring the static background, and the third group deals with tracking dynamic objects and scene reconstruction; that is, it integrates groups one and two. SLAM in dynamic environments requires further research to enable practical implementations. The current challenges are related to handling missing, noisy, and outlier data. Although statistical-based techniques can tackle some of these challenges due to their recursive sampling approach, they have to trade off accuracy for computation cost.

6.2 Planar Scenes–Based

In this category, algorithms assume the existence of targets with a planar shape such as floors, walls, tables, books, pictures, among others. With that assumption, the problem is simplified. The objective is to estimate the pose of a target object using a reference image (localized) instead of a model. It is for this reason that this group is also known as the 2D-2D estimators. This group of techniques assumes the existence of an initial pose, from which the pose of interest is estimated by concatenating the poses that allowed the camera to be brought from its initial position to its final position. A disadvantage associated with this strategy is the accumulation of errors between intermediate poses. The alternatives of this group are classified in two groups, as depicted in Fig. 2.

The first group assumes that the model that relates points between different images of the same plane is the homography matrix. This matrix is defined using the rotation matrix and the translation vector between the camera and the target object. Then, the pipeline is composed of two stages: (*i*) compute the homography matrix, (*ii*) extraction of the pose from homography matrix. The reported algorithms in the first stage are based on direct linear transformations [21], minimization of cost functions based on geometric distances [26] and deep learning approaches [8].

The second group formulates the problem as the minimization of a dissimilitude function between the reference image $I_0(x)$ and the new image $I(w(x,h))$;

where w is the model that relates the pixels of the target object in the two images and h the parameters of this model. Multiple dissimilitude functions are reported in [21], some of which are sensitive to illumination changes and must be compensated with robust parameter estimators [26].

7 Opportunities and Working Lines

This study aimed to contribute to the identification of current working lines in each of the components of AR interfaces to aid packing operations. The analysis of packing typologies in Sect. 4 allowed us to identify the absence of a taxonomy for the assembly of the PPS in the container. This taxonomy is required to guide the selection of components in the AR architecture.

The first analysis for the tracking system in Sect. 5 showed that the tracking technology that better satisfies the packing operation is marker–less. The second analysis on tracking in Sect. 6 allowed to identify enormous progress in the area of visual-based tracking for AR. The main challenges related with the packing operation were identified as simultaneous tracking of multiple targets, specifically boxes and containers, robustness under dynamic scenes, especially in the segmentation of the dynamic object and robustness in tracking with partial occlusion of the target object, where accuracy is mostly degraded. Also, there exists a need to enhance robustness and speed in response times of deep learning algorithms under simpler but efficient pipelines. The analysis of sensor systems in Sect. 5 showed a preference in solutions that operate with the AR devices' built-in sensors, usually IMU and RGB-D camera, to reduce complexity in the installation, configuration, and maintenance of the systems. However, user interfaces' analysis showed that it is necessary to combine diverse alternatives for input and output to the system as a strategy to increase usability. So, there is a challenge to resolve the compromise between complexity and usability of these systems. Lastly, the Assembly Management component analysis in Sect. 5 allowed us to identify a set of functions required for context awareness in the specific packing operation. These functions are open problems that require advances in packing of known items with non-deterministic arrival order [34] and detection of assembly states using computer vision algorithms.

8 Conclusions

The presented review identified current opportunities and working lines to implement AR interfaces that allow communicating a Physical Packing Sequence (PPS) in the context of packing operations. To the best of our knowledge, this paper is the first survey on this topic. The analysis allowed identifying ten current working lines. We highlight the absence of a taxonomy for the assembly of the PPS in the container; this taxonomy is required to guide the selection of components in the AR architecture. The analysis of displays available in the market, let us conclude that the most suitable alternative for packing requirements resides on a wireless, binocular, OST-HMD device. In this alternative, there are still open challenges related to ergonomics, the field of view, resolution, refresh rate, and display of occlusion.

The analysis in the marker-less tracking algorithms allowed formulating a taxonomy and identify a wide advance in approaches to the problem, as well as contribution opportunities in topics related to: (1) detection of objects in cluttered scenes, (2) segmentation of objects under scenes with partial occlusion and (3) pose tracking in temporary image sequences for dynamic scenes. We expect that these research lines contribute to research in specific aspects of the problem.

References

1. Araújo, L.J.P., Özcan, E., Atkin, J.A.D., Baumers, M.: Analysis of irregular three-dimensional packing problems in additive manufacturing: a new taxonomy and dataset. Int. J. Prod. Res. **57**(18), 5920–5934 (2019)
2. Berkemeier, L., Zobel, B., Werning, S., Ickerott, I., Thomas, O.: Engineering of augmented reality-based information systems: design and implementation for intralogistics services. Bus. Inf. Syst. Eng. **61**(1), 67–89 (2019)
3. Bimber, O., Raskar, R.: Spatial Augmented Reality Merging Real and Virtual Worlds (2005)
4. Bortfeldt, A., Wäscher, G.: Constraints in container loading - a state-of-the-art review. Eur. J. Oper. Res. **229**(1), 1–20 (2013)
5. Byravan, A., Fox, D.: SE3-nets: learning rigid body motion using deep neural networks. In: Proceedings - IEEE International Conference on Robotics and Automation, no. 3, pp. 173–180 (2017)
6. Chatzopoulos, D., Bermejo, C., Huang, Z., Hui, P.: Mobile augmented reality survey: from where we are to where we go. IEEE Access **5**, 6917–6950 (2017)
7. de Souza Cardoso, L.F., Queiroz, F.C.M., Zorzal, E.R.: A survey of industrial augmented reality. Comput. Ind. Eng. **139**, 106–159 (2019)
8. DeTone, D., Malisiewicz, T., Rabinovich, A.: Deep Image Homography Estimation (2016)
9. Duong, N.D., Kacete, A., Sodalie, C., Richard, P.Y., Royan, J.: XyzNet: towards machine learning camera relocalization by using a scene coordinate prediction network. In: Adjunct Proceedings - 2018 IEEE International Symposium on Mixed and Augmented Reality, ISMAR-Adjunct 2018, pp. 258–263 (2018)
10. Dyckhoff, H.: A typology of cutting and packing problems. Eur. J. Oper. Res. **44**(2), 145–159 (1990)
11. Easy Cargo. Easy Cargo (2016)
12. Garon, M., Lalonde, J.F.: Deep 6-DOF tracking. IEEE Trans. Visual Comput. Graphics **23**(11), 2410–2418 (2017)
13. Ghiani, G.: Intelligent software for logistics. In: Hanne, T., Dornberger, R. (eds.) Computational Intelligence in Logistics and Supply Chain Management, 1 edn, chap. 7, p. 472. Springer, Cham (2017)
14. Hansson, K., Hernvall, M.: Performance and Perceived Realism in Rasterized 3D Sound Propagation for Interactive Virtual Environments, June 2019
15. Kim, K., Lepetit, V., Woontack, W.: Scalable real-time planar targets tracking for digilog books. Visual Comput. **26**(6–8), 1145–1154 (2010)

16. Koulieris, G.A., Akşit, K., Stengel, M., Mantiuk, R.K., Mania, K., Richardt, C.: Near-eye display and tracking technologies for virtual and augmented reality. Comput. Graph. Forum **38**(2), 493–519 (2019)
17. Kretschmer, V., Plewan, T., Rinkenauer, G., Maettig, B.: Smart palletisation: cognitive ergonomics in augmented reality based palletising. Adv. Intell. Syst. Comput. **722**, 355–360 (2018)
18. Maettig, B., Hering, F., Doeltgen, M.: Development of an intuitive, visual packaging assistant. In: Nunes, I.L. (ed.) AHFE 2018. AISC, vol. 781, pp. 19–25. Springer, Cham (2019). https://doi.org/10.1007/978-3-319-94334-3_3
19. Maettig, B., Kretschmer, V.: Smart packaging in intralogistics: an evaluation study of human-technology interaction in applying new collaboration technologies. In: Proceedings of the 52nd Hawaii International Conference on System Sciences, pp. 739–748 (2019)
20. Majewski, M., Kacalak, W.: Human-machine speech-based interfaces with augmented reality and interactive systems for controlling mobile cranes. IEEE Commun. Mag. **54**(3), 63–68 (2016)
21. Marchand, E., Uchiyama, H., Spindler, F.: Pose estimation for augmented reality: a hands-on survey. IEEE Trans. Visual Comput. Graphics **22**(12), 2633–2651 (2016)
22. Masood, T., Egger, J.: Augmented reality in support of industry 4.0-implementation challenges and success factors. Robot. Comput.-Integrated Manuf. **58**(March), 181–195 (2019)
23. Murauer, N., Panz, N., von Hassel, C.: Comparison of scan mechanisms in augmented reality supported order picking processes. In: CEUR Workshop Proceedings, vol. 2082, pp. 69–76 (2018)
24. Payet, N., Todorovic, S.: From contours to 3D object detection and pose estimation. In: 2011 International Conference on Computer Vision, (ICCV), pp. 983–990 (2011)
25. Popple, R.: The science of Palletizing - Vol. 3. Technical report, Columbia Machine (2009)
26. Pressigout, M., Marchand, E.: Model-free augmented reality by virtual visual servoing. In: Proceedings - International Conference on Pattern Recognition, pp. 887–890 (2004)
27. Ramos, A.G., Oliveira, J.F., Lopes, M.P.: A physical packing sequence algorithm for the container loading problem with static mechanical equilibrium conditions. Int. Trans. Oper. Res. **23**(1–2), 215–238 (2016)
28. Rusinkiewicz, S., Levoy, M.: Efficient variants of the ICP algorithm. In: Proceedings of International Conference on 3-D Digital Imaging and Modeling, 3DIM, pp. 145–152 (2001)
29. Saputra, M.R.U., Markham, A., Trigoni, N.: Visual SLAM and structure from motion in dynamic environments-a survey. ACM Comput. Surv. **51**(2), 1–36 (2018)
30. Schmalstieg, D., Hollerer, T.: Augmented Reality: Principles and Practice. Pearson Education (2016)
31. Tan, D.J., Navab, N., Tombari, F.: 6D Object Pose Estimation with Depth Images: A Seamless Approach for Robotic Interaction and Augmented Reality. arXiv, pp. 1–4 (2017)
32. Tareen, S.A.K., Saleem, Z.: A comparative analysis of SIFT, SURF, KAZE, AKAZE, ORB, and BRISK. In: 2018 International Conference on Computing, Mathematics and Engineering Technologies: Invent, Innovate and Integrate for Socioeconomic Development, pp. 1–10 (2018)
33. Uchiyama, H., Marchand, E.: Object detection and pose tracking for augmented reality: recent approaches. In: Foundation in Computer Vision, pp. 1–8 (2012)

34. Wang, F., Hauser, K.: Robot packing with known items and nondeterministic arrival order. In: Proceedings of RSS, pp. 1–9 (2019)
35. Wang, X., Ong, S.K., Nee, A.Y.C.: A comprehensive survey of augmented reality assembly research. Adv. Manuf. **4**(1), 1–22 (2016). https://doi.org/10.1007/s40436-015-0131-4
36. Wang, Y., Zhang, S., Yang, S., He, W., Bai, X.: Mechanical assembly assistance using marker-less augmented reality system. Assembly Autom. **38**(1), 77–87 (2018)
37. Wäscher, G., Haußner, H., Schumann, H.: An improved typology of cutting and packing problems. Eur. J. Oper. Res. **183**(3), 1109–1130 (2007)
38. Wurll, C.: Mixed case palletizing with industrial robots summary/abstract state of the art. In: Proceedings of ISR 2016: 47st International Symposium on Robotics, pp. 682–687 (2016)
39. Zou, D., Cao, Q., Zhuang, Z., Huang, H., Gao, R., Qin, W.: An improved method for model-based training, detection and pose estimation of texture-less 3D objects in occlusion scenes. In: 11th CIRP Conference on Industrial Product-Service Systems, pp. 541–546 (2019)
40. Zou, Z., Shi, Z., Guo, Y., Ye, J.: Object Detection in 20 Years: A Survey, pp. 1–39 (2019)

Airline Workforce Scheduling Based on Multi-agent Systems

Nicolas Ceballos Aguilar, Juan Camilo Chafloque Mesia(✉), Julio Andrés Mejía Vera, Mohamed Rabie Nait Abdallah, and Gabriel Mauricio Zambrano Rey

Pontificia Universidad Javeriana, Bogotá, Colombia
{nicolasceballos,juanchafloque,julio.mejia,rnait-abdallah, gzambrano}@javeriana.edu.co

Abstract. This study addresses employee's scheduling problems of an airline company who are assigned to customer service at an airport. The problem can be defined as a dynamic scheduling problem that includes uncertainties and unexpected environmental changes. Therefore, this paper reports an agent-based simulation model that implements the workforce scheduling problem including features such as personnel skills, preferences, and certain constraints imposed by the airline. Besides the workforce scheduling problem, the proposed approach also integrates the personnel transportation problem, focused on the welfare of the employees upon certain shift conditions defined by the company. The integration of both problems seeks to offer a holistic solution, given the close relationship of both problems. By using agent-based modelling, the proposed approach allows facing the problem complexity, whilst making possible to quickly react to unexpected changes and events which is not possible to achieve with traditional optimization methods.

Keywords: Workforce scheduling · Personnel transportation · Agent-based scheduling · Optimization

1 Introduction

A major element in the development of a country is a robust airport connection system because it influences the political, socio-economic, and technological aspects. An airport system plays a vital role in responding to passengers' demands of different operations, faced by the airlines response plans which include the preparation of flights, passengers' attendance and boarding, control of departing and arriving flights, among others [8]. Over the years, there have been issues concerning the provision of airline services in airports resulting in passenger dissatisfaction due to long waiting times. On the airline personnel side, work overload, lack of control over operations and poor management of workforce scheduling have been some issues that also affect passengers service. In addition, there has been a lack of prevention of events, considering the highly dynamic behaviour of flights and other airport operations [1].

D. Trentesaux et al. (Eds.): SOHOMA 2021, SCI 987, pp. 95–107, 2021.
https://doi.org/10.1007/978-3-030-80906-5_8

This paper aims at developing an efficient and flexible workforce scheduling model for an airline company. The main issue with such volatile demand is that it leads to a mismatch between the original workforce schedule and the demands being covered in the different airport operations. Additionally, an inadequate workforce schedule can result in job dissatisfaction, work overload, schedule variability, high personnel rotation, among others. The higher uncertainty of a post-pandemic world for airlines makes it even more crucial to rely on adaptable and robust scheduling processes.

The workforce scheduling problem considers the following decisions: weekly rostering, shift scheduling for every day and task assignment during the day with the possibility of rapid changes given sudden increase in demand. In addition, the airline must ensure transportation to employees.

This paper addresses the problem with a list of features that are not usually considered all together. For the personnel scheduling, factors such as preferences, skills, and restrictions are applied with the addition of personnel transportation. To do so, the proposed approach integrates a multi-objective function that seeks to minimize the unattended demand of passengers and maximize the welfare of the employees by considering certain aspects of the schedule and adequate transportation.

2 Background and Related Works

In related literature, there have been many approaches to solve the workforce scheduling problem. Particularly, agent-based models (ABM) have been recently applied to solve optimization problems that present several interrelated components in a distributed and heterogeneous environment, such as scheduling problems, followed by transportation and logistics problems [3]. Most studies appear to feature deterministic approaches, while real-world workforce scheduling problems must deal with a variety of uncertainty sources. Therefore, these statements show that in a reactive and non-deterministic environment where uncertainty has a strong effect on the personnel schedule, such as volatile demand, last-minute changes, or problems with transportation, it could prove very beneficial to incorporate such uncertainty in the decision-making process [7]. As well, there are opportunities for ABM by providing various tools that allow the modelling, simulation, and interactions of complex adaptive systems by using agents, providing a way to simulate results graphically according to several designed scenarios [2].

For instance, Wang, Z. and Wang, C. proposed in [9] a multi-agent system focusing on scheduling based on preferences using what they called preference points. These points are used to quantify the value on each preference, so that employees can indicate their preferences before the beginning of the planning horizon. Similarly, there are reports of an agent-based system that solves both the workforce scheduling and re-rostering problems, minimizing the difference between the initial and the reconstructed list, and also reducing the negative impact on the nurses' preferences. The generated solutions avoided any reduction in the satisfaction of nursing preferences, which is vital to reduce employee dissatisfaction [6]. In various scenarios, the personnel must carry out tasks at different locations requiring transportation. These scenarios are referred to as Workforce Scheduling and Routing Problems (WSRP) as they usually involve the scheduling of personnel combined with some form of routing to ensure that employees arrive on time to the locations where tasks need to be performed [5].

In an attempt to improve operational management problems related to the programming of human resources, the authors proposed a methodology that minimizes travel times in vehicle routes during home visits, characterized by the dynamic context with uncertainties and random events. A first module performs the optimized programming for vehicle routes, and the second performs the dynamic reprogramming, which is responsible for responding to interruptions or changes in conditions. The vehicle follows the planned route but can dynamically adapt the schedule in case of interruptions through interaction with other vehicles to redirect the previously assigned visits. The proposed alternative obtained drastic improvements and decreases in travel times and journeys, by finishing the routes one hour faster than the centralized alternative [2].

3 Structure of the Agent-Based Model

3.1 Organizational Rules

The organizational rules and norms express global constraints and directives that govern the multi-agent system behaviour. It is essential to understand what kind of indicators and restrictions the system must have because it helps identify the agents, requirements, behaviours and features to accomplish the objective of the organization. The main organizational rules set up by the case study airline are described in Table 1.

Table 1. Liveness restrictions

Restriction	Description
Hours per worker	Each worker must have 9 h per shift, and a total of 4 timeslots divided as follows (TS1: 2 h, TS2: 2:30 h, TS3: 2:30 h, TS4: 2 h)
Pauses during shifts	Each worker must have one-hour break during the shift. Each break depends on the time the agent starts the shift
Number of agents	A maximum of 75 agents can be assigned and a minimum of 30 agents need to be assigned
Rest between shifts	Each worker must rest 12 h between shifts
Resting days	Each worker must have at least one resting day during the week and the resting days are pre-assigned by the airline
Transport assignation	Depending on the shift start and finish hours, agents may require a vehicle for transportation, either to the airport or from the airport
Scheduling activities	The airline has a total of three main activities: International flights (A), domestic flights (B) and various (C). Each agent will have to participate in one of the three activities for each working timeslot, for each shift during the planning horizon, considering their abilities for each activity

The main indicator and objective function for the workforce scheduling are presented in Eq. 1 and Eq. 2. These equations represent the impact of unattended demand per activity and the maximum unattended demand during the entire planning horizon.

$$OFUnattended = \sum_{d=1}^{7}\sum_{p=0}^{23} Un_{d,p} + 25 * \max_{\substack{d=1\ldots7\\ p=0\ldots.23}} (Un_{d,p}) \tag{1}$$

$$Un_{d,p} = Max(0,\ Dem_{d,p,h} - N_{d,p,h}) \tag{2}$$

where:

- $Un_{d,p}$: Number of agents missing to satisfy the total demand in period p on day d.
- $Dem_{d,p}$: The demand of agents required in period p on day d.
- $N_{d,p,h}$: Number of agents assigned to attend the demand of the activity h during the period p on day d.

In addition, the indicator and objective function for personnel transportation are presented in Eq. 3. This indicator reflects the employee's well-being with regard to the transportation routes they take. Indirect routes refer to those that include at least one pick-up/drop-off point; hence service agents do not travel directly from their homes to the airport and vice-versa, incurring in extra kilometres on their journeys. Direct routes are considered ideal and taken as reference. Thus, an ideal scenario for each service agent is to have a direct ride to maximize its wellness but at a high transportation cost.

$$OFWellness = \frac{\frac{\sum_{t\in IndRoutes} KmExtraIndirect_t}{NIndirect}}{\frac{\sum_{t\in Routes} IdealKm_t}{N}} \tag{3}$$

- $IdealKm_t$: Ideal distance a passenger would have if he/she had a direct route.
- $KmExtraIndirect_t$: Difference between the distance of a passenger and the ideal km if he/she had a direct route.
- *NIdirect*: Number of indirect routes.
- N: Number of passengers carried.

3.2 GAIA Methodology

The model proposed in this paper is based on the GAIA methodology, because it helps to represent the different roles within the airline, considering the organizational rules. Figure 1 depicts the following roles: The Airline Agent (AA) represents the organization, where the impact of solutions will be measured by the performance indicators. The Customer Service Supervisor (CSS) is responsible for ensuring that the company's requirements are met, as well as the well-being of the customer service agents. The CSS assigns activities and starting hours per shift to customer service agents, and it also reports the results obtained back to the AA. The Transport Supervisor (TS) generates the transportation needed for every shift, facilitating the vehicle arrangement. Lastly, the Customer Service Agents (CSA) represent the staff that takes care of customers and activities. CSA are defined by their abilities and preferences.

In this model, CSA are part of the decision-making process, participating in the construction of schedules and routes, aiming for their welfare and the improvement of the scheduling and routing indicators.

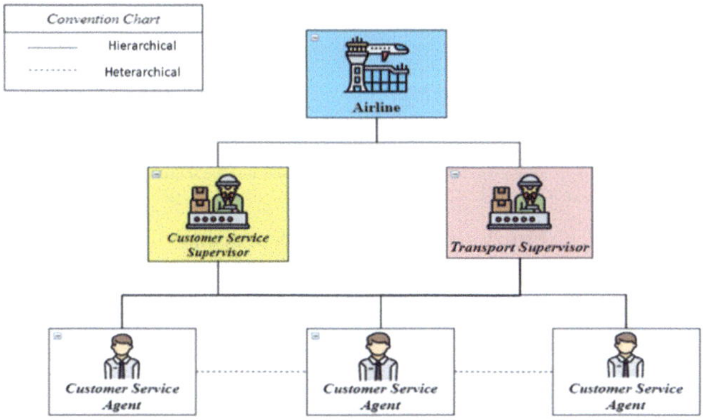

Fig. 1. Roles description

4 Decision-Making and Agent Interaction

Besides the agents and their relationships using GAIA, there are implementations of each agent that facilitate the decision-making process of these agents. The solution is generated in a sequential form where the solution generated by the CSS is sent by the Airline to the TS to generate routes for vehicles. Once the TS is finished, the result is reported back to the AA.

4.1 Decision-Making for Scheduling

To generate the workforce schedule, the CSS has embedded a Genetic Algorithm (GA) where each chromosome, as shown in Fig. 2, is implemented by a vector of n positions (each position represents a customer service agent - CSA). Each of the n positions is encoded with a random decimal number between 0 and 47 (it represents the 48 timeslots of a day). In this indirect codification, the integer part of the number represents the shift starting hour during the entire week, and the remaining decimal number (with two decimal positions) represents an index in the list of activities that the CSA has to attend in the four timeslots of each shift. Each CSA has this list of activities with all the four possible combinations (each shift has four timeslots and the three different activities can be repeated between timeslots) of activities that the agent can do. Also, each position is given a binary number (encoding mask) to help during the crossover position.

Figure 2 shows the chromosome encoding and decoding. For example, there is a chromosome with 5 genomes where the first genome has the number 8.65. The integer part of the number states that the first agent will start the shift at 8 am for every day during the planning week. The decimal part of the number, 0.65, must be multiplied with the number of possible combinations that the agent had previously generated. For instance, there are 9 possible combinations that the agent can do in this example as seen in Fig. 3. So, multiplying 9 by the decimal position 0.65, gives a result of 5.85, and rounding to the nearest whole the result is 6. This means that the CSA will do the four activities that are in the sixth combination, every day during the whole planning week.

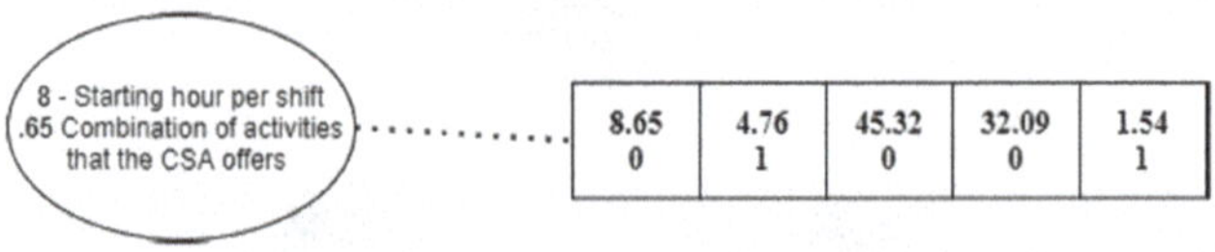

Fig. 2. Indirect coding chromosome

Combinations
AAAA
ABAB
ABBA
BBBA
BBBB
ABAA
AAAB
AABB
BBAA

Fig. 3. Possible combinations based on the activities

The crossover is done by determining the encoding mask given by the binary numbers in each genome. If the genome is 0, then the gene is inherited directly to the offspring, if not, the number is added with the genome of the other parent. To generate the second offspring, the process is similar but instead of adding the numbers, they are subtracted. Since the adding or subtracting of the possible timeslots for the new offspring can fall out of range, and a rebound operation is done for the gene to remain within the range [0, 48]. This process is represented in Fig. 4.

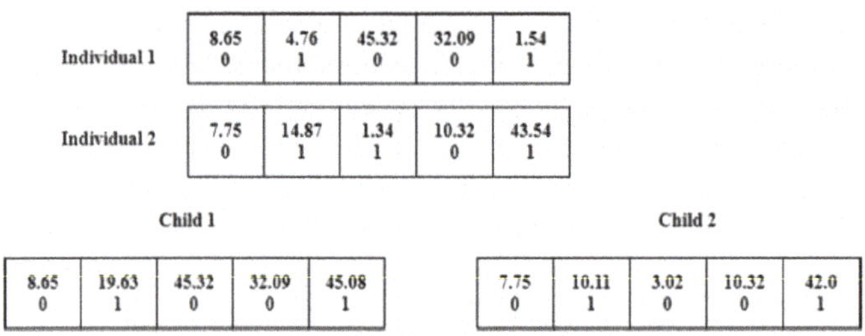

Fig. 4. Crossover representation

Finally, the mutation process is done by selecting a random position between 0 to the number of agents (genomes) to do a mirror mutation. Also, partial elitism is applied to preserve the best solutions from the population size. The partial elitism presented is guided by a threshold value. This value is established as an input to the model, usually between 0.01 and 0.05. To do this, the fitness for each chromosome is calculated with Eq. 4. The best solutions are the ones with a greater fitness.

$$Fitness = \frac{1}{OFUnattended} \tag{4}$$

4.2 Decision-Making for Transportation

The routing process starts when the TS agent determines all the CSA that need transportation, considering the transport assignation restriction in Table 1. Then, the TS generates leaders depending on the geographical position of the CSA, at each timeslot. Each vehicle has one leader and this procedure tries to imitate a clustering procedure, where each leader tries to create its own cluster to fill up a vehicle (up to 4 passengers). The negotiation follows a Contract-Net protocol. The non-leader chooses who is the best option for his/her local indicator (closeness) between all the leaders. Finally, the leader generates the best route using the Nearest Neighbour heuristic and reports the travellers and the route to the TS to calculate the global transportation indicators as presented in Eq. 3. The general decision-making process is represented in Fig. 5.

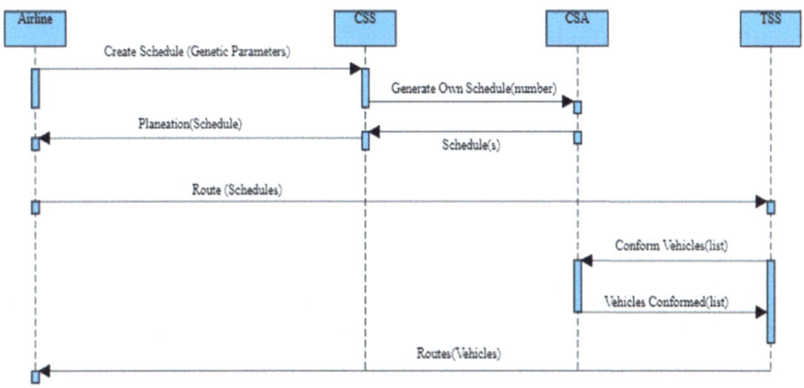

Fig. 5. Sequence diagram decision making

4.3 Reactivity to Perturbations

In an airline context, different unexpected scenarios are resolved at the moment that they occur. These are usually solved by the supervisors or by workers depending on the situation. For this reason, in this section two different perturbations scenarios are presented to evaluate the reactivity of the system to unexpected changes in the environment. The basic operation of the two perturbations is that there is a problem with the airline and that problem is reported to the relevant supervisor agent (CSS or TS), which reports the problem to the CSA. The perturbations considered are the most relevant for the airline in the case study. The procedure to tackle both types of perturbations is handled by the CSA, that makes proposals to minimize changes to the original schedule, routes, and objective function.

Peak Demand Perturbation
The process starts with the CSS detecting the extra demand and asking if the CSA can attend the demand of the concerned activity. Each agent determines if he is currently working when the peak is presented and if he can attend this request based on his skills. If any of the agents can attend the peak, a second evaluation is considered, taking into account the max demand indicator.

Absence Perturbation
This perturbation focusses on simulating the number of possible agents that could be absent in a given day during the planned week. The process starts when the CSS agent detects and absence and asks other agents whether they can work on their resting day. Once every agent responds to the CSS, the latter evaluates all the options taking into primary consideration the minimization of the variability indicator and the maximization of the activities the new agent can attend. Afterwards, the new agent checks on the TS to get information about transportation leaders and starts negotiating an inclusion into a route, if the new agent does not affect the additional kilometres indicator. Otherwise a new route is created for the new agent.

5 Results and Analysis

The implementation of the MAS is made in JADE, whose role is that of generator of high indicators of wellness and efficiency. The proposed agent-based model is compared with a deterministic model based on a metaheuristic approach using Tabu search. Besides the deterministic metaheuristic, a validator template was used to check scheduling indicators such as the unattended demand, distribution of activities and variability of schedules, as well as indicators of transportation such as number of routes, total kilometres travelled, ideal average distance and additional kilometres of agents, that include current airline indicators. Experiments were carried out with 30 replicas to compare the indicators generated by the Airline (template), deterministic metaheuristic, and the proposed MAS-based approach, withdrawing the best and worst solution from the methods developed by the authors.

5.1 Workforce Scheduling Results

Scheduling System
Table 2 reports the performance indicators for five solutions: the solution obtained with the airline method, the best and the worst solution obtained running a metaheuristic approach (Tabu Search) and the best and the worst solution obtained after running the proposed MAS-based model. In terms of unattended demand, even the worst solution of the proposed MAS-based model outbids the airline solution. Although the metaheuristic approach offers a better solution, by revising the max unattended demand by activity

the MAS-based model achieves a more homogeneous solution, avoiding important gaps between covered demand at each activity. Additionally, the distribution of activities in the MAS-based model tends to be more homogeneous because the activities are not overloaded as the other two approaches. Consequently, in the distribution of the unattended demand, there are activities that do not need additional personnel, because with the current personnel the unattended demand rate is low (see activity C). Table 2 also illustrates the indicators such as schedule variability, unattended demand, and the distribution of the three activities in each model. It is evident that the best correlation between scheduling variables is that of the agent-based solution, which reports schedule variability of zero and an unattended demand of 994 with a maximum value of 15,

Table 2. Scheduling results per method

Algorithm		VBA solution (deterministic)		MAS-based proposed model	
Indicators	Airline	Best	Worst	Best	Worst
Replica	–	13	24	11	6
Unattended demand	1434	723	1417	994	1318
Max Unattended Demand	20	17	17	15	14
%Penalties	48%	66%	70%	96%	56%
Activity A-Distribution	33%	23%	10%	16%	14%
Activity B-Distribution	60%	60%	69%	58%	55%
Activity C-Distribution	7%	16%	20%	26%	31%
Act A-Unattended demand	12%	24%	66%	52%	46%
Act B-Unattended demand	42%	58%	28%	44%	52%
Act C-Unattended demand	46%	18%	5%	4%	2%
Schedule Variability	1208.14	724.85	519.69	0	0
Objective function	**19,346,336**	**11,487,480**	**18,517,718**	**13,699,805**	**16,685,874**

which accomplish the objectives of the problem by generating wellness and reducing the unattended demand with a fixed number of workforce.

Transport System

Table 3 reports the value of the total kilometres of routes by all CSA, showing that the proposed MAS-based approach reduces this indicator by 34% compared to the airline solution. The reason of having better results with MAS is due to the algorithm which in its best and worst solution decreases the number of vehicles (no. of routes) approximately by the third of the airline solution and with a significant approach of 50 routes compared to the deterministic model (but even so, underneath the airline solution).

The average ideal distance for the MAS-based model presents higher values of additional km per agent, which represents an efficient use of vehicles, because the airline solution focused on more direct routes, using more vehicles thus higher costs. Although for each agent the average additional km is slightly higher, the MAS-based model offers a better global value in terms of Total Km of Routes.

As well, the association of variables such as the average ideal distance is the lowest for the MAS-based model, but incurring in additional kilometres which are the consequence of indirect trips, generating greater efficiency in the use of routes.

Table 3. Transport Results per method.

Algorithm		VBA solution (deterministic)		JADE solution (distributed)	
Indicators	Airline	Best	Worst	Best	Worst
Additional km	5.02	7.36	7.53	9.66	6.37
Average ideal distance	10.41	11.15	10.68	10.05	11.35
Total km of routes	5010	3362	3078	3311	4379
Max km travelled by agent	–	130.11	88.50	231.43	190.24
Min km travelled by agent	–	0.29	0.29	0.02	0.30
Number of routes	309	144	144	205	274

Perturbations Scenarios

The first scenario of perturbations is the generation of a peaks of demand for any activity. In this case, the perturbations are generated in the same model, but in different activities to determine the reactivity of the system and the impact on the solution. One of the results obtained for this perturbation is shown in Table 4.

Table 4. Peak demand perturbation

#	Description	Result	HHRR indicators before		HHRR indicators after	
1	There is a lack of 2 more agents at the time slot 21:30 of the day Monday for activity B	The CSS has 7 proposals and assign those 7 because all of them are idle and to prevent new peaks	Unattended	1212	Unattended	1214
			Max unattended	15	Max unattended	15
			Variability	0	Variability	0

The second scenarios of perturbations are the generation of absent agents for any given day. Table 5 shows one of the results of the scenarios done with this perturbation.

Table 5. Absence perturbation

#	Description	Result	HHRR indicators before		HHRR indicators before	
1	A total of 4 agents are said to be absent during the week: - Agent 11 [Fri: 18:00] - Agent 69 [Tue: 4:30] - Agent 31 [Sat 15:00] - Agent 60 [Mon 21:30]	A total of 4 agents were found that are capable to attend the missing demand for the shifts of the absent agents: - Agent 10 instead of 11 - Agent 38 instead of agent 69 - Agent 65 instead of agent 31 - Agent 13 instead of agent 60 All agents can cover exact the same activities for the previous agents For transportation, all 4 new agents will need transport and all 4 will travel alone to minimize the impact of the additional km indicator	Unattended	1300	Unattended	1300
			Max Unattended	14	Max Unattended	14
			Variability	0	Variability	10.48
			Additional Km	7.62	Additional Km	7.62
			Ideal Distance	11.08	Ideal Distance	11.08

It is worth clarifying the importance of reactivity for the results obtained in the perturbations case. The reason is that in the solutions found there was a high level of interaction, communication and cooperation between the agents and their supervisor, responding to changes that occurred, while maintaining the feasibility of the solution, in addition to reacting quickly and with low complexity. This low complexity is given by the cooperation and communication between agents, and the distributed environment in which they work. This is contrary to the deterministic approach complexity which is much higher and the whole model would be needed to be run from the start in order to accommodate changes, generating a new solution with important modifications on the whole schedule, and requiring between 15–20 min to report a new solution.

As general observations and managerial implications that can be established and concluded with the results obtained, it can be recommended to the airline to take into consideration the number of passengers per vehicle when transporting the agents. Currently the airline is seeking to reduce costs without taking into account the well-being of employees during transportation; nevertheless, as seen in the results obtained, having fewer people in each vehicle increases the well-being of the agents, reduces the total kilometres travelled while maintaining transportation costs low and reducing the number of vehicles used.

For the scheduling section, the airline can be advised to start implementing a system of low variability-schedules, since with the obtained results it is evident that the well-being and satisfaction of the employees increased considerably. As well, not only the well-being is increasing, but the activity distribution percentages are also increased, which means that leaving a weekly-fixed plan in terms of the activities that are going to be attended by each agent could also be taken into consideration.

6 Conclusion and Future Work

Several positive aspects can be highlighted by the use of MAS that allows the airline to be more competitive in the market and improve its customer service. First, the algorithmic complexity to find a feasible solution of the proposed MAS model is reduced considerably compared to a deterministic system, due to the characteristics of agents that facilitates the solution of a complex model in addition to its capability to react to unexpected events. Secondly, the proposed system focuses on the well-being of service agents, both in scheduling and transportation.

For this reason, it was decided that all agents had a fixed entry time and the same distribution of activities for every day, which, compared with the information provided by the airline and the deterministic solution generates better solutions in terms of well-being and therefore could generate a positive impact to employees. As above mentioned, the system also focuses on the welfare of agents for their transportation. It was decided to give higher priority to the welfare of agents than to the efficiency of the routes, so that agents do not have to make so many indirect routes and travel fewer kilometres.

Also, the system allows simulating unexpected changes. For this reason, the agents perceive these changes and decide rapidly without significant impact on the indicators; an example: non-attended demand and ideal kilometre increasing in less of 1% due to the design of agents that focus on distributing the decision-making process.

As future work, the system could be better used by coupling it to an information system that provides data to agents to make better decisions for the organization. As well, the airline could add more perturbed scenarios that focus more on their needs, so the system can adapt and give more flexibility to the changes in the environment.

References

1. Aeropuerto El Dorado: Statistics of passengers and aerial operations, second quarter of 2019 (2019). https://eldorado.aero/wp-content/uploads/2019/07/Consolidado-2-Trimestre-de-2019.pdf
2. Alves, F., Pereira, A.I., Barbosa, J., Leitão, P.: Scheduling of Home Health Care Services Based on Multi-agent Systems. In: Bajo, J., et al. (eds.) PAAMS 2018. CCIS, vol. 887, pp. 12–23. Springer, Cham (2018). https://doi.org/10.1007/978-3-319-94779-2_2
3. Barbati, M., Bruno, G., Genovese, A.: Applications of agent-based models for optimization problems: a literature review. Expert Syst. Appl. **39**(5), 6020–6028 (2012). https://doi.org/10.1016/j.eswa.2011.12.015
4. Billari, F.C., Fent, T., Prskawetz, A., Scheffran, J. (Eds.): Agent-based computational modelling: applications in demography, social, economic and environmental sciences. Taylor & Francis (2006)
5. Castillo-Salazar, J.A., Landa-Silva, D., Qu, R.: Workforce scheduling and routing problems: literature survey and computational study. Ann. Oper. Res. **239**(1), 39–67 (2014). https://doi.org/10.1007/s10479-014-1687-2
6. Chiaramonte, M., Caswell, D.: Re-rostering of nurses with intelligent agents and iterated local search. IIE Trans. Healthc. Syst. Eng. **6**(4), 213–222 (2016). https://doi.org/10.1080/19488300.2016.1226211
7. Van den Bergh, J., Beliën, J., De Bruecker, P., Demeulemeester, E., De Boeck, L.: Personnel scheduling: a literature review. Eur. J. Oper. Res. **226**(3), 367–385 (2013). https://doi.org/10.1016/j.ejor.2012.11.029
8. Warren, D.E.: The regional economic effects of airport infrastructure and commercial air service: quasi-experimental evaluation of the economic effects of commercial air service near smaller airports, working paper, University of Illinois at Urbana-Champaign Dept. of Agricultural and Consumer Economics (2007). http://www.ace.illinois.edu/reap/Warren_PP_071108.Pdf
9. Wang, Z., Wang, C.: Automating nurse self-rostering: a multiagent systems model. In: 2009 IEEE International Conference on Systems, Man and Cybernetics, pp. 4422–4425. IEEE (2009). https://doi.org/10.1109/ICSMC.2009.5346925
10. Zhao, J., DeLaurentis, D. Issues and opportunities for agent-based modeling in air transportation. In: 2008 First International Conference on Infrastructure Systems and Services: Building Networks for a Brighter Future (INFRA), pp. 1–6. IEEE (2008). https://doi.org/10.1109/INFRA.2008.5439611

Digitalization of Design and Planning Activities

A Novel Analysis Framework of 4.0 Production Planning Approaches – Part I

Estefania Tobon Valencia[1(✉)], Samir Lamouri[2], Robert Pellerin[3], and Alexandre Moeuf[4]

[1] Square Flow & Co., Neuilly-sur-Seine, France
e.tobon@flowandco.fr
[2] LAMIH, Arts et Métiers ParisTech, Paris, France
samir.lamouri@ensam.eu
[3] Polytechnique Montréal, Montréal, Canada
robert.pellerin@polymtl.ca
[4] Exxelia, Paris, France
alexandre.moeuf@exxelia.com

Abstract. The emergence of the Fourth Industrial Revolution has brought enterprises to review their production planning processes. Characterized by important technologies, this revolution provides managers and planners with multiple means to increase productivity, get added value from data mining processes and become more agile. This paper, divided into two parts, aims at proposing an analysis framework for conducting a literature review of the production planning approaches developed during the Fourth Industrial Revolution. This first part of the paper presents the analytical framework, the research methodology and the results of the systemic review. The proposed framework characterizes the approaches in terms of the addressed production planning activity, the planning horizon, the company size, the dimension of agility and the employed means.

Keywords: Production planning · Agility · Analysis framework · SME · Large company · Planning horizon · Resolution approach

1 Introduction

Nowadays enterprises dispose of a growing amount of available information that comes from the production process, the supply chain, the logistics department, the products and from the market [1]. The amount of information has increased during the Fourth Industrial Revolution. This Revolution has been defined as a new approach for controlling production processes by providing real time synchronization of flows and by enabling the unitary and customized fabrication of products [2].

Three main strategies for the transition to the Fourth Industrial Revolution have been proposed for industrial enterprises. These strategies are based on processes, products and services [1]. Process strategy is based on the transition from mass production to individualized production. Real-time data analysis allows production decisions to be

D. Trentesaux et al. (Eds.): SOHOMA 2021, SCI 987, pp. 111–132, 2021.
https://doi.org/10.1007/978-3-030-80906-5_9

adjusted in real time. Product strategy is based on the principle of intelligent objects. The collection of data through products makes it possible to innovate on the development of new products or to increase the autonomy of the product in its environment. The last strategy is based on the concept of product-driven automation powered by data gathered during the user's experience with products.

Among these strategies, the production planning activity could be classified as part of the Process Strategy. The production planning systems have looked so far to balance priority and capacity [3], i.e., is satisfying the market demand by establishing priorities while ensuring an optimal use of the available resources. Recent research in this area has pointed that current production planning systems are using data analytics to support real-time decision-making, adapt planning to the real-time conditions of the production system, and allow decentralization [4–10].

The integration of the technologies of the Fourth Industrial Revolution in the production planning activity has been a subject of recent research. In fact, the number of publications on these topics has increased since the year appearance of the term Industry 4.0 (2011) Hanover, see Fig. 1.

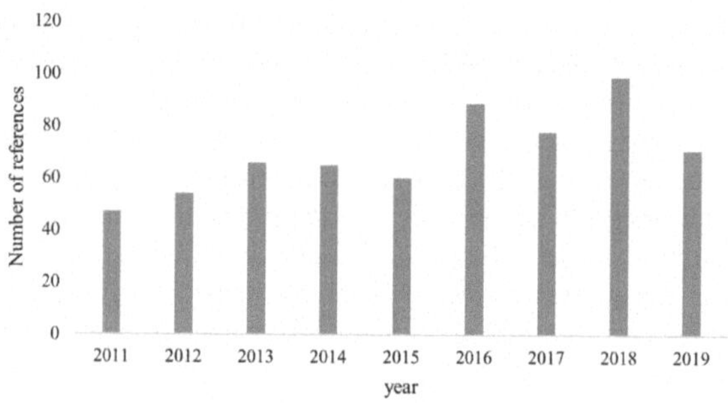

Fig. 1. Number of publications per year in production planning and 4.0 technologies

The results presented in Fig. 1 come from a research performed in Scopus on 24 August 2019 with the research string: TITLE-ABS-KEY ("Production Scheduling" OR "Production Planning" OR "Line Balancing" OR "Manufacturing Planning" OR "Operation Planning" OR "Production Sequency" OR "Capacity Planning") AND TITLE-ABS-KEY ("Big Data" OR "Simulation" OR "Internet of Things" OR "IoT" OR "Cyber Physical Systems" OR "CPS" OR "Cloud Computing" OR "Collaborative Robots" OR "Machine-to-Machine Communication" OR "comM2M"). In addition, the following filters have been applied to limit the results: only publications from 2011 concerning articles from journals and conference proceedings being part of the JCR list and having as publication language English have been kept.

The research in this field has mainly been carried in Germany, the United States and China. Figure 2 shows the top ten countries that have published the most in this area since 2011. Some researchers argue that agility is the main driver of the Fourth Industrial Revolution [1–11]. In fact, the technologies of this revolution provide enterprises with more flexibility in processes, more personalized production and improvement or even reinvention of the commercial offer.

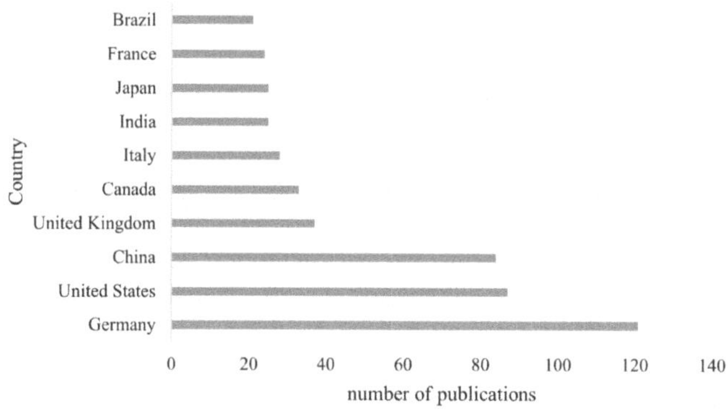

Fig. 2. Top ten-country number of publications since 2011 in production planning and 4.0 technologies

Nevertheless, the objectives of using the technologies of the Fourth Industrial Revolution in production planning activities have not clearly been identified in the literature. Therefore, this paper aims to propose an analysis framework for conducting a literature review and determine whether the production planning approaches within the Fourth Industrial Revolution look for some agility improvement. The proposed framework segments the analysis in five different elements: *activity*, *time frame*, *objectives*, *scope* and *means*.

This paper has been divided in two parts. The remainder of this first part is organized as follows. Section 2 presents the analytical framework, and the research methodology employed to review the literature is presented in Sect. 3. Section 4 summarizes the results obtained from the application of the proposed framework on the selected references. In the second part of the paper, we describe the contributions of the selected references, as well as research gaps and opportunities.

2 Framework for Literature Review Analysis

This paper intends to characterize the production planning approaches in the light of the Fourth Industrial Revolution. Our three main research objectives are:

- Identify what forms of agility improvements are pursued in new production planning approaches developed in the 4th Industrial Revolution
- Identify the scope of these approaches and
- Identify the means employed by these approaches.

We propose an analysis framework based on the following perspectives: production planning activity, planning horizon, agility, company size, Industry 4.0 technology and solution approach. Each one of the proposed perspectives has a relationship with one of the research objectives.

- *General characterization of the production planning approach*

Two perspectives have been proposed for the production planning approach. The first one concerns the production planning activity. Based on the works of [2], we propose to classify these activities in the following four categories: definition of quantities to be produced in groups of products in each period (DL: for demand levels), precision of the desired inventory levels (IL: for inventory levels), setting of the required resources (RR: as required resources) and determination of the availability of the needed resources (RA: resource availability). The proposed approach contributes to multiple activities.

For the second perspective, planning horizon, the three classic planning levels have been kept: operational, tactical and strategic. The strategic planning level determines actions to support the mission, goals, and objectives of an organization [12]. The strategic business plan defines the product lines and markets that the company wants to have in the future [3]. The tactical planning level synchronizes activities across functions that specify production levels, capacity levels, staffing levels, funding levels, and other levels, for achieving the intermediate goals and objectives and supporting the strategic plan [12]. The operational planning level sets short-range plans and schedules detailing specific actions to achieve the goals set in the tactical level [12]. The activities carried on this level ensure the implementation and control of the production planning and the flow of raw materials into the factory according to the objectives set in the tactical level [3].

Each planning level concerns different periods as follows: short-term (O: operational, from one week to a month), mid-term (T: tactical, from three to 18 months) and long-term (S: strategic, from two to five years). Some approaches look to cover multiple planning horizons.

- *Agility*

The first research objective of this study is to identify what form of agility is pursued when implementing production planning approaches in the 4th Industrial Revolution. In the research carried by [13], and based on the formal definition of agility used in military operations [14], agility in organizations has been defined as a synergic combination of the following dimensions:

- *Robustness*: the ability to maintain effectiveness across a range of tasks, situations and conditions.
- *Resilience*: the ability to recover from or adjust to misfortune, damage, or a destabilizing perturbation in the environment.
- *Responsiveness*: the ability to react to a change in the environment in a timely manner.
- *Flexibility*: the ability to employ multiple ways to succeed and the capacity to move seamlessly between them.
- *Innovation*: the ability to do new things and to do old things in new ways.
- *Adaptation*: the ability to change work processes and to change the organization.

Pellerin [11] pointed out that these dimensions are analytically distinct and often interdependent. Therefore, a single production planning approach could be addressed to multiple agility objectives.

- *Scope of the production planning approach*

In order to identify the scope of the production planning approach three categories were defined: large companies, small and middle enterprises (SME), and not specified. SME were defined by the European Commission in 2008 as firms having max. 250 employees and a total turnover less than 50 million euros (EU Commission 2008). The not specified category refers to experimental applications.

- *Means of the production planning approach*

Means were studied in the basis of two perspectives: Industry 4.0 technologies and resolution approaches. The definition of the technologies of the 4th Industrial Revolution is given in [15], which defines the following nine methods and technologies: Big Data, Simulation, Internet of Things, Cyber Physical Systems, Cloud Computing, Virtual Reality, Cyber Security, Collaborative robots and Machine-To-Machine communication. For the analysis of the literature review, this research focused on the technologies:

- Big Data analytics concerns the collection and comprehensive evaluation of data from many different sources and systems (enterprise and customer) that support real-time decision-making.
- Simulation has multiple applications from engineering to the production process and the plant operations. This technology allows exploiting real-time data to mirror the physical world in a virtual model that includes machines, products and humans. Tests and optimizations can be carried before implementing change.
- Internet of Things (IoT) consists of devices provided with embedded computing and connections that allow them to communicate and interact with other devices and more centralized controllers. This technology allows to decentralize analytics, decision-making and enables real-time responses.
- Cyber physical Systems (CPS) connect the physical objects and processes with virtual objects and processes on open, partly global and constantly connected IT networks [16, 17].
- Cloud Computing provide access to a large pool of data and computational resources through distribution computing techniques and multiple interfaces [18].
- Virtual Reality allows stereoscopic representing an environment. Users can the visualize real products and simulated objects, study the geometrical structure of the products and analyze the possibility to change the design and the manufacturing sequence [19].

Some production planning approaches employed multiple technologies; further information about these technologies is given in the Research Methodology section.

The second perspective of the means employed for production planning concerns the resolution approach, i.e., the mechanism that calculates the production plan. Three categories are defined for this perspective: Artificial Intelligence (AI), optimization and simulation. AI is a computing domain that programs machines to solve problems, learn, plan and manipulate objects. For this category we have focused on machine learning which aims to generate output data from the identification of trends by classifying input data and digital regressions.

Taking into consideration optimization three main groups were defined: exact methods, meta-heuristics and heuristics. Exact methods include linear programming, constraint programming and branch and bound algorithm. Meta-heuristics are stochastic resolution approaches that aim to find and approximation of the optimal solution through iterative processes. These algorithms take as basis natural systems, as genetic algorithm, physical phenomena like the simulated annealing algorithm, animal behaviour and ant colony mode. The inspiration in natural phenomena is the main characteristics of heuristics that tries to find feasible solutions that are not necessarily optimal.

Finally, the simulation category includes discrete event simulation, multi-agent systems and digital twins.

3 Research Methodology

The literature review was carried out in two online databases: Science Direct and Scopus. The accepted document types were journal publications and conference papers. Taking into consideration the first appearance of the term "Industry 4.0" only documents that had been published since 2011 were kept. Only references that had a source being part of the JCR 2019 list were considered, all published in English.

Multiple research strings were employed in both databases and integrated in two main parts. The first part refers to Production Planning; multiple synonyms and activities of this term were employed to build the first part of the research strings. The second part considered some of the technologies of the 4th Industrial Revolution. The following research strings were employed in both databases:

- TITLE-ABS-KEY ("Production Scheduling" OR "Production Planning" OR "Line Balancing" OR "Manufacturing Planning" OR "Operation Planning" OR "Production Sequency" OR "Capacity Planning") AND TITLE-ABS-KEY ("Simulation").
- TITLE-ABS-KEY ("Production Scheduling" OR "Production Planning" OR "Line Balancing" OR "Manufacturing Planning" OR "Operation Planning" OR "Production Sequency" OR "Capacity Planning") AND TITLE-ABS-KEY ("Cloud Computing").
- TITLE-ABS-KEY ("Production Scheduling" OR "Production Planning" OR "Line Balancing" OR "Manufacturing Planning" OR "Operation Planning" OR "Production Sequency" OR "Capacity Planning") AND TITLE-ABS-KEY ("IoT" OR "Internet of Things").
- TITLE-ABS-KEY ("Production Scheduling" OR "Production Planning" OR "Line Balancing" OR "Manufacturing Planning" OR "Operation Planning" OR "Production Sequence" OR "Capacity Planning") AND TITLE-ABS-KEY ("CPS" OR "Cyber Physical Systems").
- TITLE-ABS-KEY ("Production Scheduling" OR "Production Planning" OR "Line Balancing" OR "Manufacturing Planning" OR "Operation Planning" OR "Production Sequency" OR "Capacity Planning") AND TITLE-ABS-KEY ("Big Data").

- TITLE-ABS-KEY ("Production Scheduling" OR "Production Planning" OR "Line Balancing" OR "Manufacturing Planning" OR "Operation Planning" OR "Production Sequency" OR "Capacity Planning") AND TITLE-ABS-KEY ("Virtual Reality").

Using the five research strings allowed finding 777 references with 29 cross-references. The references from simulations were selected from a previous literature review carried out by the research team [20]. For Cloud Computing, IoT, CPS, Big Data and Virtual Reality the following subjects were excluded: control programming, energy management, industrial automation, information system communication and robotic automation. Taking into consideration this selection 53 references out of 153 were kept.

Based on the four main production planning activities proposed by [3], a full-text analysis was conducted. This analysis reduced from 53 to 33 the group of selected references. This group was completed with 13 references of simulation. Finally, 46 references were analysed through the proposed framework.

Figure 3 describes the reference selection process. Detailed information about the reference selection process for each Industry 4.0 technology is presented in Table 1.

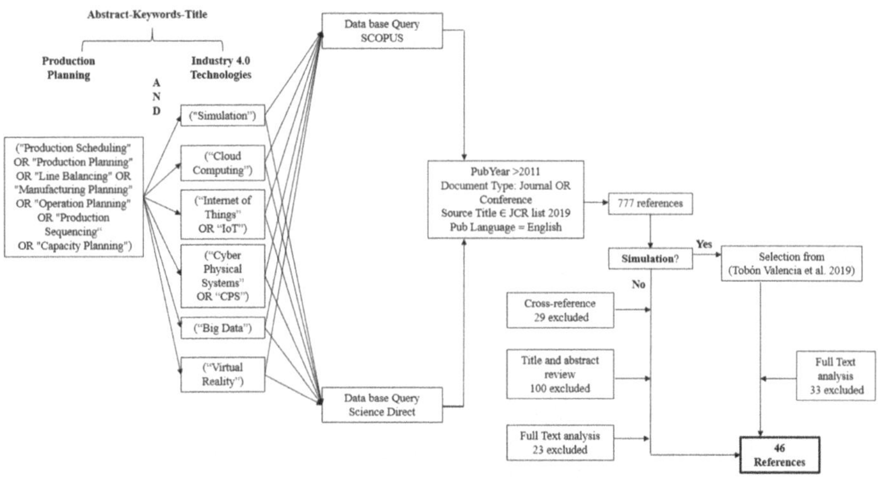

Fig. 3. Mapping reference selection process

The selection of the five technologies employed to build the research strings is based on an analysis carried in VOSviewer. Research strings similar as the ones presented before were built for eight of the nine technologies and means defined by Ruessmann et al. [15]. Only Cyber-Security was not included in the analysis. Then, the same filters (publication year, source type and source title selection) were applied to collect all the publications in Science Direct and Scopus. A cross-reference and a duplicate analysis were then carried to leave apart doubled references (113 references were excluded). Finally, the resulting list of references was imported to VOSviewer (664 references).

The study carried in VOSviewer allowed us to find among the 664 references the existent relationship between the keywords employed by the authors. It was also possible

Table 1. References selection

Industry 4.0 technology	No. of Scopus references	No. of references in Science Direct	Cross reference	Abstract selection	Full text analysis selection
Simulation	388	207	-	-	13
Cloud comput.	14	16	2	8	5
IoT	23	16	8	15	10
CPS	33	23	8	20	16
Big data	11	10	5	3	1
Virtual reality	31	5	6	7	1

to identify the technologies from the 4th Industrial Revolution that the authors exploited the most in references dealing with production planning.

The following technologies had the greatest number of references: big data, cloud computing, cyber physical systems, IoT, simulation and virtual reality. Based on this analysis the research team decided to focus only in references that exploited one of the aforementioned technologies.

The results of this analysis is presented in Fig. 4. Technologies have been identified with rectangles.

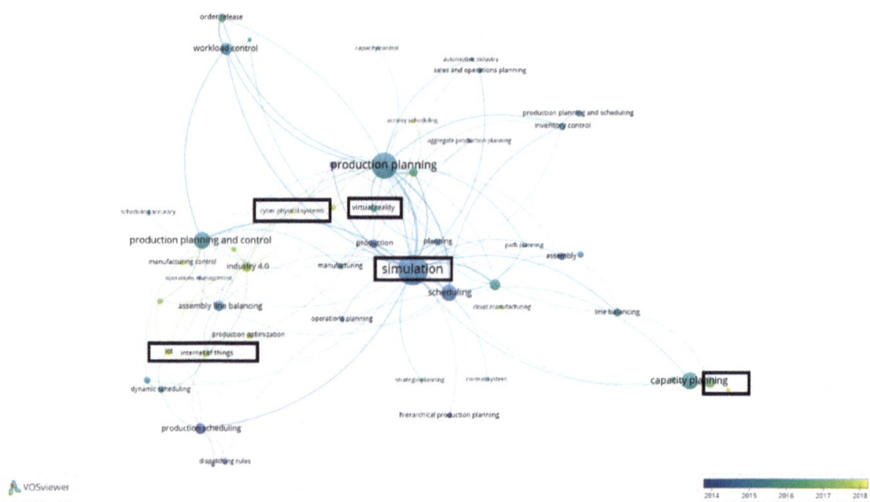

Fig. 4. Selection of the 6 technologies of 4th Industrial Revolution for literature review Part 1

An expansion on the right side of Fig. 4 is presented in Fig. 5 to show the appearance of the two technologies, which are not legible:

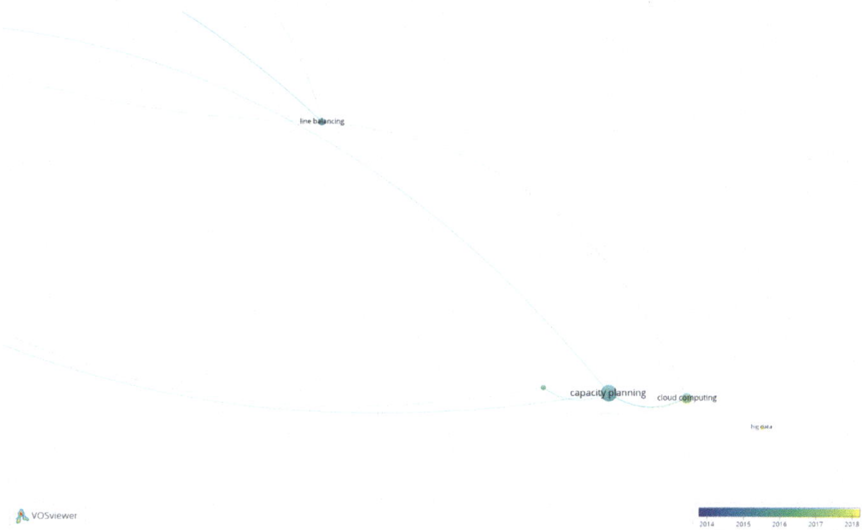

Fig. 5. Selection of the 6 technologies of 4th Industrial Revolution for literature review Part 2

4 Results for the Systematic Review of Production Planning and Some Technologies of the 4th Industrial Revolution

The database-driven literature review that was conducted in Scopus and Science Direct using the research strings selected 46 references. Table 2 presents the number of references by source that were used in the literature review.

Table 2. Distribution of articles by source

Source	*Selected Articles*
CIRP	9
IEEE	9
International Journal of Production Research	6
Journal of Manufacturing Systems	3
FAIM- International Conference on Intelligent Computation in Manufacturing Engineering	2
IFAC	2
Computers and Industrial Engineering	1
European Journal of Operational Research	1
International Journal of Simulation, Systems, Science & Technology	1
CASE- International Conference in Automation Science and Engineering	1
International Journal of Advanced Manufacturing Technology	1
Journal of Open Innovation	1
Journal of Engineering Manufacture	1
Advances in Mechanical Engineering	1
Knowledge Based Systems	1
Computers and Operations Research	1
DET- Conference on Digital Enterprise Technology	1
Production Planning and Control	1
International Journal of Computer Integrated Manufacturing	1
International Journal of Operations Research	1
12th conference on Intelligent Computation in Manufacturing Engineering	1
Total number of articles	46

The rest of this section is structured in four parts that develop the three research objectives. The first part characterizes production planning in the following dimensions of the proposed framework: production planning activity and planning horizon. In the second part the agility objectives sought by the approach are analysed. Then, the results about the scope of each approach are presented. The fourth part presents the classification of the employed means based on the 4.0 technology and the resolution approach.

- *Characterization of the production planning approach*

Production Planning Activity: Taking into consideration the production planning activities, the determination of the availability of the required resources was the main activity of the study. From the references studied, 37 out of 45 were addressed to this production planning activity. The results shown in Fig. 6 reveal that the distribution of the current production planning approaches in the proposed production planning activities is not uniform at all. A detail of the classification of each production planning approach in the production planning activity is presented in Table 3.

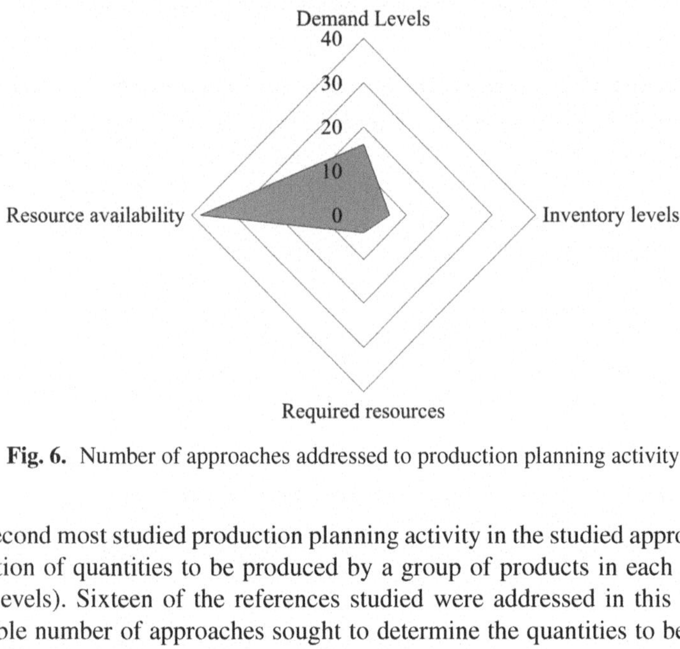

Fig. 6. Number of approaches addressed to production planning activity

The second most studied production planning activity in the studied approaches was the definition of quantities to be produced by a group of products in each period (D: Demand levels). Sixteen of the references studied were addressed in this activity. A considerable number of approaches sought to determine the quantities to be produced in short-term period [7, 22–25, 31, 34, 39].

With six references the precision of the desired inventory levels, was positioned in the third place of the studied production planning activities [4, 6, 7, 43, 45, 46].

Finally, the production planning activity that had the smaller number of approaches was the setting of the required resources [18, 37, 38, 44].

Planning Horizon-Perspective: From this perspective, almost all the approaches considered a short-term planning level (43 out of 45 studied references). The results shown in Fig. 7 reveal that in the 4th Industrial Revolution the production planning approaches

Table 3. References according to the characterization of the production planning approach

Perspective	***Production Planning Activity***				***Planning Horizon***		
References	DL	IL	RR	RA	O	T	S
(Asadzadeh, 2015)				x	x		
(Berger *et al.*, 2019)				x	x		
(Biesinger *et al.*, 2019)				x	x		
(Dallasega *et al.*, 2017)		x			x		
(Denno, Dickerson and Anne, 2018)	x			x	x		
(Engelhardt and Reinhart, 2012)				x	x		
(Erol and Sihn, 2017)				x	x	x	
(Fang *et al.*, 2016)	x	x			x	x	x
(Georgiadis and Michaloudis, 2012)		x			x		
(Graessler and Poehler, 2017)				x	x		
(Gräler and Pöhler, 2018)			x	x	x		
(Grundstein, Freitag and Scholz-Reiter, 2017)				x	x		
(Helo, Phuong and Hao, 2019)				x	x		
(Iannino *et al.*, 2018)				x	x		
(Illmer and Vielhaber, 2018)			x			x	
(Jiang *et al.*, 2017)				x	x		
(Klein *et al.*, 2018)				x	x		
(Kubo *et al.*, 2016)			x	x	x		
(Kuhnle, Röhrig and Lanza, 2019)				x	x		
(Kumar *et al.*, 2018)	x	x			x		
(Lachenmaier, Lasi and Kemper, 2017)				x	x		
(Lanza, Stricker and Peters, 2013)	x			x	x		
(Lanza, Stricker and Moser, 2014b)				x	x		
(Leusin *et al.*, 2018)				x	x		
(Li *et al.*, 2017)	x				x	x	
(Lin *et al.*, 2018)	x			x	x		
(Meyer, Hans Wortmann and Szirbik, 2011)		x			x		
(Mourtzis and Vlachou, 2018)				x	x		
(Nonaka *et al.*, 2015)	x			x	x		
(Qu *et al.*, 2016)	x			x	x		
(Rajabinasab and Mansour, 2011)				x	x	x	
(Rossit, Tohmé and Frutos, 2019)				x	x		
(Ruppert and Abonyi, 2018)				x	x		
(Saif *et al.*, 2019)	x			x	x	x	
(Shamsuzzoha *et al.*, 2016)	x			x	x	x	
(Shim, Park and Choi, 2018)	x			x	x		
(Shpilevoy *et al.*, 2013)	x			x	x		
(Tang *et al.*, 2017)			x	x	x		
(Um, Choi and Stroud, 2014)		x				x	
(Wandt et al. 2012)				x	x		
(Wang *et al.*, 2019)	x			x	x		
(Waschneck *et al.*, 2018)				x	x		
(Xu and Chen, 2016)				x	x		
(Xu and Chen, 2017)	x			x	x		
(Zhou *et al.*, 2019)	x			x	x		
(Zhu, Qiao and Cao, 2017)	x			x	x		

DL: demand levels, IL: inventory levels, RR: required resources, RA: resource availability, O: operational or short-term, T: tactical or mid-term, S: strategic or long-term.

focused on the operational planning level. For details of the planning horizon to which the production approach addresses see Table 3.

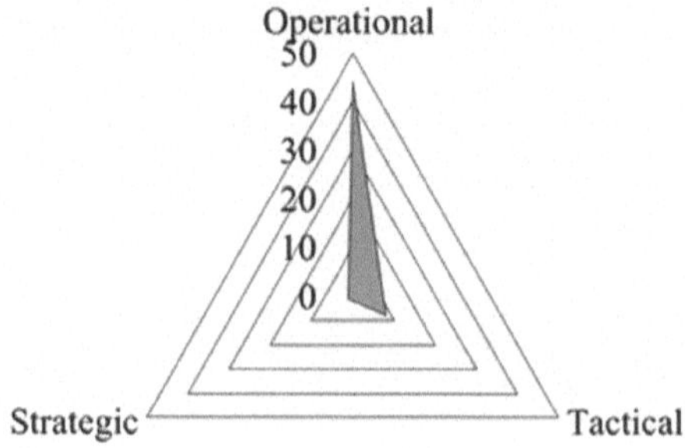

Fig. 7. Number of approaches by planning horizon

In the operational planning level, a certain number of approaches were addressed to the scheduling [7, 9, 18, 27–30, 32, 33] and the control activities [35, 36] which have a short-term planning. The mid-term planning horizon was considered exclusively in two approaches [44, 46]. The rest of the approaches that considered mid-term planning horizon presented a proposition for the short-term planning as well [5, 10, 26, 31, 40], or for the three planning horizons [43].

- *Agility*

Five dimensions of agility have been proposed to determine whether the production planning approaches look for some agility improvement. It was possible to classify all the references studied in almost one of the defined dimensions. Responsiveness is the main dimension of agility pursued by the selected production planning approaches (32 references out of 45 looked for this agility dimension). Figure 8 shows the distribution of the approaches in the proposed dimensions of agility. For specific classification of each production planning approach in the dimension of agility, see Tables 4 and 5.

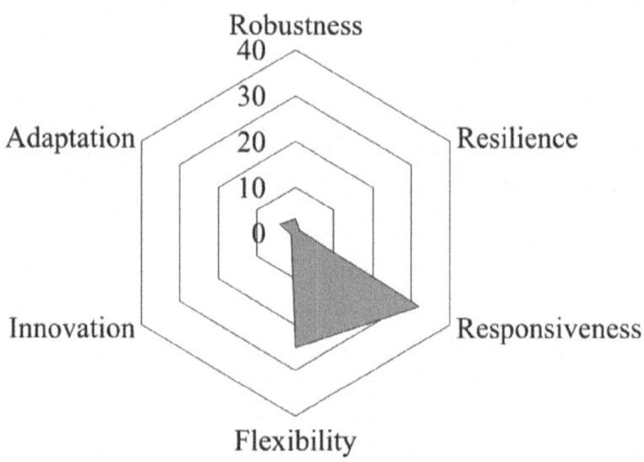

Fig. 8. Number of approaches addressed to agility dimension

Table 4. References according to the agility objectives Part I

Perspective	***Agility Objectives***					
References	Robustness	Resilience	Responsiveness	Flexibility	Innovation	Adaptation
(Asadzadeh, 2015)				x		
(Berger *et al.*, 2019)			x			
(Biesinger *et al.*, 2019)			x			
(Dallasega *et al.*, 2017)	x					
(Denno, Dickerson and Anne, 2018)			x			
(Engelhardt and Reinhart, 2012)			x			
(Erol and Sihn, 2017)	x					
(Fang *et al.*, 2016)				x		
(Georgiadis and Michaloudis, 2012)			x	x		
(Graessler and Poehler, 2017)				x		
(Gräler and Pöhler, 2018)			x	x		
(Grundstein, Freitag and Scholz-Reiter, 2017)	x			x		
(Helo, Phuong and Hao, 2019)			x	x		
(Iannino *et al.*, 2018)			x			
(Illmer and Vielhaber, 2018)					x	
(Jiang *et al.*, 2017)			x			
(Klein *et al.*, 2018)				x		
(Kubo *et al.*, 2016)			x	x		
(Kuhnle, Röhrig and Lanza, 2019)			x	x		
(Kumar *et al.*, 2018)				x		
(Lachenmaier, Lasi and Kemper, 2017)			x	x		
(Lanza, Stricker and Peters, 2013)		x	x			
(Lanza, Stricker and Moser, 2014b)			x	x		x
(Leusin *et al.*, 2018)			x			
(Li *et al.*, 2017)			x			
(Lin *et al.*, 2018)			x			
(Meyer, Hans Wortmann and Szirbik, 2011)			x			
(Mourtzis and Vlachou, 2018)			x			
(Nonaka *et al.*, 2015)			x			
(Qu *et al.*, 2016)				x		
(Rajabinasab and Mansour, 2011)				x		x
(Rossit, Tohmé and Frutos, 2019)			x	x		
(Ruppert and Abonyi, 2018)				x		
(Saif *et al.*, 2019)			x	x		
(Shamsuzzoha *et al.*, 2016)			x	x		x
(Shim, Park and Choi, 2018)			x			
(Shpilevoy *et al.*, 2013)			x			
(Tang *et al.*, 2017)			x	x		
(Um, Choi and Stroud, 2014)						x

Table 5. References according to the agility objectives Part II

Perspective	*Agility Objectives*					
References	Robustness	Resilience	Responsiveness	Flexibility	Innovation	Adaptation
(Lachenmaier, Lasi et Kemper, 2017)			x	x		
(Lanza, Stricker et Peters, 2013)		x	x			
(Lanza, Stricker et Moser, 2014b)			x	x		x
(Leusin *et al.*, 2018)			x			
(Li *et al.*, 2017)			x			
(Lin *et al.*, 2018)			x			
(Meyer, Hans Wortmann et Szirbik, 2011)			x			
(Mourtzis et Vlachou, 2018)			x			
(Nonaka *et al.*, 2015)			x			
(Qu *et al.*, 2016)				x		
(Rajabinasab et Mansour, 2011)				x		x
(Rossit, Tohmé et Frutos, 2019)			x	x		
(Ruppert et Abonyi, 2018)				x		
(Saif *et al.*, 2019)			x	x		
(Shamsuzzoha *et al.*, 2016)			x	x		x
(Shim, Park et Choi, 2018)			x			
(Shpilevoy *et al.*, 2013)			x			
(Tang *et al.*, 2017)			x	x		
(Wandt et al. 2012)			x			
(Wang et al., 2019)			x	x		
(Waschneck et al., 2018)			x			
(Xu et Chen, 2016)			x	x		
(Xu et Chen, 2017)			x	x		
(Zhou et al., 2019)			x	x		
(Zhu, Qiao et Cao, 2017)				x		

- *Scope of the production planning approach*

Three categories classify the scope of the production planning approach. From the company size perspective, the approaches could be addressed to large companies or to SME. For the approaches that did not specify the company size a third category was proposed (Not Specified). Figure 9 shows the general distribution of approaches by scope category. Tables 6 and 7 present the detailed classification of each production approach.

Not Specified: About 76% of the analysed approaches did not specify the company size to which they were addressed. Two options drove the analysis to this category: a) the approach had not been validated; b) the authors did not specify details of the company size in which they had carried the validation tests. The following approaches had not yet been validated: [4, 8, 17, 21, 22, 27, 28, 34, 37, 41, 42].

Large Company: Around 13% of the studied approaches had been proposed or tested in large companies.

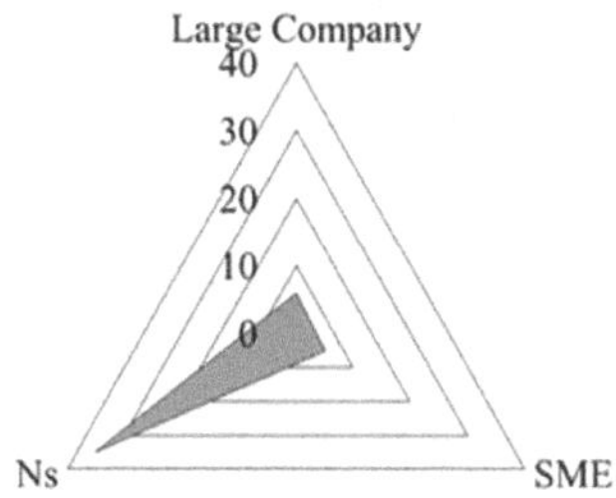

Fig. 9. Number of approaches by company size

Table 6. References related to the scope of production planning approach Part I

Perspective	*Company Size*		
References	Large Company	SME	Not Specified
(Asadzadeh, 2015)			x
(Berger *et al.*, 2019)			x
(Biesinger *et al.*, 2019)			x
(Dallasega *et al.*, 2017)			x
(Denno, Dickerson and Anne, 2018)			x
(Engelhardt and Reinhart, 2012)			x
(Erol and Sihn, 2017)		x	
(Fang *et al.*, 2016)			x
(Georgiadis and Michaloudis, 2012)		x	
(Graessler and Poehler, 2017)			x
(Gräler and Pöhler, 2018)			x
(Grundstein, Freitag and Scholz-Reiter, 2017)		x	

SME: This company size was targeted by 11% of the selected approaches.

- *Means of the production planning approach*

Two perspectives have been proposed to study the means of the production planning approach: technology 4.0 and resolution approach. Figure 10 shows the general distribution of production planning approaches for 4.0 technologies. A detailed classification of each approach and the employed 4.0 technologies are presented in Tables 8 and 9.

CPS: From the Industry 4.0 technologies, cyber physical systems were the most employed ones in the analysed approaches.

Simulation: This technology was the second most employed one of the references studied. Around 30% of the approaches employed simulation.

IoT: The Internet of Things was employed as a mean in 20% of the studied approaches.

Cloud Computing: Only five of the 45 studied approaches employed Cloud Computing as a technological resource.

Table 7. References according to the scope of the production planning approach

Perspective	*Company Size*		
References	Large Company	SME	Not Specified
(Helo, Phuong and Hao, 2019)			x
(Iannino *et al.*, 2018)			x
(Illmer and Vielhaber, 2018)			x
(Jiang *et al.*, 2017)			x
(Klein *et al.*, 2018)			x
(Kubo *et al.*, 2016)			x
(Kuhnle, Röhrig and Lanza, 2019)			x
(Kumar *et al.*, 2018)	x		
(Lachenmaier, Lasi and Kemper, 2017)	x		
(Lanza, Stricker and Peters, 2013)			x
(Lanza, Stricker and Moser, 2014b)	x		
(Leusin *et al.*, 2018)			x
(Li *et al.*, 2017)			x
(Lin *et al.*, 2018)	x		
(Meyer, Hans Wortmann and Szirbik, 2011)			x
(Mourtzis and Vlachou, 2018)		x	
(Nonaka *et al.*, 2015)			x
(Qu *et al.*, 2016)			x
(Rajabinasab and Mansour, 2011)			x
(Rossit, Tohmé and Frutos, 2019)			x
(Ruppert and Abonyi, 2018)			x
(Saif *et al.*, 2019)			x
(Shamsuzzoha *et al.*, 2016)		x	
(Shim, Park and Choi, 2018)			x
(Shpilevoy *et al.*, 2013)	x		
(Tang *et al.*, 2017)			x
(Um, Choi and Stroud, 2014)	x		
(Wandt et al. 2012)			x
(Wang *et al.*, 2019)			x
(Waschneck *et al.*, 2018)			x
(Xu and Chen, 2016)			x
(Xu and Chen, 2017)			x
(Zhou *et al.*, 2019)			x
(Zhu, Qiao and Cao, 2017)			x

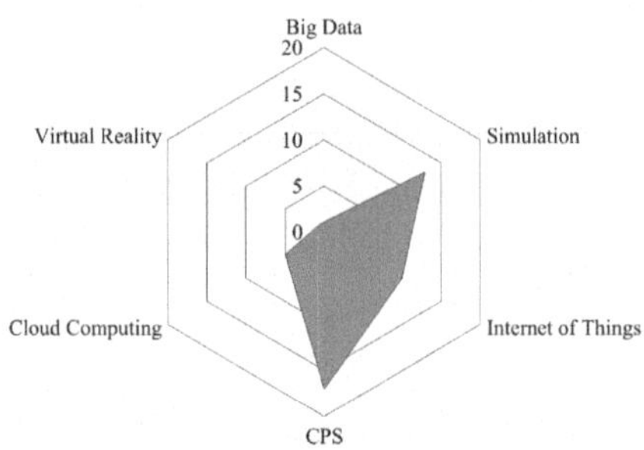

Fig. 10. Number of approaches addressed to 4.0 technology

For the other two technologies, *Big Data* and *Virtual Reality*, only one approach employed each of them. The second category of the perspective is the resolution approach; the results are presented in Fig. 11. A detailed classification for each approach in the resolution categories is presented in Tables 8 and 9.

AI: Artificial Intelligence was the resolution approach the most employed in the references studied. Ten approaches out of 45 employed AI.

All optimization methods (exact methods, heuristics and meta-heuristics) that were defined to analyze the resolution approach were employed in an equal number of approaches (seven approaches for each optimization method).

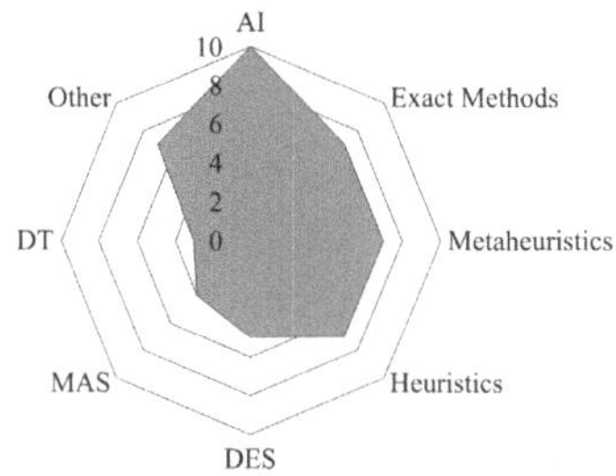

Fig. 11. Number of production planning approaches addressed to resolution approach

Simulation as resolution approach was studied through the DES, DT, and MAS methods. These methods presented a similar number of approaches that employed them (around four approaches per method).

Table 8. References according to Means of the production planning approach Part I

Perspective	**Industry 4.0 Technology**						**Resolution Approach**							
References	*Big Data*	*Simulation*	*IoT*	*CPS*	*Cloud Comp.*	*Virtual Reality*	*AI*	*Exact Methods*	*Meta-heuristics*	*Heuristics*	*DES*	*MAS*	*DT*	*Other*
(Asadzadeh, 2015)		x					x		x	x				
(Berger *et al.*, 2019)				x									x	
(Biesinger *et al.*, 2019)				x									x	
(Dallasega *et al.*, 2017)		x									x			
(Denno, Dickerson and Anne, 2018)		x							x					
(Engelhardt and Reinhart, 2012)			x											x
(Erol and Sihn, 2017)					x		x							
(Fang *et al.*, 2016)			x					x						
(Georgiadis and Michaloudis, 2012)														
(Graessler and Poehler, 2017)				x									x	

IoT: Internet of Things, CPS: Cyber Physical System, AI: Artificial Intelligence, DES: Discrete Event Simulation, MAS: Multi-Agent Systems, DT: Digital Twin.

Table 9. References according to means of the production planning approach Part II

Perspective	**Industry 4.0 Technology**						**Resolution Approach**						
References	*Big Data*	*Simulation*	*IoT*	*CPS*	*Cloud Comp.*	*Virtual Reality*	*AI*	*Exact Methods*	*Meta-heuristics*	*Heuristics*	*DES*	*MAS*	*DT*
(Gräler and Pöhler, 2018)				x									
(Grundstein, Freitag and Scholz-Reiter, 2017)				x				x					
(Helo, Phuong and Hao, 2019)					x				x				
(Iannino *et al.*, 2018)		x						x					
(Illmer and Vielhaber, 2018)				x									
(Jiang *et al.*, 2017)				x								x	
(Klein *et al.*, 2018)				x								x	
(Kubo *et al.*, 2016)				x							x		
(Kuhnle, Röhrig and Lanza, 2019)				x			x				x		
(Kumar *et al.*, 2018)		x							x				
(Lachenmaier, Lasi and Kemper, 2017)				x									
(Lanza, Stricker and Peters, 2013)				x						x			
(Lanza, Stricker and Moser, 2014b)				x						x			
(Leusin *et al.*, 2018)		x					x						
(Li *et al.*, 2017)	x							x					
(Lin *et al.*, 2018)			x										
(Meyer, Hans Wortmann and Szirbik, 2011)			x									x	
(Mourtzis and Vlachou, 2018)				x	x					x			
(Nonaka *et al.*, 2015)		x					x				x		
(Qu *et al.*, 2016)		x					x						
(Rajabinasab and Mansour, 2011)		x							x				
(Rossit, Tohmé and Frutos, 2019)				x									
(Ruppert and Abonyi, 2018)			x							x			
(Saif *et al.*, 2019)			x							x			
(Shamsuzzoha *et al.*, 2016)			x				x						
(Shim, Park and Choi, 2018)		x								x			
(Shpilevoy *et al.*, 2013)		x										x	
(Tang *et al.*, 2017)					x				x				
(Um, Choi and Stroud, 2014)					x			x					
(Wandt et al. 2012)						x					x		
(Wang *et al.*, 2019)			x						x				
(Waschneck *et al.*, 2018)		x					x						
(Xu and Chen, 2016)			x										
(Xu and Chen, 2017)			x					x					
(Zhou *et al.*, 2019)		x					x						
(Zhu, Qiao and Cao, 2017)				x			x						

In the second part of this paper we present a discussion of the results obtained.

5 Conclusions

This paper characterized the production planning approaches in the context of the 4th Industrial Revolution. A data metric analysis (carried in VOSViewer software), based on the keywords relationship between the production planning and the 4.0 technologies, allowed selecting the research strings to conduct the queries and choose the papers. A framework for analyzing the selected production planning approaches based on the production planning activity, the planning horizon, the targeted dimension of agility, the scope and the means, was proposed; this framework could be applied in other works analyzing different production planning activities. A more detailed analysis of the contributions of the studied production planning approaches is presented in the second part of this paper.

References

1. CEFRIO PME 2.0: CEFRIO: Prendre part à la révolution manufacturière? Du rattrapage technologique à l'industrie 4.0 chez les PME, Analysis Report (2016)
2. Kohler, D., Weiz, J.: Industry 4.0: Les défis de la transformation numérique du modèle industriel allemand, Industry, 176 éd. La Documentation française, Paris (2016)
3. Arnold, J. T., Chapman, S.N., Clive, L.M.: Introduction to materials Management. New Jersey (2008). https://doi.org/10.1017/mdh.2014.75
4. Dallasega, P., Rojas, R.A., Rauch, E., Matt, D.T.: Simulation based validation of supply chain effects through ICT enabled real-time-capability in ETO production planning. Procedia Manuf. **11**, 846–853 (2017). https://doi.org/10.1016/j.promfg.2017.07.187
5. Erol, S., Sihn, W.: Intelligent production planning and control in the cloud – towards a scalable software architecture. In: 10th CIRP Conference on Intelligent Computation in Manufacturing Engineering, pp. 571–576. https://doi.org/10.1016/j.procir.2017.01.003 (2016)
6. Georgiadis, P., Michaloudis, C.: Real-time production planning and control system for job-shop manufacturing: a system dynamics analysis. Eur. J. Oper. Res. **216**, 94–104 (2012). https://doi.org/10.1016/j.ejor.2011.07.022
7. Kumar, S., Purohit, B.S., Manjrekar, V., Singh, V., Kumar Lad, B.: Investigating the value of integrated operations planning: a case-based approach from automotive industry. Int. J. Prod. Res. **7543**, 1–22 (2018). https://doi.org/10.1080/00207543.2018.1424367
8. Lanza, G., Stricker, N., Moser, R.: Concept of an intelligent production control for global manufacturing in dynamic environments based on rescheduling. In: IEEE International Conference on Industrial Engineering and Engineering Management, pp. 315–319. Malaysia (2014). https://doi.org/10.1109/IEEM.2013.6962425
9. Jiang, Z., Jin, Y., Mingcheng, E., Li, Q.: Distributed dynamic scheduling for cyber-physical production systems based on a multi-agent system. IEEE Access **6**, 1855–1869 (2017). https://doi.org/10.1109/ACCESS.2017.2780321
10. Rajabinasab, A., Mansour, S.: Dynamic flexible job shop scheduling with alternative process plans: an agent-based approach. Int. J. Adv. Manuf. Technol. **54**, 1091–1107 (2011). https://doi.org/10.1007/s00170-010-2986-7
11. Pellerin, R.: The contribution of Industry 4.0 in creating agility within SMEs. In: Proceedings of the 2018 IRMBAM Conference, pp. 1–10. Nice, France (2018)
12. APICS: The Association of Operations (2011). http://www.apics.org/docs/default-source/industry-content/apics-ombok-framework.pdf?sfvrsn=c5fce1ba_2

13. Lemieux, A.A., Pellerin, R., Lamouri, S., Carbone, V.: A new analysis framework for agility in the fashion industry. Int. J. Agil. Syst. Manag. **5**, 175–197 (2012). https://doi.org/10.1504/IJASM.2012.046904
14. Alberts, F., Hammond, J., Weil, D.: Power to the Edge: Command and Control in the Information Age. Washington DC Report (2003). http://edocs.nps.edu/dodpubs/org/CCRP/Alberts_Power.pdf
15. Ruessmann, M., et al.: Industry 4.0: The Future of Productivity and Growth in Manufacturing (2015). https://inovasyon.org/images/Haberler/bcgperspectives_Industry40_2015.pdf
16. Geisberger, E., Broy, M.: Agenda CPS: Integrierte Forschungsagenda Cyber Physical Systems. Springer Acatech (2012)
17. Berger, C., Zipfel, A., Braunreuther, S., Reinhart, G.: Approach for an event-driven production control for cyber-physical production systems. In: 12th CIRP Conference on Intelligent Computation in Manufacturing Engineering, pp. 349–354. Elsevier (2019). https://doi.org/10.1016/j.procir.2019.02.085
18. Tang, L., et al.: Online and offline based load balance algorithm in cloud computing. Knowl.-Based Syst. **138**, 91–104 (2017). https://doi.org/10.1016/j.knosys.2017.09.040
19. Wandt, R., Friedewald, A., Lödding, H.: Simulation aided disturbance management in one-of-a-kind production on the assembly site. In: IEEE International Conference on Industrial Engineering and Engineering Management, pp. 503–507 (2012). https://doi.org/10.1109/IEEM.2012.6837790
20. Tobon Valencia, E., Lamouri, S., Pellerin, R., Forget, P., Moeuf, A.: Planification de la production et Industrie 4.0: Revue de la littérature. In: CIGI-Qualita, Montréal (2019)
21. Xu, Y., Chen, M.: Improving just-in-time manufacturing operations by using internet of things based solutions. I. In: 9th International Conference on Digital Enterprise Technology, pp. 326–331. Elsevier (2016). https://doi.org/10.1016/j.procir.2016.10.030
22. Xu, Y., Chen, M.: An internet of things based framework to enhance just-in-time manufacturing. J. Eng. Manuf. **232**, 2353–3263 (2017). https://doi.org/10.1177/0954405417731467
23. Lin, P., Li, M., Kong, X., Chen, J., Huang, G.Q., Wang, M.: Synchronization for smart factory – towards IoT-enabled mechanisms. Int. J. Comput. Integr. Manuf. **31**, 624–635 (2018). https://doi.org/10.1080/0951192X.2017.1407445
24. Nonaka, Y., Suginishi, Y., Lengyel, A., Nagahara, S., Kamoda, K., Katsumura, Y.: The S-model: a digital manufacturing system combined with autonomous statistical analysis and autonomous discrete-event simulation for smart manufacturing. In: International Conference on Automation Science and Engineering, CASE 2015. pp. 1006–1011. Gothenburg, Sweden (2015). https://doi.org/10.1109/CoASE.2015.7294230
25. Qu, S., Wang, J., Govil, S., Leckie, J.O.: Optimized adaptive scheduling of a manufacturing process system with multi-skill workforce and multiple machine types: an ontology-based, multi-agent reinforcement learning approach. In: 49th CIRP Conference on Manufacturing Systems, CIRP-CMS 2016, pp. 55–60. Elsevier Stuttgart, Germany (2016). https://doi.org/10.1016/j.procir.2016.11.011.
26. Shamsuzzoha, A., Toscano, C., Carneiro, L.M., Kumar, V., Helo, P.: ICT-based solution approach for collaborative delivery of customized products. Prod. Plan. Control. **27**, 280–298 (2016). https://doi.org/10.1080/09537287.2015.1123322
27. Zhu, X., Qiao, F., Cao, Q.: Industrial big data-based scheduling modeling framework for complex manufacturing system. Adv. Mech. Eng. **9**, 1–12 (2017). https://doi.org/10.1177/1687814017726289
28. Asadzadeh, L.: A local search genetic algorithm for the job shop scheduling problem with intelligent agents. Comput. Ind. Eng. **85**, 376–383 (2015). https://doi.org/10.1016/j.cie.2015.04.006

29. Helo, P., Phuong, D., Hao, Y.: Cloud manufacturing – Scheduling as a service for sheet metal manufacturing. Comput. Oper. Res. **110**, 208–219 (2019). https://doi.org/10.1016/j.cor.2018.06.002
30. Waschneck, B., et al.: Optimization of global production scheduling deep reinforcement learning. In: 51st CIRP Conference on Manufacturing Systems. pp. 1264–1269. Stockholm, Sweden (2018). https://doi.org/10.1016/j.procir.2018.03.212
31. Saif, U., Guan, Z., Wang, C., He, C., Yue, L., Mirza, J.: Drum buffer rope-based heuristic for multi-level rolling horizon planning in mixed model production. Int. J. Prod. Res. **57**, 3864–3891 (2019). https://doi.org/10.1080/00207543.2019.1569272
32. Shim, S., Park, K., Choi, S.: Sustainable production scheduling in open innovation perspective under the fourth industrial revolution, J. Open Innov., 4(4), 42; https://doi.org/10.3390/joitmc4040042
33. Klein, M., et al.: A negotiation-based approach-based production International for scheduling. In: 28th Int. Conference on Flexible Automation and Intelligent Manufacturing, FAIM 2018, pp. 334–341, Elsevier Columbus, USA (2018). https://doi.org/10.1016/j.promfg.2018.10.054
34. Denno, P., Dickerson, C., Anne, J.: Dynamic production system identification for smart manufacturing systems. J. Manuf. Syst. **48**, 1–11 (2018). https://doi.org/10.1016/j.jmsy.2018.04.006
35. Leusin, M.E., Kück, M., Frazzon, E.M., Maldonado, M.U., Freitag, M.: Potential of a multi-agent system approach for production control in smart factories. IFAC-Papers on-line **51**(11), 1459–1464 (2018). https://doi.org/10.1016/j.ifacol.2018.08.309
36. Kuhnle, A., Röhrig, N., Lanza, G.: Autonomous order dispatching in the semiconductor industry using reinforcement learning, In: 12th CIRP Conference on Intelligent Computation in Manufacturing Engineering, pp. 391–396, Elsevier, Naples, Italy (2019). https://doi.org/10.1016/j.procir.2019.02.101
37. Gräler, I., Pöhler, A.: Intelligent Devices in a Decentralized Production System Concept. In: 11th CIRP Conference on Intelligent Computation in Manufacturing Engineering. pp. 116–121 (2018). https://doi.org/10.1016/j.procir.2017.12.186
38. Kubo, R.H., Asato, O.L., Dos Santos, G.A., Nakamoto, F.Y.: Modeling of allocation control system of multifunctional resources for manufacturing systems. In: 12th IEEE Int. Conf. on Industry Applic., INDUSCON (2016). https://doi.org/10.1109/INDUSCON.2016.7874596
39. Lanza, G., Stricker, N., Peters, S.. Ad-hoc rescheduling and innovative business models for shock-robust production systems. In: Forty Sixth CIRP Conference on Manufacturing Systems, pp. 121–126. Elsevier (2013). https://doi.org/10.1016/j.procir.2013.05.021
40. Li, Q., Wang, L., Shi, L., Wang, C.: A data-based production planning method for multi-variety and small-batch production. In: IEEE 2nd Int. Conf. on Big Data Analysis, ICBDA, pp. 420–425. Beijing (2017). https://doi.org/10.1109/ICBDA.2017.8078854
41. Biesinger, F., Meike, D., Kraß, B., Weyrich, M.: A digital twin for production planning based on cyber-physical systems: A case study for a cyber-physical system-based creation of a digital twin. In: 12th CIRP Conference on Intelligent Computation in Manufacturing Engineering, pp. 355–360. Elsevier (2019). https://doi.org/10.1016/j.procir.2019.02.087.
42. Graessler, I., Poehler, A.: Integration of a digital twin as human representation in a scheduling procedure of a cyber-physical production system. In: IEEE International Conference on Industrial Engineering and Engineering Management, pp. 289–293. Singapore (2017). https://doi.org/10.1109/IEEM.2017.8289898
43. Fang, C., Liu, X., Pei, J., Fan, W., Pardalos, P.M.: Optimal production planning in a hybrid manufacturing and recovering system based on the internet of things with closed loop supply chains. Oper. Res. Int. J. **16**(3), 543–577 (2015). https://doi.org/10.1007/s12351-015-0213-x
44. Illmer, B., Vielhaber, M.: Virtual validation of decentrally controlled manufacturing systems with cyber-physical functionalities, 51st CIRP Conf. on Manufacturing Systems Virtual, pp. 509–514. Elsevier Stockholm (2018). https://doi.org/10.1016/j.procir.2018.03.195

45. Meyer, G.G., Hans Wortmann, J.C., Szirbik, N.B.: Production monitoring and control with intelligent products. Int. J. Prod. Res. **49**, 1303–1317 (2011). https://doi.org/10.1080/00207543.2010.518742
46. Um, J., Choi, Y.C., Stroud, I.: Factory planning system considering energy-efficient process under cloud manufacturing. In: 47th CIRP Conf. on Manufacturing Systems, pp. 553–558. Elsevier, Windsor, Canada (2014). https://doi.org/10.1016/j.procir.2014.01.084

A Novel Analysis Framework of 4.0 Production Planning Approaches – Part II

Estefania Tobon Valencia[1(✉)], Samir Lamouri[2], Robert Pellerin[3], and Alexandre Moeuf[4]

[1] Square Flow&Co, Neuilly-sur-Seine, France
e.tobon@flowandco.fr
[2] LAMIH, Arts et Métiers ParisTech, Paris, France
samir.lamouri@ensam.eu
[3] Polytechnique Montréal, Montréal, Canada
robert.pellerin@polymtl.ca
[4] Exxelia, Paris, France
alexandre.moeuf@exxelia.com

Abstract. The emergence of the Fourth Industrial Revolution has brought enterprises to review their production planning processes. Characterized by many technologies this revolution provides managers and planners with multiple means to increase productivity, get an added value from data mining processes and become more agile. This paper, divided into two parts, proposes an analysis framework to conduct a literature review of the production planning approaches developed during the 4th Industrial Revolution. This second part of the paper presents a summary of the contributions, a discussion of results, the gaps in literature and opportunities for further research. The results show that current production planning approaches do not exploit all the 4.0 tools and technologies; researchers usually employ CPS and Simulation. The results demonstrate that all the approaches followed some form of agility even though not all its dimensions have been pursued equally. Our results also indicate that production planning approaches are mainly focused on balancing resource utilization at operational planning level. Finally, the literature review showed that there are not real-case validations.

Keywords: Production planning · Agility · Analysis framework · SME · Large company · Planning horizon · Resolution approach

1 Introduction

The first part of this paper presented the analytical framework, the research methodology and the results of the systemic review. In this second part we first present the contributions of the selected references based on the analytical framework. A discussion of the contributions and gaps in the literature is then exposed. The paper concludes with opportunities for further research in this area.

D. Trentesaux et al. (Eds.): SOHOMA 2021, SCI 987, pp. 133–150, 2021.
https://doi.org/10.1007/978-3-030-80906-5_10

2 Detailed Description of the Contributions Based on the Framework for Systematic Review of 4.0 Production Planning Approaches

The structure of this section is based on the framework for literature review analysis that we presented in the first part of the paper. We describe the research contributions considering the production planning activity, the planning horizon, the dimension of agility, the scope and the means.

- *Characterization of the production planning approach.*

Production Planning Activity-Perspective: Seeking to better classify the approaches addressed to *determine the availability of the required resources*, an additional classification has been proposed for this activity. The following subcategories have been proposed from activities found in the references studied: scheduling problems, support decision making, resource allocation in reconfigurable production lines and others.

Scheduling Problem: The scheduling problem was the core of 22 approaches. Multiple approaches looked at proposing schedules when unforeseen changes, interruptions and disturbances are encountered [5, 6, 9, 12–22]. Disruptions varied from one reference to another. Changes with customer orders, machine breakdowns, shortage of materials and labour strike were some of the studied disruptions by the authors. Multiple solutions were proposed to make the system adaptable to disturbances.

An approach based on event-driven production control was proposed by Berger *et al.* [9]. Nonaka *et al.* [19] proposed to add temporal machines or workers into a bottleneck process to recover production. A strategy of local and global actions was proposed by Lanza *et al.* [5] to face disruptions. Locally, the system was able to choose between processing the part in a different resource with available capacity, change the sequence of products if the subsequent product does not require the broken resource or change the sequence of the production process if no machine is available. At a global scale, the task could be assigned to a different production site. The objective was to minimize the effect of the disruption in the system in terms of due dates and productivity.

Xu and Chen [13] presented a dynamic scheduling framework that allows identifying the required resources to process one order and the required capacity (in hours). The same authors proposed to gather data from the manufacturing process in real time to adjust production schedules taking into consideration variations of the production environment [14]. Real-time data was also exploited by Zhou *et al.* [15], Zhu *et al.* [24] and Asadzadeh [25]. The first group of authors proposed an optimization model that generates a production schedule that is adaptable to the changing status of resources. Zhou *et al.* [15] exploited real-time data as input of a simulation that allows generating an optimal scheduling plan by optimizing the resource utilization. Asadzadeh [25] proposed a scheduling approach for production systems that gathers data through CPS.

Lin et al. [17] presented a model to has traceability and visibility of the production process in real time looking to minimize the holding time, the setup time and the tardiness. Mourtzis and Vlachou [18] proposed only to reschedule the pending tasks of the plan when disturbances are detected. The ongoing tasks remain as they are in the plan.

Other approaches were addressed to improve the scheduling activity itself. Tang *et al.* [10], Asadzadeh [25], Helo *et al.* [26] employed meta-heuristics while Waschneck *et al.* [27] used machine learning techniques for planning function of resource availability and processing time. Rajabinasab and Mansour [7], Saif *et al.* [28], Shim *et al.* [29] presented scheduling approaches that best use capacity and minimize lead-time. A novel scheduling method in which products are scheduled incrementally has been proposed by Klein *et al.* [30]. Grundstein *et al.* [31] proposed an exact method (APC) that balances capacity and workload of the workstations relative to due delivery dates.

Support Decision Making: The approaches presented by Denno *et al.*, Engelhardt and Reinhart and Leusin *et al.* [32–34] use decision making in order release, capacity control and sequencing activities. An autonomous order dispatching system was proposed by Kuhnle *et al.* [35] using CPS and reinforcement learning. The decision-making process was linked to workload, throughput or tardiness optimization.

Local decision-making regarding resource availability and product requirements was proposed in the models presented by Lachenmaier *et al.* [36]. Decentralized decision-making was implemented by using a Cyber Physical Production System (CPPS) [37].

A method that combines simulation and virtual reality to support the decision-making for planners when dealing with a disturbance was proposed by Wandt *et al.* [11]. The simulation allowed evaluating the capacity utilization of resources.

Resource Allocation in reconfigurable production lines: Erol and Sihn [2] presented a model for reconfigurable production lines with efficiency and an optimal resource allocation objectives; according to customer orders, the architecture finds a process model that schedules the operations. Kubo *et al.* [38] proposed a methodology for modelling resource allocation in a system having the capacity to regenerate the manufacturing system. An ad-hoc model for reconfiguring and rescheduling production plants when resource or supplier disturbances affect the internal and external environment was proposed in the research of Lanza *et al.* [39].

Others: A production planning method for mid-term and short-term planning horizons was presented in the works of Li et al. [40]. The authors intend to minimize the number of delay products and the production cycle, and balance production capacity.

Referring to resource allocation in a short-term planning horizon the approaches presented in the works of Biesinger *et al.*, Graessler and Poehler and Ruppert and Ahonyi [41–43] proposed solutions that update in real time the cycle times of processes in the production line. The objective is to maximize productivity and avoid production stops.

The second most studied production planning activity in the analysed approaches was the definition of quantities in which groups of products are made in each period (D: Demand levels). Shpilevoy *et al.*, Qu *et al.*, Zhu *et al.*, Saif *et al.*, Shim *et al.* Klein *et al.* developed in their research [12, 20, 24, 28, 29] the decision to define the quantities to be produced with similar objectives. Qu *et al.* [20] considered lateness minimization for time delivery. Shim *et al.* [28] considered makespan minimization and reduction of set-up times when different types of products require to be processed. Klein *et al.* [29] had as objective to minimize the lead time. Shpilevoy *et al.* [12] proposed a system that allows to increase customer service level. Zhu *et al.* [23] sought to minimize cycle time.

Some references proposed the definition of quantities to be produced taking into account unexpected events. The approach proposed by Wang *et al.* [23] considered new jobs arrivals to define a new schedule. Zhou *et al.* [15] included the analysis of materials' availability in their scheduling process. Fang *et al.* [44] extended production planning to closed-loop supply chains. They proposed to consider returns and their value deterioration during storage, by defining the required quantities of components to satisfy demand. Only one of the references studied considered the definition of demand levels in a tactical planning horizon [22].

With six references the precision of the desired inventory levels was placed third in the studied production planning activities. Dallasega *et al.* [1] presented a production planning approach for optimizing inventory levels on-site and off-site in the construction industry. The objectives of this approach were to avoid buffers on-site in order to decrease the risk of damages and to prevent construction interruptions due to unavailable components on site. Georgiadis *et al.* [3] defined the optimal values of WIP inventory. Meyer *et al.* [46] extended the definition of the inventory levels to the procurement, production and shipment activities. The platform proposed in Um *et al.* [47] defined the quantities that must be completed by subcontractors to reduce inventory holding costs.

Finally, the production planning activity that had the smallest number of approaches was the setting of the required resources. A model-based approach that integrates CPS and design in the production planning to determine the capacity and resource requirements was presented in [45]. Tang et al. [10] reported an online and offline balancing method allowing to identify the required resources. The works of Kubo *et al.* [38] describe a method that reconfigures the manufacturing team by identifying alternative resources when an unexpected event causes a resource failure.

Planning Horizon - Perspective: On the operational planning level, a certain number of approaches were addressed to the scheduling [4, 6, 10, 24–27, 29, 30] and the control activities [34, 45] which have a short-term planning.

New concepts have emerged in production planning approaches addressed to short-term planning. One of them was the generation of plans to face disturbances or disruptions in real time. The following approaches focused on this concept [3, 9, 11, 12, 16, 20, 21, 23, 33, 38, 39, 46]. Another concept was the generation of short-term plans, based on real time data collection from the production environment and data analysis. Real time planning was studied in the works [13–15, 17, 18, 32, 37, 41–43]. On a short-term planning horizon five approaches were proposed to define inventory plans [1, 3, 4, 44, 46]. The mid-term planning horizon was considered exclusively in two approaches. Illmer *et al.* [45] proposed a model-based approach integrating CPS and CAD design for mid-term capacity planning. Um *et al.* [47] did not specify the planning horizon to which the model addresses; instead, when analyzing the production planning, the quantities to be produced per subcontractor were specified on mid-term horizon.

The rest of the approaches that considered mid-term planning horizon presented a proposition for the short-term planning as well [2, 7, 22, 28, 40], or for all three planning horizons [44].

- *Agility*

Five dimensions of agility have been proposed to determine whether the production planning approaches consider some agility improvement.

Responsiveness: Based on the identification of disturbances in the production environment the following approaches described alternatives that combine different methods and respond in a timely manner. Berger *et al.* [9] proposed an event-controlled system to identify disturbances and take corrective actions (using closed-loop control) to carry on with the production plan. Denno *et al.* [32] developed a model responding to disturbances in workstation capacity and detecting resources needing more processing time for a specific job. Gräler and Pöhler [37] describe a system that reacts when malfunctions or missing materials are detected in the CPPS. A CPPS was also referred in the work of Biesinger *et al.* [41], who used a digital twin to minimize the integration time of a new product in the production process. Iannino *et al.* [16] described a MAS in which agents were able to able to exchange information in order to decide the best working strategy or process parameter adaptation.

Jiang et al. [6] developed a rescheduling algorithm that minimizes the processing time. The approach presented by Lin *et al.* [17] is based on vertical and horizontal synchronization of planning processes to improve the timeliness of the due delivery date.

Kubo et al. [38] refer to an automated guided vehicle that, using a regenerated schedule, directs parts to the available resource when machine breakdowns occur. Lanza *et al.*, Mourtzis *et al.* and Kuhnle *et al.* [5, 18, 35] decide upon rescheduling at variations of the available capacity. Xu and Chen, Leusin *et al.*, Meyer *et al.* [14, 34, 46] created adaptable planning systems that take corrective actions when disruptions are detected. Similarly, Nonaka *et al.* [19] proposed a system that is able to detect machine failures and to predict their impact on production on a short-term planning horizon. Once the failure detected, temporal machines or workers could be affected to the bottleneck processes in order to recover production. Based on resource availability analysis that uses machine learning techniques, Waschneck *et al.* [27] design a real time planning algorithm that ensures continuity of the production flow.

Tang *et al.*, Wandt *et al.*, Shpilevoy *et al.*, Zhou *et al.*, Rossit *et al.*, Shamsuzzoha *et al.*, Helo *et al.*, Shim *et al.* and Engelhardt *et al.* configure respectively in [10–12, 15, 21, 22, 26, 29, 33] planning approaches that reduce the response time to unexpected events while trying to increase the number of orders completed for the requested customer date on short-term horizons. The approach presented by Li *et al.* [40] has the same objectives but also covers mid-term planning.

Saif *et al.*, Leusin *et al.*, Lachenmaier *et al.* [28, 34, 36] describe adaptable planning systems that allow to meet new demands from markets by incorporating them in short-term strategies. Saif *et al.* considered in [28] the acceptance of new orders on a mid-term planning horizon. The approaches presented by Georgiadis *et al.* and Wu *et al.* [3, 13] considered a new schedule in case of order cancellation.

Flexibility: A number of approaches studied refer to flexibility as defined in this analysis. Asadzadeh *et al.* and Saif *et al.* [25, 28] proposed optimal short-term plans with multiple objectives: minimize makespan, maximize on time delivery and minimize overdue dates. Georgiadis *et al.* [3] considered other objectives to adapt the schedule, like the desired system's work in process level, the finished product inventory level and the desired customer order backlog. The integration of different functions (production scheduling, maintenance and inventory) that calculate different plans by changing process parameters was proposed in [4]. Rajabinasab *et al.* [7] considered alternative process plans using simulation and varying the job-shop parameters. Different plan propositions came from machine selection and job sequencing.

An adaptable planning system was proposed by Klein *et al.* [30]; plans were adapted according to the values of the available capacity, the processing times and the production costs. In a similar way, Ruppert and Ahonyi [43] conceived an algorithm that continuously updates the cycle time to maximize productivity; this ensured that all operators finish their job in every cycle step.

Other approaches were addressed to identify alternative suppliers to meet customer demands [44]. Shamsuzzoha *et al.* and Lachenmaier *et al.* [22, 36] considered in their approaches collaborative product design. Wang *et al.*, Helo *et al.* and Graessler *et al.* [23, 26, 42], developed methods to select and allocate resources based on information about process times, real time disturbances and skills of the employees. Similarly, Gräler and Pöhler [37] allocated resources by anticipating the consequences of disturbances. Grundstein *et al.* [31] presented a solution that aligns load and capacity. Their systems allowed increasing capacity at stations to compensate demand deviations; also, when the utilization of resources decreases, orders are released as early as possible. Lanza *et al.*, Tang *et al.*, Xu *et al.*, Zhou *et al.*, Rossit *et al.*, Zhu *et al.* and Kubo *et al.* proposed respectively in [5, 10, 13, 15, 21, 24, 38] to generate updated resource plans to adapt the load to the available capacity. An autonomous order dispatching system designed to adapt the production plan to the available capacity of resources was proposed by Kuhnle *et al.* [35]. Qu *et al.* defined in [20] an ontology based on multi-stage, multi-machine and multi-product elements that to provide optimal schedules when allocating multi-skilled employees.

Adaptation: Following the definition adopted of this dimension of agility, four of the studied approaches proved the ability to change work processes. A solution for rescheduling in a global manufacturing environment was presented by Lanza *et al.* in [5]. Based on network collaboration new schedules were proposed by changing the production site to which the product is allocated. Similarly, Shamsuzzoha *et al.* [22] conceived an ICT platform that facilitates industrial collaboration to provide new products in a network.

Industrial collaboration on the supplier side was proposed in the works of Um *et al.* [47]. Their model allows selecting among the existing suppliers the one that has the best machine to perform an operation, which allows at the same time minimizing energy consumption.

The system proposed by Rajabinasab *et al.* [7] evaluates the possibility of performing an operation on alternative machines and also of manufacturing the product with alternative operations or sequences of operations.

Robustness: Three of the studied research works included this dimension of agility. A system that allows to simplify the material flow between the off-site and the on-site construction was proposed by Dallasega *et al.* in [1]. This approach aims at preventing construction interruptions due to unavailable components on site. Three additional objectives were pursued by this method: to reduce overproduction, to avoid not necessary materials' movements and handling and to reduce components lead times.

In order to maintain effectiveness when facing demand variations Erol and Sihn [2] employed a machine learning method to analyse and predict changes in the demand structure. Grundstein *et al.* proposed in [31] an adaptable short-term planning that compensates demand deviations.

Innovation: This dimension of agility was considered to the analysis of the capability to innovate (the ability to do new things and, and old things in a new way). Only one of the studied approaches could be classified in this dimension; the model proposed by Illmer et Vielhaber [45] identifies the resource that could best process a specific task from the commissioning process design. On a mid-term planning horizon this approach intends to set the basis for capacity planning.

Resilience: From the studied methods, only one presented the ability to recover from a destabilizing perturbation in the environment. The model proposed by Lanza *et al.* [5] provides solutions to handle external disruptions such as market variations and short economic cycles.

- *Scope of the production planning approach*

Three categories were defined to classify the scope of the production planning approach: Large Companies, SME and Not Specified (the company's size is not specified).

Not specified: A group of research works have been validated in real production environments, but the authors did not present details about the company size. For example, the configuration methodology proposed by Engelhardt and Reinhart [33] was implemented in a German automotive manufacturer of vehicle seats. Leusin *et al.* [34] tested their MAS system in a production line for automotive transmission forks, with 11 machines and 5 product types. Saif *et al.* [28] validated their heuristic approach in a workshop of a Chinese company that has 8 workstations and produces 15 different types of products. The model proposed by Fang *et al.* [44] was applied in a context of closed loop supply chains with a very large number of returns.

The rest of the approaches that presented a real-case validation involved multiple industrial sectors: sheet metal manufacturing [26], steel coating [16], vehicle production lines [45], semiconductor industry [35], aerospace manufacturing [40], high-precision mould-making industry [46], plant pipe fabrication [19], and a boat constructor with a one-of-a-kind production [11].

Another group of approaches developed experimental case studies [6, 7, 10, 15, 20, 21, 23, 26, 29, 30, 38, 43].

Large Company: Kumar et al. [4] tested their approach in a company that produces engines and transmission sets for automotive manufacturers. A large car manufacturer provided the case-study presented by Lin *et al.*, Lachenmaier *et al.* and Um *et al.* [17, 36, 47]. Kumar et al. [4] exploited their approach in an aircraft manufacture, while Shpilevoy *et al.* [12] tested their model in a jet engine manufacturing environment.

SME: Erol and Sihn, Georgiadis and Michaloudis and Grundstein *et al.* [2, 3, 31], tested their approaches in SMEs from the engineering sector. Mourtzis *et al.* [18] validated their proposition on a real case in a mould-making SME. Shamsuzzoha *et al.* [22] employed three real cases of different industrial sectors: textile and apparel, footwear, and machine tools to apply their platform.

- *Means of the production planning approach*

Two perspectives were proposed to study the employed means for the production planning approach: technology 4.0 and resolution. We first present the results for the studied 4.0 technologies: CPS, Simulation, IoT, Cloud Computing, Big Data and Virtual Reality.

CPS: Georgiadis *et al.*, Mourtzis and Vlachou, Grundstein *et al.*, Gräler, and Pöhle and Kubo *et al.*, [3, 18, 31, 37, 38], employed this technology as a data gathering tool for the production process. In the research works [9, 21, 24, 30, 41] respectively of Berger *et al.*, Rossit *et al.*, Zhu *et al.*, Klein *et al.* and Biesinger *et al.*, this technology was employed to collect information about the state of each resource, the cycle times of each station, the idle time of resources and the load level. Jiang *et al.* [6] employed CPPS not only to collect information but also to perform actions after rescheduling.

In the research carried out by Graessler and Poehler [42], CPS were employed to select the best resource according to their availability and their performance to execute the task. The employees interact with the CPS by indicating whether they could execute the task. A Digital Twin (DT) checks if the assigned task is suitable for the employee in terms of availability and performance. Once the employee is selected, the information of task execution is transferred to the DT. Illmer *et al.* [45] integrated CPS, CAD modelling and simulation to determine the required resource to process a specific task from the design phase. Lanza *et al.* and Li *et al.* [5, 39] use in their works technology to collect information about disruptions in real time; thus, the decision-making and planning processes were aligned with real conditions of the production system. In the study made by Lachenmaier *et al.* [36], the authors defined the requirements, needs and opportunities for applying simulation in CPS environments.

Simulation: Rajabinasab *et al.*, Shpilevoy *et al.*, Iannino *et al.*, Qu *et al.*, Asadzadeh *et al.* founded their approaches in [7, 12, 16, 20, 25] on MAS. Leusin *et al.* [34] employed MAS in a model that loads the demand data from the ERP system. Rajabinasab [7] proposed a multi-agent scheduling system in which the agents are coordinated based on a pheromone approach. The approach of Waschneck *et al.* [27] combines MAS and DT, the latter concept assured the interaction of the employees with the model.

Dallasega *et al.* [1] used data from the ERP and the Manufacturing Execution System (MES) as input for the simulation. Also, information gathered in real time from the production process, the resources and the WIP inventory was used in the simulation-based approach of by Zhou *et al.* [15].

Autonomous DES (discrete event simulation) modules were proposed by Nonaka *et al.* and Denno *et al.* [19, 32]. In the approach of Nonaka *et al.* [19], the values of the DES parameters and from the real manufacturing environment were compared; when a measure performed better than the current production conditions, the modified process parameters were employed in the MES. Denno *et al.* [32] automated the simulation by using genetic algorithms and Petri Nets. Qu *et al.* [20] defined an ontology of multi-stage, multi-machine and multi-product with a multi-agent learning approach that seeks optimal schedules for multi-skilled workforce. The use of simulation allowed Kumar *et al.* [4] to integrate multiple functions (production, maintenance and inventory).

IoT: The main reason for using this technology was to increase visibility and traceability (Lin *et al.* [17]). Some works employed a product-centric approach. Saif *et al.* and Meyer *et al.* [28, 46], used the concept of intelligent product that makes possible the exchange of information with resources and information systems in order to coordinate purchasing, production and shipment. In the approach of Engelhardt and Reinhart [33] the products (backrests and seats) were able to communicate the order number and specific quality data (failure code and tolerance data). In case of problems with order processing, the system directs the order to the rework process and accelerates its treatment. Data from end-of-life products, defective products and products in the manufacturing process are collected through IoT in a closed-loop supply chain environment in the research of Gang *et al.* [44]. Ruppert and Ahonyi [43] connected information about products and assembly lines.

Other approaches use the IoT technology to gather information about resources. In the research of Shamsuzzoha *et al.* [22], IoT was used to monitor the resources of the partner organizations. Real time data from resources transferred to users allowed Xu and Chen, and Wand *et al.* [13, 14, 23] to optimize the production schedules.

Cloud Computing: A cloud-based Software as a Service (SaaS) model for intelligent production planning and control was proposed by Erol and Sihn [2]. The model proposed by Helo *et al.* [26] is based on cloud manufacturing service. The cloud technology allowed to monitor and coordinate all the production lines in real time. Um *et al.* [47] evaluated in a cloud computing platform the best option among multiple subcontractors to develop a specific aircraft project. Tang *et al.* and Mourtzis and Vlachou [10, 18], developed short-term planning systems integrated in a cloud-based CPS.

Big Data: The approach of Li *et al.* [40] was the only one that referred to this technology. A big data set containing information of the product, processing steps, inventory levels, work centres, and resources represent the input of the model.

Virtual Reality: The model proposed by Wandt *et al.* [11] used virtual reality to evaluate the geometry and positioning of critical assembly pieces in a boat constructor environment. These evaluations were made after simulating the disturbance and the countermeasures. If all assembly sequences are valid, then the production plan is accepted.

The second perspective category was the resolution approach. Three categories were considered: AI (Artificial Intelligence), Optimization and Simulation.

AI: In the studies of Erol *et al.* and Kuhnle *et al.* [2, 35], the actions of the agents are performed according to the reinforcement learning and neural network analysis. Neural networks were also employed in the optimization approach of Zhu *et al.* [24]. Leusin *et al.* [34] defined an AI agent that is responsible for optimizing the production sequence. Waschneck *et al.* [27] present an application of machine learning with agent algorithms (DQN – Deep Q-Networks) and reinforcement learning for production scheduling.

Zhou *et al.* and Shamsuzzoha *et al.* [15, 22] applied knowledge extracted from data to intelligent manufacturing. Nonaka *et al.* [19] extracted knowledge from cyber models by help of data mining that provide the input for a DES module. An autonomous statistical module parameterizes through data mining the net processing time according to the production strategy (stock-to-order, engineer-to-order). In the proposed scheduling approach formulated by Qu *et al.* [20], a multi-agent approximate Q-learning algorithm updates in real time the production schedule function of the changes in the production system. This algorithm, inspired by the depth-limited search method, allows distributing the load between employees and machines..

Optimization - Exact Methods: Fang et al. used in [44] and Um *et al.* in [47] the Mixed-integer linear programming (MILP) model, while Iannino *et al.* [16] used the Integer linear programming model (ILP). The System Dynamics (SD) approach was used by Georgiadis and Michaloudis in [3]. Grundstein *et al.* [31] named their resolution approach APC, after Autonomous Production Control. Xu and Chen [14] proposed a dynamic production planning framework that includes an optimization module. The base of these approaches resided in exact methods. Li *et al.* [40] presented the HTN (Hierarchical Task Network) modelling method. An optimization algorithm named by the authors HTN-CPM was used to determine the number of parts to be produced in a defined time interval and to minimize the delay in due delivery dates.

Optimization Meta-heuristic: Genetic algorithms were developed in the research works [4, 25, 26, 32] carried out respectively by Kumar *et al.*, Asadzadeh *et al.*, Helo *et al.* and Denno *et al.* Rajabinasab et al. applied in their work [7] an ant colony algorithm, while Tang *et al.* [10] developed a bacterial foraging optimization algorithm and Wang *et al.* designed in [23] a particle swarm optimization algorithm.

Optimization Heuristic: Lanza *et al.* and Asadzadeh *et al.* [5, 25], proposed local search algorithms as resolution approaches. In the research of Lanza *et al.* [5], a reverse branch and bound algorithm allowed successive evaluation of affected assembly lines in a network. The algorithm compared the costs of rescheduling, of tardiness and of allocating products to other production sites. Mourtzis and Vlachou [18] developed a new heuristic named ASA (adaptive scheduling algorithm); they compare the efficiency of the resolution approach (in terms of mean flow time, mean capacity utilization and mean tardiness) with different dispatching rules, like FIFO (first in first out) and EDD (earliest due date).

Production plans for tactical and operational planning horizons were computed by Saif *et al.* [28] with a heuristic based on the drum buffer rope (DBR) theory.

The algorithm described by Ruppert and Ahonyi in [43] sets the cycle time based on the difference between the theoretical and the measured activity times. Shim *et al.* [29] proposed a scheduling heuristic algorithm based on five-lot-selection inventory rules.

Simulation – DES: In the planning methods presented by Dallasega *et al.*, Kuhnle *et al.* and Kubo *et al.* respectively in [1, 35, 38], the analysis and design of the manufacturing system control is based on the concept of Discrete Event Dynamic Systems (DEVS). Nonaka *et al.* [19] used DEVS to support the decision-making of planners when detractors generate fluctuations in the manufacturing process. Similarly, Wandt *et al.* [11] worked with DEVS to simulate firstly the disturbance and its impact in the production plan, and secondly the reactions at disturbance to evaluate whether the requirements on due delivery dates and resource capacity are still respected.

Simulation – MAS: Multi-agent systems were used in 4 of the 45 studied production planning developments. In the research reported by Jiang *et al.* [6] the negotiation mechanisms of the agents were based on the maximization of the processing quality and the minimization of the processing and delay costs. Klein *et al.* [30] associated an agent to each production process; agents can participate in actions and direct negotiations based on the production price, the processing time and CO_2 emission. In the work of Meyer *et al.* [46], an agent was assigned to each product, and to the following activities of the manufacturing process: purchase, shipment, production and sales. A MAS that performs resource management in a smart factory was described by Shpilevoy *et al.* in [12]. The negotiations of the system are carried by the agents taking into account the order priority, the structure or products, the technological processes, the particularities of operators and machines, the operation sequence and the time needed to execute each operation.

Simulation - DT: Digital twins were used in three of the studied approaches: Berger et al. [9] used production DT to simulate machine failure, longer unpredicted maintenance processes, and resource malfunctions not yet discovered. A DT was integrated in the research of Biesinger *et al.* [41] in order to reduce the existing gap between production planning and shop floor processes. A solution to involve human operators in the short-term planning process was proposed by Graessler and Poehler in [42] by developing a DT for the communication of humans with the production system in CPS.

Other: The works of Lachenmaier *et al.*, and Illmer *et al.* [36, 45], employed simulation as resolution approach. An inverse optimization approach was used by Shamsuzzoha et al. in [21]; the optimization process computes the values of the parameters that generate an imposed schedule. Xu and Chen [13] also presented an optimization approach the category of which could not be specified. Engelhardt *et al.* [33] developed a methodology for implementing an IoT framework with real time data collection in a production planning perspective. Gräler and Pöhler, [37] proposed a methodology for the implementation of computational intelligence in a manufacturing process. For the two last analysed methodologies, a resolution approach was presented. A system framework of IoT automatic synchronization with APS was developed by Lin *et al.* [17]. This synchronization was developed in RAPShell.

3 Discussion

The results of the characterization of the production planning approaches show a great disparity (see Fig. 1 in Part I). The evaluation of the availability of the necessary resources is the production planning activity the most targeted by researchers. This result is aligned with the dimension of agility that was present in the greatest number of approaches - responsiveness (see Fig. 3 in Part I). The novel production planning approaches focus on detecting disturbances or unexpected events and reacting in a timely manner by help of adaptable production plans.

Fewer approaches have been addressed to define the quantities to be produced by group of products in each period. This is surprising as it has been shown that in the 4th Industrial Revolution the demand is characterized by big variations, high standards for delivery on time and individual customer needs [5, 6, 23, 48, 49]. Very few researches have proposed solutions to specify the inventory levels and set the required resources.

The literature review clearly demonstrated that current production planning approaches have a huge preference for the short-term planning level (see Fig. 2 - Part I). This result is completely coherent with the dimensions of agility that have got special attention from the researchers. The two dimensions of agility that are the core of the studied production planning approaches – responsiveness and flexibility – were related by Pellerin [8] short-term horizons. Few approaches incorporated multiple planning levels in their solution.

The results of the studied dimensions of agility prove that the production planning approaches of the 4th Industrial Revolution look for some agility improvement. However, there is a great disparity between the agility objectives (see Fig. – Part I). Responsiveness is the most observed agility dimension targeted by production planning approaches. The researchers are interested in systems that adjust themselves in real time to the current internal and external conditions of the manufacturing environment. Similarly, flexibility is the second most targeted dimension of agility in the studied approaches; in fact researchers develop solutions that plan better production on short term.

Fewer approaches have proposed a solution for adaptation and robustness. Research in this area seeks to incorporate industrial collaboration for supplier selection and production processing. The approaches also provide means to face demand variations. Resilience and innovation do not appear as priorities of the studied production planning approaches. Both agility dimensions were related to long and mid-term planning horizons by Pellerin [8]. Tactical and strategical planning levels were not considered in this study.

The results of the study concerning production planning proved that there is a clear need to differentiate them by specific company size. Moeuf *et al.* [50] showed that SMEs have a completely different managerial structure compared to Large Companies. They observed that SMEs have a lack of expert support functions, and compared with large companies they present lower productivity, higher costs and less on time delivery. Additionally, it would be important for researchers to validate production planning approaches in real production planning scenarios. Many approaches presented only experimental validations. A real case validation could encourage practitioners to effectively apply the research propositions.

The analysis of the means used in the developed approaches showed that there is also a disparity in the use of the analysed group of technologies (see Fig. 5 - Part I). The most employed technologies in the context of the 4th Industrial Revolution are CPS and Simulation. Unfortunately, the literature review did not report systematic use of Big Data and virtual reality technologies; it is still necessary to bring these technologies closer to production planning scenarios. The approaches studied to determine the availability of resources operated in principal with CPS as 4.0 technology. It was also shown that Cloud Computing has mainly been used to define resource availability. The precision of inventory levels is the activity for which researchers have integrated more informational and operational technologies (IT – OT). At least one use case per technology (except for Big Data) was found in the literature review.

Simulation and IoT were found in applications in all the production planning activities except for the definition of the necessary resources. For this activity, the researchers employed in most of the cases CPS; the only case that was reported on Big Data was related to the demand levels.

Referring to the other category of means – the resolution approach – the results in Fig. 6 – Part I indicate that the studied production planning approaches are mainly based on AI and all optimization methods (exact methods, meta-heuristics and heuristics). Simulation as a resolution approach is clearly not the most preferred by research. While AI, meta-heuristics and heuristics are more employed in the definition of resource availability, exact methods are involved in an equal number of applications for demand level definition, inventory level precision and resource availability determination. DES has been employed as a resolution approach of all the studied production planning activities. DT was only found in approaches addressing the definition of resource availability. MAS appear in approaches of all the production planning activities except for the definition of the required resources.

Among the selected references, no case of artificial intelligence as a resolution approach for the precision of inventory levels nor the setting of the required resources was found. Similarly, heuristics were not applied as means of resolution of these two activities. An analysis of mean perspectives, technologies and resolution approaches allowed to specify that in the current production planning research works AI is used in principal together with simulation as 4.0 technology. For the optimization techniques, exact methods are the only ones that have been used with all the five analysed 4.0 technologies. Meta-heuristics have been applied with simulation as a technology in IoT and Cloud Computing frameworks. Heuristics was found in at least one application case with each technology except for Big Data. For the simulation techniques, researchers use DES when employing simulation as a technology and CPS. while MAS was used to develop an application with the latter two technologies and also with IoT. Finally, DT are integrated in principal with CPS. There appear to be few production planning approaches that use AI when Internet of Things, CPS and Cloud Computing are the selected technologies.

At present, an analysis of the agility dimensions with technological means is proposed. The approaches that employed CPS looked mainly for responsiveness and flexibility. Similarly, these two dimensions of agility were targeted when using simulation as a technology and IoT. There appear to be few approaches that use Cloud Computing when robustness and adaptation are the targeted objectives. The two use cases of Big Data and Virtual Reality point at responsiveness.

Referring to robustness, a great disparity of technology use was identified. Only one use case of the simulation, IoT, CPS and Cloud Computing technologies, were found. Similarly, when adaptation was the target one use case for each technology except was found except for Big Data. The only works addressing resilience and innovation employed CPS. In the few cases related to large companies, the approaches had as agility objectives responsiveness, innovation and/or adaptation. These objectives were also sought by the approaches that targeted SMEs, for which two cases of robustness analysis were found.

Finally, taking into consideration the production planning activity and the dimensions of agility it was found that the approaches defining the demand levels mainly targeted responsiveness, while flexibility was the second most targeted objective. Resilience and innovation were rarely targeted in this planning activity.

The main objective of the approaches that addressed the precision of inventory levels was flexibility, while responsiveness was on the second position among the targeted objectives. The research works with approaches for setting the required resources targeted responsiveness and flexibility equally. The only approach that focused on innovation as the main agility dimension considered the planning activity.

Responsiveness was the most targeted objective in the approaches that focus on determine the availability of the needed resources,. With no significant difference, flexibility was the second dimension of agility targeted by these approaches. A few number of approaches addressing this activity looked for robustness, adaptation and resilience. In all works focusing on innovation the production planning approach was present.

4 Conclusions. Perspectives and Limitations

This paper characterized the production planning approaches in the context of the 4th Industrial Revolution. The literature review identified 46 production planning approaches that employ the following 4.0 technologies: Big Data, Simulation, IoT, CPS, Cloud Computing and Virtual Reality. The reviewed publications revealed that the production planning approaches in the 4th Industrial Revolution are mainly oriented towards determining resource availability on a short-term planning horizon. The conducted review has also shown that CPS, Simulation and IoT are the most exploited technologies, that is around 30% of the nine tools and technologies that have been identified as pillars of Industry 4.0.

According to the conducted study, it seems that there are multiple ways of combining technological and resolution means to improve different dimensions of agility. However, the review identified gaps in the current production planning approaches. In fact, there is a breach in the research in the dimensions of resilience, innovation, robustness and adaptation.

In line with the scope of the production planning approaches, few research works have been addressed to a specific company size. From this gap, several issues emerge. Which agility objectives are more within the reach of SMEs from a production planning perspective? Which technological means are best suited to the industrial and organizational context of SMEs? How should large companies address agility in production planning processes?

The conducted literature review has some limitations. First, the selection of papers for the review was done taking into consideration four general production planning activities: the definition of demand levels, the precision of the inventory levels, the set of required resources and the evaluation of resource availability). An extension through a more precise specification of such activities may certainly influence the number of selected papers. Second, there are many approaches that have not been applied in a real industrial context. Around 50% of the studied approaches (22 out of 46 approaches) did not present a real-case validation. It would be important to implement these proposals in real industrial cases to ensure the sharing of knowledge. Third, artificial intelligence has been here studied as a single resolution approach. An extension of this perspective on the multiple methods that it covers would be pertinent.

Further research will be conducted to identify if the production planning approaches of the Fourth Industrial Revolution look also for other objectives than agility. The definition of the implementation process of 4.0 technologies in the production planning processes requires to be further investigated.

References

1. Dallasega, P., Rojas, R.A., Rauch, E., Matt, D.T.: Simulation based validation of supply chain effects through ICT enabled real-time-capability in ETO production planning. In: FAIM, pp. 846–853 (2017). https://doi.org/10.1016/j.promfg.2017.07.187
2. Erol, S., Sihn, W.: Intelligent production planning and control in the cloud – towards a scalable software architecture. In: 10th CIRP Conf. on Intell. Comput. in Manufacturing Engineering, pp. 571–576. Ischia, Italy (2016). https://doi.org/10.1016/j.procir.2017.01.003
3. Georgiadis, P., Michaloudis, C.: Real-time production planning and control system for job-shop manufacturing: a system dynamics analysis. Eur. J. Oper. Res. **216**, 94–104 (2012). https://doi.org/10.1016/j.ejor.2011.07.022
4. Kumar, S., Purohit, B.S., Manjrekar, V., Singh, V., Kumar Lad, B.: Investigating the value of integrated operations planning: a case-based approach from automotive industry. Int. J. Prod. Res. **7543**, 1–22 (2018). https://doi.org/10.1080/00207543.2018.1424367
5. Lanza, G., Stricker, N., Moser, R.: Concept of an intelligent production control for global manufacturing in dynamic environments based on rescheduling. In: IEEE International Conference on Industrial Engineering and Engineering Management, pp. 315–319. Malaysia (2014). https://doi.org/10.1109/IEEM.2013.6962425
6. Jiang, Z., Jin, Y., Mingcheng, E., Li, Q.: Distributed dynamic scheduling for cyber-physical production systems based on a multi-agent system. IEEE Access **6**, 1855–1869 (2017). https://doi.org/10.1109/ACCESS.2017.2780321
7. Rajabinasab, A., Mansour, S.: Dynamic flexible job shop scheduling with alternative process plans: an agent-based approach. Int. J. Adv. Manuf. Technol. **54**, 1091–1107 (2011). https://doi.org/10.1007/s00170-010-2986-7
8. Pellerin, R.: The contribution of Industry 4.0 in creating agility within SMEs. In: Proceedings of the 2018 IRMBAM Conference, pp. 1–10. Nice, France (2018)
9. Berger, C., Zipfel, A., Braunreuther, S., Reinhart, G.: Approach for an event-driven production control for cyber-physical production systems. In: 12th CIRP Conference on Intelligent Computation in Manufacturing Engineering, pp. 349–354. Elsevier B.V., Naples, Italy (2019). https://doi.org/10.1016/j.procir.2019.02.085
10. Tang, L., et al.: Online and offline based load balance algorithm in cloud computing. Knowl.-Based Syst. **138**, 91–104 (2017). https://doi.org/10.1016/j.knosys.2017.09.040
11. Wandt, R., Friedewald, A., Lödding, H.: Simulation aided disturbance management in one-of-a-kind production on the assembly site. In: IEEE Int. Conf. on Industrial Engineering and Eng. Management, pp. 503–507 (2012). https://doi.org/10.1109/IEEM.2012.6837790
12. Shpilevoy, V., Shishov, A., Skobelev, P., Kolbova, E., Kazanskaia, D., Shepilov, Y., Tsarev, A.: Multi-agent system "Smart Factory" for real-time workshop management in aircraft jet engines production. In: 11th IFAC Workshop on Intelligent Manufacturing Systems, pp. 204–209. IFAC, São Paulo, Brazil (2013). https://doi.org/10.3182/20130522-3-BR-4036.00025
13. Xu, Y., Chen, M.: Improving just-in-time manufacturing operations by using Internet of Things based solutions. In: 9th Int. Conference on Digital Enterprise Technology, pp. 326–331. Elsevier B.V. (2016). https://doi.org/10.1016/j.procir.2016.10.030
14. Xu, Y., Chen, M.: An Internet of Things based framework to enhance just-in-time manufacturing. J. Eng. Manuf. **232**, 2353–3263 (2017). https://doi.org/10.1177/0954405417731467

15. Zhou, G., Zhang, C., Li, Z., Ding, K., Wang, C.: Knowledge-driven digital twin manufacturing cell towards intelligent manufacturing. Int. J. Prod. Res. **7543**(4), 1034–1051 (2019). https://doi.org/10.1080/00207543.2019.1607978
16. Iannino, V., Vannocci, M., Vannucci, M., Colla, V., Neuer, M.: A multi-agent approach for the self-optimization of steel production. Int. J. Simul. Syst. Sci. Technol. **19**, 1–7 (2018). https://doi.org/10.5013/IJSSST.a.19.05.20
17. Lin, P., Li, M., Kong, X., Chen, J., Huang, G.Q., Wang, M.: Synchronization for smart factory – towards IoT-enabled mechanisms. Int. J. Comput. Integr. Manuf. **31**, 624–635 (2018). https://doi.org/10.1080/0951192X.2017.1407445
18. Mourtzis, D., Vlachou, E.: A cloud-based cyber-physical system for adaptive shop-floor scheduling and condition-based maintenance. J. Manuf. Syst. **47**, 179–198 (2018). https://doi.org/10.1016/j.jmsy.2018.05.008
19. Nonaka, Y., Suginishi, Y., Lengyel, A., Nagahara, S., Kamoda, K., Katsumura, Y.: The S-model: a digital manufacturing system combined with autonomous statistical analysis and autonomous discrete-event simulation for smart manufacturing. In: International Conference on Automation Science and Engineering, CASE 2015, pp. 1006–1011. Gothenburg, Sweden (2015). https://doi.org/10.1109/CoASE.2015.7294230
20. Qu, S., Wang, J., Govil, S., Leckie, J.O.: Optimized adaptive scheduling of a manufacturing process system with multi-skill workforce and multiple machine types: an ontology-based, multi-agent reinforcement learning approach. In: 49th CIRP Conference on Manufacturing Systems, CIRP-CMS, pp. 55–60. Elsevier B.V., Stuttgart, Germany (2016). https://doi.org/10.1016/j.procir.2016.11.011
21. Rossit, D.A., Tohmé, F., Frutos, M.: Industry 4.0: smart scheduling. Int. J. Prod. Res. **57**, 3802–3813 (2019). https://doi.org/10.1080/00207543.2018.1504248
22. Shamsuzzoha, A., Toscano, C., Carneiro, L.M., Kumar, V., Helo, P.: ICT-based solution approach for collaborative delivery of customized products. Prod. Plan. Control. **27**, 280–298 (2016). https://doi.org/10.1080/09537287.2015.1123322
23. Wang, X., Yew, A.W.W., Ong, S.K., Nee, A.Y.C.: Enhancing smart shop floor management with ubiquitous augmented reality. Int. J. Prod. Res. **7543**(8), 2352–2367 (2019). https://doi.org/10.1080/00207543.2019.1629667
24. Zhu, X., Qiao, F., Cao, Q.: Industrial big data-based scheduling modeling framework for complex manufacturing system. Adv. Mech. Eng. **9**, 1–12 (2017). https://doi.org/10.1177/1687814017726289
25. Asadzadeh, L.: A local search genetic algorithm for the job shop scheduling problem with intelligent agents. Comput. Ind. Eng. **85**, 376–383 (2015). https://doi.org/10.1016/j.cie.2015.04.006
26. Helo, P., Phuong, D., Hao, Y.: Cloud manufacturing – scheduling as a service for sheet metal manufacturing. Comput. Oper. Res. **110**, 208–219 (2019). https://doi.org/10.1016/j.cor.2018.06.002
27. Waschneck, B., Reichstaller, A., Belzner, L., Altenmüller, T., Bauernhasl, T., Knapp, A., Kyek, A.: Optimization of global production scheduling deep reinforcement learning. In: 51st CIRP Conference on Manufacturing Systems, pp. 1264–1269. Stockholm, Sweden (2018). https://doi.org/10.1016/j.procir.2018.03.212
28. Saif, U., Guan, Z., Wang, C., He, C., Yue, L., Mirza, J.: Drum buffer rope-based heuristic for multi-level rolling horizon planning in mixed model production. Int. J. Prod. Res. **57**, 3864–3891 (2019). https://doi.org/10.1080/00207543.2019.1569272
29. Shim, S., Park, K., Choi, S.: Sustainable production scheduling in open innovation perspective under the fourth industrial revolution. J. Open Innovation **4**(4), 42 (2018). https://doi.org/10.3390/joitmc4040042

30. Klein, M., et al.: A negotiation-based approach-based production International for scheduling. In: 28th Int. Conf. on Flexible Automation and Intelligent Manufacturing, FAIM 2018, pp. 334–341. Elsevier, Columbus, OH, USA (2018). https://doi.org/10.1016/j.promfg.2018.10.054
31. Grundstein, S., Freitag, M., Scholz-Reiter, B.: A new method for autonomous control of complex job shops – integrating order release, sequencing and capacity control to meet due dates. J. Manuf. Syst. **42**, 11–28 (2017). https://doi.org/10.1016/j.jmsy.2016.10.006
32. Denno, P., Dickerson, C., Anne, J.: Dynamic production system identification for smart manufacturing systems. J. Manuf. Syst. **48**, 1–11 (2018). https://doi.org/10.1016/j.jmsy.2018.04.006
33. Engelhardt, P., Reinhart, G.: Approach for an RFID-based situational shop floor control. In: IEEE International Conference on Industrial Engineering and Engineering Management, pp. 444–448 (2012). https://doi.org/10.1109/IEEM.2012.6837778
34. Leusin, M.E., Kück, M., Frazzon, E.M., Maldonado, M.U., Freitag, M.: Potential of a multi-agent system approach for production control in smart factories. In: IFAC – Paper Online, 51, issue 11, pp. 1459–1464. Elsevier (2018). https://doi.org/10.1016/j.ifacol.2018.08.309
35. Kuhnle, A., Röhrig, N., Lanza, G. Autonomous order dispatching in the semiconductor industry using reinforcement learning. In: 12th CIRP Conf. on Intell. Comput. in Manuf. Eng., pp. 391–396. Elsevier, Naples (2019). https://doi.org/10.1016/j.procir.2019.02.101
36. Lachenmaier, J.F., Lasi, H., Kemper, H.G.: Simulation of production processes involving cyber-physical systems. In: 10th CIRP Conf. on Intelligent Computation in Manufacturing Engineering, pp. 577–582, Ischia (2017). https://doi.org/10.1016/j.procir.2016.06.074
37. Gräler, I., Pöhler, A.: Intelligent devices in a decentralized production system concept. In: 11th CIRP Conference on Intelligent Computation in Manufacturing Engineering, pp. 116–121. Naples (2018). https://doi.org/10.1016/j.procir.2017.12.186
38. Kubo, R.H., Asato, O.L., Dos Santos, G.A., Nakamoto, F.Y.: Modeling of allocation control system of multifunctional resources for manufacturing systems. In: 12th IEEE Int. Conf. on Ind. App. INDUSCON (2016). https://doi.org/10.1109/INDUSCON.2016.7874596
39. Lanza, G., Stricker, N., Peters, S.: Ad-hoc rescheduling and innovative business models for shock-robust production systems. In: 46th CIRP Conference on Manufacturing Systems, pp. 121–126. Elsevier (2013). https://doi.org/10.1016/j.procir.2013.05.021
40. Li, Q., Wang, L., Shi, L., Wang, C.: A data-based production planning method for multi-variety and small-batch production. In: IEEE 2nd Int. Conf. on Big Data Analysis, ICBDA, pp. 420–425. Beijing, China (2017). https://doi.org/10.1109/ICBDA.2017.8078854
41. Biesinger, F., Meike, D., Kraß, B., Weyrich, M.: A digital twin for production planning based on cyber-physical systems: a case study for a cyber-physical system-based creation of a digital twin. In: 12th CIRP Conf. on Intell. Comput. in Manufacturing Engineering, pp. 355–360. Elsevier, Naples (2019). https://doi.org/10.1016/j.procir.2019.02.087
42. Graessler, I., Poehler, A.: Integration of a digital twin as human representation in a scheduling procedure of a cyber-physical production system. In: IEEE International Conference on Industrial Engineering and Engineering Management, pp. 289–293. Singapore (2017). https://doi.org/10.1109/IEEM.2017.8289898
43. Ruppert, T., Abonyi, J.: Industrial internet of things-based cycle time control of assembly lines. In: 2018 IEEE International Conference on Future IoT Technologies, Future IoT, pp. 1–4 (2018). https://doi.org/10.1109/FIOT.2018.8325590
44. Fang, C., Liu, X., Pei, J., Fan, W., Pardalos, P.M.: Optimal production planning in a hybrid manufacturing and recovering system based on the internet of things with closed loop supply chains. Oper. Res. Int. J. **16**(3), 543–577 (2015). https://doi.org/10.1007/s12351-015-0213-x
45. Illmer, B., Vielhaber, M.: Virtual validation of decentrally controlled manufacturing systems with cyber-physical functionalities. In: 51st CIRP Conf. on Manufacturing Systems Virtual, pp. 509–514. Elsevier, Stockholm (2018). https://doi.org/10.1016/j.procir.2018.03.195

46. Meyer, G.G., Hans Wortmann, J.C., Szirbik, N.B.: Production monitoring and control with intelligent products. Int. J. Prod. Res. **49**, 1303–1317 (2011). https://doi.org/10.1080/00207543.2010.518742
47. Um, J., Choi, Y.C., Stroud, I.: Factory planning system considering energy-efficient process under cloud manufacturing. In: 47th CIRP Conf. on Manufacturing Systems, pp. 553–558. Elsevier, Windsor, Canada (2014). https://doi.org/10.1016/j.procir.2014.01.084
48. Autenrieth, P., Lörcher, C., Pfeiffer, C., Winkens, T., Martin, L., Overview, A.: Current significance of IT-infrastructure enabling industry 4.0 in large companies. In: IEEE International Conference on Engineering, Technology and Innovation (ICE/ITMC 2018) (2018). https://doi.org/10.1109/ICE.2018.8436244
49. Sousa, R.A., Varela, M.L.R., Alves, C., Machado, J.: Job shop schedules analysis in the context of industry 4.0. In: 2017 International Conference on Engineering, Technology and Innovation ICE/ITMC, pp. 711–717 (2018). https://doi.org/10.1109/ICE.2017.8279955
50. Moeuf, A., Pellerin, R., Lamouri, S., Tamayo-Giraldo, S., Barbaray, R.: The industrial management of SMEs in the era of Industry 4.0. Int. J. Prod. Res. **56**, 1118–1136 (2018). https://doi.org/10.1080/00207543.2017.1372647

A Survey About BIM Interoperability and Collaboration Between Design and Construction

Léa Sattler[1,3], Samir Lamouri[1,3], Robert Pellerin[2], Thomas Paviot[1], Dominique Deneux[3](✉), and Thomas Maigne[4]

[1] Arts et Métiers ParisTech, Paris, France
{lea.sattler,samir.lamouri,thomas.paviot}@ensam.eu
[2] Polytechnique Montréal, Montréal, Canada
robert.pellerin@polymtl.ca
[3] LAMIH, Université Polytechnique Hauts-de-France, Valenciennes, France
dominique.deneux@uphf.fr
[4] Treegram, Paris, France
tmaigne@treegram.fr

Abstract. BIM interoperability has been recognized as a strong brake to BIM collaboration and is a very active research field. This paper systematically reviews 80 applicative articles about BIM collaboration and interoperability recently published, in order to analyze the different trends and to create an understandable map of the subject for researchers. The subject is analysed in three analytical frameworks: (1) the AEC context in which collaboration and interoperability issues are raised, (2) the collaborative goal of BIM interoperability, and (3) BIM collaboration and interoperability suggested solutions. The main findings of this paper are three research gaps: (1) BIM collaboration and interoperability are not often conjointly addressed; (2) solving approaches rarely consider the problem though the trivial angle of data querying and retrieval; (3) interoperability for geometrically complex BIM models could receive more attention. Research perspectives are consequently deducted.

Keywords: BIM · Interoperability · Collaboration · Parameterization · Queries · Automation · Integration

1 Introduction

Building Information Modelling (BIM) aims to improve data reliability and data-based workflows in the AEC industry (Architectural, Engineering and Construction). Among its various goals, the ultimate step of BIM is the deployment of integrated processes [1]. Integration modes, such as IPD (Integrated Project Delivery), are characterized by the fact that all participants "are involved early in the design phase of a project and work collaboratively to reduce work repetition and waste to enhance the merit of a project by considering the demand at each phase such as the design phase and the construction

D. Trentesaux et al. (Eds.): SOHOMA 2021, SCI 987, pp. 151–179, 2021.
https://doi.org/10.1007/978-3-030-80906-5_11

phase" [2]. However, integration is still in its infancy in AEC in comparison to industries such as automotive or aerospace; the fragmented nature of the AEC industry, composed by a majority of SMEs (small and medium-sized enterprises) [1, 3], may be the cause of this lateness, not to mention the difficulty of implementing standards and changes in an ever-prototyping industry where each product is singular [4]. Nevertheless, mutualizing project data via the implementation of a BIM model is the first step towards an integrative process. A second step will be the smooth interconnection between BIM models from different trades. This implies for several domain-specific BIM models to work together: in other words, to interoperate. This interoperation is still one of the biggest unresolved topics of BIM. According to a report on BIM business values edited in 2012 [5], "better interoperability" is identified as the second most important need in the BIM field. In this same report, 79% of BIM users identify an "improved interoperability between software applications" as a key factor to increase BIM benefits. Lack of interoperability is also suspected of creating an "economic burden" [6]. Its cost has been estimated to $15.8 billion in 2004 by the US National Institute of Standards and Technologies (NIST) [7], where in comparison, it only costs the auto industry $1 billion [8]. Not surprisingly, interoperability has been identified as a trending research topic by [9], with the highest number of published papers in the BIM field between 2005 and 2015.

To summarize, to date, the lack of BIM interoperability has been identified as an obstacle for more than 15 years. This obstacle directly affects one of most important aspects of BIM: collaboration. If a BIM actor is not able to import or export clean and reliable information from a given BIM model, he cannot base his collaboration on this BIM model: the two subjects are closely related. To address this obstacle, the present paper analyses the different ways BIM collaboration and interoperability are addressed in the literature.

The objective of this review is the creation of an understandable map of the subject for researchers and the identification of research trends, gaps and perspectives. It investigates how the question of BIM interoperability, in a collaborative context, is treated: how do AEC actors collaborate through BIM models today? How do they exchange information via a BIM model? What specific interoperability problems do they experience? How do they solve them? The next section describes the background and methodology of the present review. In the third, fourth, and fifth sections, three distinct analytical frameworks are developed, namely: (1) the AEC context, in which collaboration and interoperability issues are raised (2) the collaborative goals of BIM interoperability, and (3) the suggested methodologies to solve interoperability and collaboration issues. The sixth section is a discussion, followed by conclusions.

2 Methodology

2.1 Interoperability and Collaboration in the Field

The lack of BIM interoperability can be explained by: (1) BIM data heterogeneousness - for a given project, BIM data is made of several domain-specific BIM models, which contain diverse sets of data in terms of semantic, data structure, modelling logic, granularity, formats; (2) BIM practice heterogeneousness: there are very unequal levels of BIM adoption, and different AEC teams have various BIM knowledge, skills, methodologies and use different applications. However, BIM interoperability is addressed by:

1. *A neutral exchange format*: the IFC (Industry Classes Foundation, ISO 16739), maintained by Building Smart International. Created in 1997, it has become the de-facto standard for the building industry. However, the AEC trades diversity implies a large number of IFC concepts (around 800, one of the largest EXPRESS-based data model [10]), and this richness comes with a price: it creates redundancy. Hence, there are different ways to describe one piece of information in the IFC schema, which can lead to data ambiguity and data loss during the exchange.
2. *Collaborative BIM Platforms*: web-based repositories allowing project users to share BIM models within a project team. Besides storage functionality, additional collaborative tools are usually available such as file revisioning, status tracking, model annotation and access rights management for different project members according to their role and status. Unfortunately, these platforms are generally not open-access and use proprietary formats, which is paradoxical for a technology that is thought to support collaborative and integrative approaches, because it can enforce a monopoly of software vendors [11] which would tend to reinforce interoperability issues.

For practitioners, the word "interoperability" is classically used to refer to "data-oriented technical interoperability" and is computer-science oriented. It is frequently said that BIM collaboration is limited by lack of BIM interoperability.

2.2 Interoperability and Collaboration in Theory

In the research field, interoperability is classically defined as "the ability of two or more systems or components to exchange information and to use the information that has been exchanged" [12]. As mentioned in [13, 14], interoperability can be reached at three levels: technical (exchanged data), semantic (shared information) and organizational (interactions involving processes, people, organization and contractual agreements). Another distinction is made in [8]: BIM interoperability can be reached at technological level or at business level. Figure 1 shows the correspondence between these viewpoints.

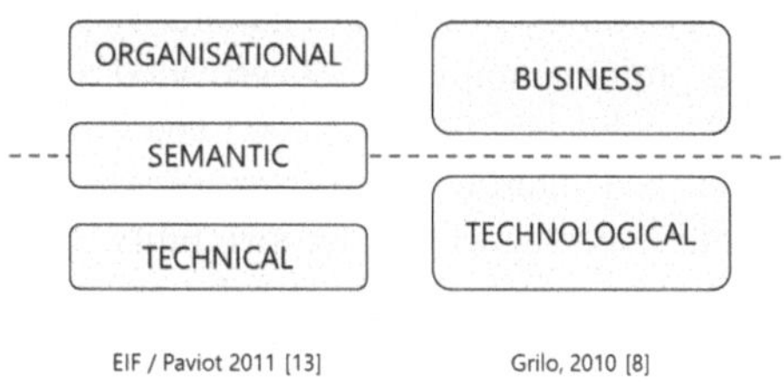

Fig. 1. Levels of interoperability

In Fig. 1 one can speculate that the semantic level would be a mixed technological and business layer. These two faces are mirrored in the essential behaviour of any BIM model, which is to be interpreted by both human and computer [15]. In [8], the business interaction layer is decomposed in five value levels of interoperability for BIM, from the most simple to the most interweaved: communication, coordination, cooperation, collaboration, channel.

In this paper, in order to create a bridge between practitioners' definitions and theoretical definitions, the word "collaboration" will be used to refer to business interoperability, and the word "interoperability" alone will actually refer to technological interoperability.

2.3 Background

The BIM field is a hot research topic; therefore a myriad reviews have been written about it, in which the interoperability and collaborative problematic are brought up. Some of them address the interoperability issues via a specific AEC angle, like for instance BIM data exchanges for precast concrete BIM models [16], or BIM and GIS integration [17]. The works [18–21] review the topic of BIM for sustainability, relations between BIM models and energy-simulation tools, energy retrofitting; [22–26] raise the problematic of interoperability between BIM and facilities management, operation and maintenance systems. Some reviews adopt a technological angle and focus on specific interoperability ways or tools, like for instance the IFC standard which history and development is addressed in [27], while [28] focuses on early IFC studies and questions the relations between research and BIM standardization. [29] reviews the literature about the semantic web technologies in the AEC and addresses the subject of BIM web-based data exchanges; [30] addresses the BIM and IoT (Internet of Things) integration. Finally, some reviews were found that have a more holistic approach to our question.

In [31], the questions of links between technological interoperability and business interoperability is addressed in the context of BIM implementation for SME, and some enablers for digital collaboration are raised at management level, such as "defining collaborative technology responsibilities, ensuring top-level management commitment, common convention, and intellectual property rights". In [32], the authors study BIM protocols for large construction clients, and reveal that interoperability modalities and collaboration modes are key topics, mentioned in 100% of the project standards guides. [33] surveyed the various ways digital collaboration is implemented into the BIM field for major projects. It reveals that it is almost never achieved in an integrated manner, and even when it is, the integration remains partial, limited to a phase or a specific topic. It was written in 2013, in the pre-infancy of BIM collaborative platforms, and the present review could be seen as a prolongation of it. [9] provided a bibliometric analysis of research articles about BIM published between 2005 and 2015. The subsection about "Collaborative environment and Interoperability" exposes a growing interest for this topic. It is divided into four subcategories: interoperability & IFC, Semantic BIM & Ontology, Collaborative environment, Knowledge & Information management.

This present paper could be an exploration of this specific article's section. In [34], the authors create a bibliometric mapping of the BIM knowledge, and interoperability is cited as one of the most important research topic of the knowledge base of the BIM, and "universal interoperability" is identified as a key milestone in 2012–2013. Similarly, in a review of global BIM research [35], a citation burst for the word "interoperability" between 2010 and 2012 is mentioned. In [36], "collaboration in design, engineering, and construction stakeholders", and "enhancing exchange of information and knowledge management", that is business and technological interoperability are identified as two of the five common major sets of critical success factor to measure the efficiency and accuracy of a BIM implementation. Similarly, in [37], collaboration and interoperability, as well as accurate data exchanges are considered as key subjects for construction and transportation industry.

To sum up, to the authors' knowledge, to date, no literature reviews exclusively focus on BIM interoperability and collaboration, although this subject is partially raised in BIM general reviews. In addition, three recent systematic reviews about BIM mention a milestone for this subject around 2013 [9, 34, 35], which seems to be coincident with the inception date of BIM collaborative platforms in the field. As a consequence, the present literature review will concern BIM interoperability and collaboration from 2013 to nowadays.

2.4 Corpus Constitution

In order to investigate the link between interoperability and collaboration in BIM, the subject was circumscribed with two types of key words: the first set is relative to interoperability on the technological side – *conversion, IFC, MVD (Model View Definition)* – and the second is linked to collaboration (business aspect of interoperability) – *exchange, share, cooperation, coordination, framework, collaboration*; finally some words refers to both business and technological levels, such as *integration* and of course, *interoperability*. The research has been limited to domains (engineering and computer science), time (2013 to nowadays, for the reasons mentioned previously) and language (English).

Figure 2 summarize the corpus constitution process. The research string has given a selection of 124 papers, on which have been processed two manual exclusions (as opposed to algorithm-based): (1) thematic and (2) methodological.

1. 39 papers were excluded for two main reasons (Table 1): (1) the paper was outside the scope of this subject, either too far or too general, and (2) the paper addressed the question of interoperability and collaboration at the extremes of the AEC industry, such as before or at the very beginning of the design like the BIM & Geographic information system (GIS), and at the very end of construction, the operational phase, and facilities management, with the exception of cases of refurbishment or HBIM (Historical BIM, requiring renovation). This was done on purpose, as the goal was to investigate how actors collaborate through BIM inside the AEC industry, i.e. during the design and construction phases.
2. Only the applicative papers were kept from the 93 papers that were left. The 13 reviews were excluded from the corpus (Table 2), because the goal is to investigate applicative case-scenarios. In addition, these reviews have been studied for the introduction of this paper.

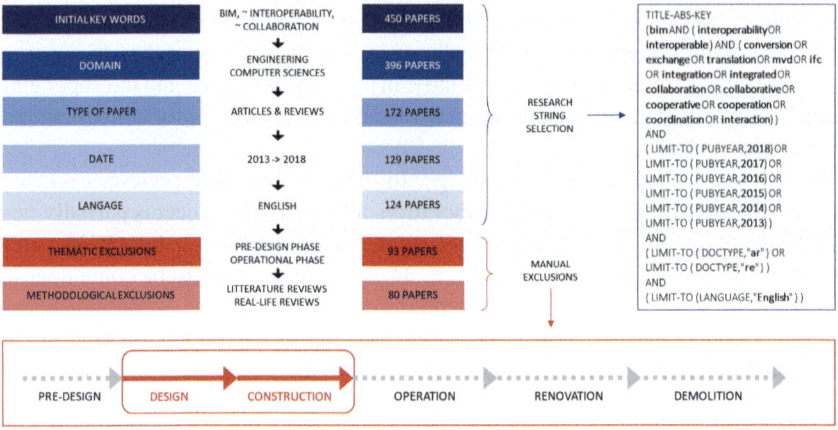

Fig. 2. Data collection process

Table 1. Excluded communications

Cause of exclusion	Number
About BIM and GIS	10
About operational phase	17
Not focused enough AEC	2
Not focused enough on interoperability	2
Total	31

Table 2. Selected communications

Type of article	Number
Article	80
Framework	8
Framework + Application	68
Tools survey	4
Review	13
BIM Body of knowledge (BOK)	2
Literature review	6
Real-life review	5
Total	**93**

2.5 Analytical Frameworks Implementation

A very small number of articles were left: a total of 80 papers have been carefully studied. Figure 3 shows the increase in publications about BIM interoperability and collaboration between 2013 and 2018. This corpus has been analysed through three analytical frameworks. Three sets of categorizations have been implemented, which, again, has been done via a manual process (as opposed to an algorithm-based process). The categories that were set up are globally mutually exclusive, but when an article could be assigned to several categories, the most salient categories were selected. The categorization process can be further detailed.

The three analytical frameworks imply three basic questions that can be applied to any problem: (1) what is the context in which the problem is raised? In our case: in which particular AEC context is the BIM collaboration and interoperability challenge raised? (2) What is the problem, how is it characterized? In our case, for which collaboration purpose do BIM actors need interoperability? (3) How is the problem solved? In our case, what is the range of solution suggested in the literature to improve BIM collaboration through better interoperability?

Fig. 3. Number of publications per year

3 Survey of AEC Contexts in Which Collaboration and Interoperability Issues are Raised

Each of the reviewed papers captures the interoperability and collaborative challenge in a specific AEC context, i.e. building types, construction phase, discipline – or in a generic context, addressing the question on the AEC general level. With regards to methodology, it is worth specifying how an AEC context category is inferred: *directly* (the paper explicitly addresses an AEC context) or *indirectly* (the paper addresses the challenge of the AEC generic level but takes a specific context as a test case study). The first general findings of this review are the following (1) the challenge of BIM collaboration and interoperability has not been raised equally in these different contexts, and this section details this distribution (2) almost 20% of the papers are not based on a precise AEC context and target to solve the challenge at a generic level (Table 3).

Table 3. Categorization of articles according to AEC context.

AEC CONTEXT	Number	%
1. Energy analysis	21	26%
2. All disciplines and regular buildings	18	23%
3. Structure	9	11%
4. Precast Concrete Structure	9	11%
5. Multi-criteria Performance Assessment	7	9%
6. On site operations - Construction management	6	8%
7. Transportation infrastructure	5	6%
8. HBIM	2	3%
9. Building code & regulation	1	1%
11. Modular Building	1	1%
10. Complex geometry	1	1%
Total	**80**	**100%**

3.1 Most Significant Categories

Energy Analysis

The first salient category is the energy performance domain: 21 articles are concerned (26%). Interestingly, one of the perspectives raised in [33] in 2013 was that more studies were needed about "issues around the use of integrated digital technologies for green design and construction," and it seems that this wish has been strongly accomplished. The papers raise the question of how to base an energy performance assessment on BIM data [15–27], how to assign a green score to a building on the base of its BIM data [51, 52] and concerns embodied energy as well as operational energy [43]. The question of energy modelling is raised [53], and a specialized cousin of BIM is rising: "BEM", for "Building Energy Modelling" [54]. The obvious reason for this high representation of the energy subject is the actuality of the environmental problem: energy simulations are required earlier in the design-construction phases. Another explanation might be inferred: building energy analysis requires digging into very diverse data sets in terms of granularity (dealing with abstract objects: zones, envelopes, and detailed objects like electrical terminal), and in terms of discipline (weather data, GIS data, MEP data), and a strong engineering is required to solve the connection of this heterogeneous data sets.

Structure and Precast Concrete Structure

The second salient category, with 18 articles (22%) is the structural domain. In this category, half of the papers are about the general structure domain, questioning data syntactic and semantic interoperability between BIM models and structural models [4, 6, 55], automatic translation of BIM models into structural models [56–58], BIM collaboration from the structural engineering perspective [59], and structural health monitoring (SHM) [60, 61]. The other half is about precast concrete structure. Three articles explicitly address this domain, via the questions of BIM data representation for precast concrete elements

[16], precast component BIM integration via catalogues [62] and a precast component BIM driven laser-scanning quality control process [63]. Six other articles use precast concrete data as case studies, to test precast concrete MVD checking processes [64, 65], semantic enrichment processes [66] or web-based representation [67–69]. The predominance of precast in BIM interoperability studies is explained by the complexity of precast concrete BIM modelling: it deals with both discrete elements and continuous materials, different levels of granularity and assembling rules (adjacency, overlapping, etc.) and low tolerances. As the pure description of a precast model is a challenge itself in BIM, transferring it without losing data is challenging as well. The US Precast Concrete Institute (PCI) and National BIM Standards (NBIMS) have made strong efforts to standardize the BIM interoperability approach in this domain that is very prone to automation.

3.2 Emerging Categories

Multi-criteria Performance Assessment

The multi-criteria performance assessment category contains 7 articles (9%). Three of them target specific goals, such as the evaluation of instance refurbishment scenarios [70, 71] or cost and energy performance optimization [72]. Four articles are theoretical insights addressing the challenges of interoperability for multi-disciplinary design optimization [73], concurrent engineering [74], design and construction phase optimization [75], or predictive building performance checking [76]. These papers mark an emerging trend in AEC: concurrent engineering, with the goal to incorporate very diverse domain knowledge earlier in the conception phase to evaluate and enhance the building performance. This emerging field involves combining very heterogeneous data sets, in terms of semantic, level of detail (LOD) and granularity, hence the setup of very strong interoperability systems between various BIM data sets.

On-Site Operations - Construction Management

This category includes 7 articles (8%) about the quality inspection and evaluation of construction [77, 78], supply-chain management [79], data flow reliability on construction sites [80], and emerging technologies like the Internet of Things [81] or augmented and virtual reality [82]. As in concurrent engineering, the use of BIM to monitor construction activities implies the ability to connect very diverse sets of data, with different semantics, LOD and granularity, and therefore, a strong level of interoperability.

Transportation Infrastructure

This category includes 5 articles (6%), about bridges [11, 83], tunnels [84, 85], and airport terminals [86]. The use of BIM for this type of project is a rising trend, as evidenced in [87], and implies dealing with larger sets of data. The volume of the required data exchanges and their various LOD justify a need for robust collaboration processes.

3.3 Less Significant Categories

Some AEC contexts are uncommon, such as historical BIM (HBIM), with 2 articles [88, 89], or modular building [90], and under-represented in the analysed corpus. Nonetheless, HBIM could be considered as a BIM laboratory, because it has to deal with an inverted challenge: a traditional project BIM model evolves from design to construction by an increase in the level of detail, whereas HBIM starts with a high LOD and has to deal with model simplification. New concepts have been brought about in this domain, such as the level of geometry (LOG), level of accuracy (LOA), level of information (LOI). It opens the road to model transformation and simplification, which are needed in every project for cross-domain communication, not only HBIM or refurbishment. BIM could learn a lot from HBIM. Only one article has been found on BIM interoperability for building codes and regulation compliance checks [15], which indicates that this AEC subdomain is still in its infancy. Likewise, only one article deals with interoperability for complex geometry buildings which is both understandable, given the small proportion of such building types in AEC, and surprising, given the fact that complex geometry is an emerging trend in AEC, enabled by technologies such as 3D printing, 3D scanning, and since most common BIM software tools do not support such geometry. This is the origin, for instance, of a strong need for translation between NURBS tools and classic BIM tools.

To conclude this section, the interoperability and collaboration subject is addressed through diverse trades and subjects in the AEC industry, with an emphasis on the energy assessment domain and structural domains; multi-criteria performance assessment, on site operations and transportation infrastructure are rising trends, and it seems that the complex geometry building category could gain more attention. Now that the industrial context in which the challenge is raised is made clear, the "why" question can be addressed: what do BIM actors need interoperability for, and more specifically, for which collaborative goals?

4 Survey of Collaborative Goals of BIM Interoperability

This second analytical framework analyses the reviewed articles through the following prism: what do BIM actors need to improve BIM interoperability for? In other words: for a given set of BIM data, what collaborative operation do they plan to process that justifies a need for interoperability? As shown in Table 4, seven major categories of collaborative needs have been set up, following a logical gradient categorization, from the most basic exchange (conversion) to the most complex (integrative approach): convert data, re-use data, check data, retrieve data, link data, combine data and combine data hubs.

Table 4. Categorization of articles according to the interoperability need.

COLLABORATIVE GOALS OF INTEROPERABILITY	Number	%
1. Convert Data	7	9%
2. Re-use Data	9	11%
3. Check Data	8	10%
4. Retrieve Data	4	5%
5. Link Data	25	31%
6. Combine Data	21	26%
7. Combine Data hubs	6	8%
Total	**80**	**100%**

4.1 Convert Data: Access BIM Data

In this 1st category, actors need to access a BIM data set and open it in a format they have access to, generally an open one. The industrial challenge of this is how can a set of data be syntactically transferred from one format to another? The focus is on a format, and almost exclusively on IFC. These papers address subjects such as IFC data serialization for web format [69], IFC file compression [91], IFC reliability for structural modelling [55], and technical interoperability improvement measurement [4]. Three articles [16, 67, 68] focus on the improvement of MVD robustness and reliability. According to BuildingSmart, an MVD (model view definition) "defines a subset of the IFC schema that is needed to satisfy one or many exchange requirements of the AEC industry"[1]. An MVD can be neutral (like the IFC Coordination view, to be understood by any AEC actors) or domain-specific (like the IFC Structural Analysis View that targets the structural domain). With MVD subsets, the IFC format inherently addresses the question of BIM exchange purposes and, with trade-oriented MVDs, the question of BIM data-exchanges between various AEC domains.

4.2 Reuse Data: Prevent Recreation of BIM Data

In this 2nd subcategory, actors need to prevent BIM data re-creation and avoid rework. The first sub-challenge is how to make BIM models suitable for collaboration. In [59], the question of how to reinforce the use of a structural model in the collaboration process is raised. In [83], the question of which items in a BIM model are more likely to be collaborative vectors is raised. The second sub-challenge is how a specific set of data can be transformed from one trade-semantic to another. Eight articles [6, 44, 56–58, 92] address the question of how to transform a BIM non-trade-specific model into a BIM domain-specific model with a higher LOD. Most of these papers imply "BIM coordination model" or "BIM architectural model" when they mention "BIM model".

By default, a "BIM model" generally designates the "original" BIM model, i.e. the BIM model made by an architect in the design phase or the coordination model on which

[1] http://www.buildingsmart-tech.org/specifications/ifc-view-definitions.

every trade base their specific models, in the construction phase. It suggests that: (1) a BIM model, by default, is non-trade specific – which can be discussed; (2) a BIM model, in its essence, is made to be understood by all trades, as a collaborative base. Beyond the trade translation level, the challenge is addressed with a more generic point of view in [66], through the following semantic enrichment question: how can any given received model from trade no. 1 be enriched with implicit semantic in order to be translated into a given trade no. 2? How can information contained in a model be used to deduct new sets of information? This very generic expectation corresponds to the need to prevent iterative data re-inputting between BIM actors: actor no. 1 sends a BIM model to actor no. 2, who needs to manually enrich the model with trade-specific knowledge, which is: (1) time-consuming, (2) error-prone (3) subject to a future iteration in the next exchange or project phase. Such semantic enrichment processes could significantly minimize rework in the AEC industry.

4.3 Check Data: Validate BIM Data

In this 3rd category, actors need to validate a set of BIM data. Two kinds of checks are involved: a data-structure check, and a content check. The former is about detecting misclassified IFC elements [93], reviewing MVD checkers [94], checking an IFC file against its MVD [10, 64, 65]. It corresponds to a need, for a given BIM received file, to check its internal structuration and to detect if it is compliant with predefined expectations. The latter is about validating the content of a BIM file against building code and regulation [15] or against clients expectation [76] and environmental performance [49]. It concerns only three papers, suggesting that BIM-based content checking is still in its infancy.

4.4 Retrieve Data: Querying BIM Data

In this 4th category, actors need to retrieve a precise set of BIM data throughout a bigger set. The industrial challenge is, how can a set of BIM data be queried efficiently? Only four articles address this question: about performing queries on a BIM model [95, 96], about retrieving data throughout a rich multiple LOD HBIM data set [89], and about transforming the exogenous data query results into an IFC file [97]. The underlying question is on how to access and manipulate a sharp targeted BIM data set without knowing the BIM parent software or being an IFC schema expert. It seems that easing BIM data retrieval, identification and isolation would democratize BIM collaboration.

4.5 Connect Data - Linking a BIM Model to an External Trade

In this 5th category, actors need to create persistent links between a BIM model and a domain specific set of data. The difference between this 5th subsection and the 2nd subsection is the following: in the former, the question is passively asked to avoid rework. Here, the question is asked actively, with the perspective to create a robust and reusable bridge between two AEC domains. Two modes of connection between a BIM model and a specific domain are addressed: (1) the upstream mode, with questions such as, "how to embed structural health monitoring into a BIM model?" [60, 61], or "how to represent pedestrian routes in IFC?" [98], i.e., "how to structure a model upstream to make it fit a specific domain?"; (2) the downstream mode, with questions like "how to base an energy assessment on the base of a BIM model" [38–43, 47, 48, 53, 99, 100], i.e. "how to extract and transform, from a BIM model, the appropriate information for a domain-specific model"? This category is the most salient, with 25 papers (31%). It suggests a strong need to use BIM models as collaborative vectors between several domains, and it raises the question of domain knowledge encapsulation in BIM models.

4.6 Combine Data – Merging Multiple BIM Domains

In this 6th section, actors need to combine multiple heterogeneous data sources throughout a BIM process. It is a scaled version of the question addressed by the 5th category: linking upstream and downstream data from more than two sources. The question is addressed on three types of levels.

1. **Conception level**: in this sub-category, diverse data sets need to be combined to assess a performance like cost or energy [70, 72, 101] during the design phase.
2. **Construction level**: in this subcategory, data combination happens in the construction phase, with examples like BIM data feeding self-inspection processes [78], supply-chain management [79], surface subsidence risk [84], quality surface assessment for precast concrete [63], VR/AR on construction site [82] and Internet of Things [81]. In this phase, in addition to being very heterogeneous in terms of semantics and granularity, the data sets have higher LOD and are heavier than in the conception phase, which increases the data combining challenge.
3. **Generic AEC Level**: in this subcategory, data combination is questioned through the angle of business modalities and common data structuration in the AEC industry. The focus is on the conceptual and structural locks that impede smooth data combining [85], and on the challenge of implementing BIM cross-domain models [73, 102, 103]: the question of various domain-specific AEC data-structuration alignment through a shared core is raised in [104]. In [105, 106], the question of combining BIM data and informal social interaction about a project is raised. A step back is made in [107], by addressing the business context allowing data combining, with the creation of a business interoperability quotient measurement. It is clear that data combination, which can be considered as the highest level of data exchange, is not a purely technical question, but also depends on interactions allowed by the disposition of contracts and human dispositions.

With 26% of the articles, it seems that the data-combining subject is well addressed on multiple granularities, which suggest a strong need in the AEC not only for setting up cross-domain collaboration, but also for automatic cross-domain data confrontation and analyses.

4.7 Combining Data Hubs

In this 7th section, actors need to navigate between different data-hubs and to connect different BIM collaborative platforms or cloud tools together. This could be understood as a scaled version of the 6th category: the interoperability is addressed on the level of the collaborative tool itself, and is no longer on the data level. This raises a new question: can, and should, collaborative platforms be interoperable? Diverse angles are adopted, such as the trivial but necessary question of a robust Internet connection on a construction site to allow the collection of live and synchronous as-built data via Wi-Fi [80], or the possibility of creating multi-user interfaces to review multiple-domain digital mock-ups [74]. Two papers question the BIM web-based collaboration via a technical cloud angle: on a private cloud context [86] and on a federated cloud context [11]. A new question is raised in [108]: what must be done to allow different web-based BIM collaborative platforms to be interoperable? Such platforms were expected to solve the interoperability and collaborative problems by allowing different BIM files to be shared, viewed and edited online. As they are often proprietary tools, the BIM technical interoperability problem has just been moved from the file level to the platform levels. This section contains a low number of articles, all very recent (2017–2018), which suggests that it is a new research topic.

To conclude this section, distinct data exchange modalities justify a need for interoperability, from the simplest use such as "convert data" to more complex ones like "combine data". The challenge here is not on whether actors will be able to "read" each other's data, but on how they will be able to actually use and process it. While a strong effort is being made to link data from one trade to another and combine data from different sources, it seems that the basic use of retrieving data is not addressed in the scientific literature. Now that the "why" question has been addressed, the "how" question can follow: how does the scientific community suggest solving the interoperability and collaboration issue?

5 Survey of Methodologies to Solve Interoperability and Collaboration Issues

This section focuses on the solutions that are suggested to solve interoperability and collaboration issues. Three categories of methodologies for problem resolution are implemented: (1) the **a priori approach**, in which the responsibility of the exchange integrity is on the sender level, (2) the **a posteriori approach**, in which the responsibility of the exchange integrity is placed on the receiver side, and (3) the **bidirectional approach**, which addresses the data flux itself, rather than the sender or the receiver perspective. The corpus is analysed through this specific framework (Table 5).

Table 5. Categorization of articles according to the interoperability solution.

SUGGESTED METHODOLOGIES	Number	%
1. A priori	**7**	**9%**
1.1 Data model	2	3%
1.2 IFC/MVD Extension	4	5%
1.3 MVD implementation enhancement	1	1%
2. A posteriori	**22**	**28%**
2.1 Data automatic deduction & transformation	12	15%
2.2 IFC / MVD validation	5	6%
2.3 Conversion enhancement	3	4%
2.4 Easy queries and extraction	2	3%
3. Bidirectional	**51**	**64%**
3.1 Data Articulation	18	23%
3.2 Data Centralization & Connection	14	18%
3.3 Workflows Orchestration	10	13%
3.4 Semantic Web & Ontologies	9	11%
Total	**80**	**100%**

5.1 A Priori Approach

In this category, the responsibility of the exchange integrity is placed at the sender level, who must ensure that the delivered set of data can be accessed and used by the receiver. The solutions focus on upstream data structuration, in order to enable robust and reliable downstream exchange, and the industrial challenge is: "how should the data ideally be structured"? The IFC being the de facto AEC standard, the idea of creating a new data schema is almost non-existent, with the exception of a new data-structuration suggestion for structural domain modelling [55]. The article focuses on optimal IFC use to represent the circulation in a building [98], and IFC schema extensions (new concepts and class definition) for IFC-based product catalogues [109], and for structural sensor modelling [60, 61]. Other papers focus on MVD enhancements, with, for instance, the creation of a new MVD to support specific domains like modular buildings [90], or a suggestion to modularize MVD development [16]. As MVD are based on specific exchange requirements, their implantation and maintenance is expensive, time consuming, and not very well documented. By setting up reusable semantic exchange modules (SEM) as modular bricks (the basis) for the creation of various MVD for different AEC domains, the effort of MVD implementation could be mutualized and distributed throughout the AEC, so its cost could be reduced [16]. The a priori approach is almost exclusively focused on open standards and addresses the ideal representation and structuration of BIM data, or ideal ways BIM actors should create IFC model views. However, the number of articles adopting this angle is decreasing with time, suggesting that this approach, if necessary, is not sufficient to solve AEC interoperability challenge.

5.2 A Posteriori Approach

In this category, the responsibility of the exchange integrity is on the receiver level, and the perspective is inversed from the first one. The solutions are optimizing the downstream use of a given set of received BIM data. The onus is placed on the output data.

Data Automatic Deduction and Transformation

The most salient category is the data automatic deduction and transformation approach. The principle is to infer implicit semantic content from explicit BIM data, as stated and developed in [66], in which precast joints and slab aggregations are detected in a BIM model by inferring rules as a test for the general process of semantic enrichment. This article insists on the paradigm change that is made by placing the onus on the receiving application and its pragmatism. It is also developed in [56, 57], with the setup of interpreted information exchanges (IIE) to capture information that could be considered as latent into a BIM model, such as geometrical wireframe inputs (points and lines) needed for a structural analysis model.

The process is two-fold: (1) extraction of the appropriate data, i.e. wireframe geometrical model, and (2) transformation of this data in order it to fit to a structural analysis geometrical schema, i.e. merge points and recompose lines to obtain network of connected vertical and horizontal lines. In this case, the process consists of a trade-oriented geometric transformation of a model, and a similar process is stated in [44] with geometry simplification for the energy domain.

Similar cases of BIM model transformation for specific trades are found: for the structural domain [6, 58], the energy analysis domain [40, 53, 99], and from CAD models to BIM models: in the research work [92] a tool using a visual programming language (VPL) is developed to automate the transformation of a geometrical 3D CAD model into a BIM model.

IFC/MVD Validation

Another angle is IFC and MVD validation. A strong effort has been made to validate IFC against MVDs, with the proposal of a robust and modularized MVD validation process [64] and the development of logic rules and a validation framework to syntactically and semantically check an IFC instance against a specific MVD via a user friendly interface [65]. The idea here is to check if an IFC is syntactically correct according to its MVD. A machine learning approach has been developed [93], with the creation of a novel detection learning algorithm: training individual one-class SVMs on multiple BIM models to detect IFC misclassified elements. The use of artificial intelligence seems to be promising for this use case.

Conversion Strategies

This challenge is still addressed via classical methodologies such as conversion enhancements, with, for instance an evaluation of the evolution of technical data interoperability through time [4], the proposal of a content-based compression algorithm, deleting redundant information of a big IFC received file [91], or the implementation of JSON as language to translate IFC into a web-compatible format.

Easy Queries and Extractions

Two articles focus on resolving BIM interoperability issues by facilitating BIM data queries: by setting up a prototype tool that automates the generation of MVD-based query code on BIM models, and that can be set up by a BIM end user with no programming skills [95] or by demonstrating that informal queries can be performed on BIM data using fuzzy logic [96]. As mentioned earlier in the previous section, BIM data retrieval corresponds to a strong need in the field, and the fact that only two articles address BIM model queries suggests a gap in the research question.

5.3 Bidirectional Approach

In this category, solving of interoperability and collaborative issues is not placed on the sender or receiver side, but rather on the data-exchange process itself: the data flow is questioned rather than the input or output data.

Data Articulation

The main approach for this bidirectional methodology with 18 articles is data articulation, i.e. creation of tools to automate the bridges between different data sets. Technological ways to articulate data imply various vectors, as for instance, a 3D CAD parametric reference curve to federate data exchanges on a steel bridge [83], a Modelica-based BIM physical library to facilitate semi-automatic translations from BIM to BEM [42], a building-regulation-specific semantically rich object model as a base for a building-regulation compliance checking process [15], a standard classification system used as a mediator to implement performance checking on a BIM model [76], or the use of a visual programming (VPL) tools to create a connection between 3D-CAD data and BIM models. Some tools are explicitly based on open-BIM, as for instance an IFC-based evaluation framework for construction quality evaluation [77], while some are explicitly based on BIM commercial tools' API, as for instance a plugin that automates the transformation of BIM data to feed an energy simulation engine [46]. Other applications question the use of IFC vs API as interoperability vectors, with for instance mapping typical two-way exchanges between a structural model and BIM coordination models via IFC or API [59].

The ultimate challenge in a data articulation framework is the roundtrip scenario. The question of bilateral exchanges between two sets of BIM data are raised in only 2 papers, highlighting the broken feedback loop between IFC and gbXML data in a tested framework [71], and the creation of a framework that bilaterally connects a multi-objective optimization module to a BIM model, which allows the result to be re-imported back into the BIM input model [43]. Inferring some data from a BIM model no. 1 to enrich a BIM model no. 2 is addressed, but sending back the inferred information to the model no. 1 is not. It seems that roundtrip scenarios and two-way exchanges between several BIM domain-specific models could receive more attention.

Data Centralization and Connection

A second approach, with 14 articles, is based on data centralization and distribution frameworks, i.e. tools that collect and distribute different BIM data sets. A first subcategory is the technical angle. Three articles address the question of data centralization and distribution via workspaces and BIM user data access rights, with the setup of a multi-server on a private cloud [86], or the setup of a cloud federated framework: a distributed collaborative environment in which each team maintains its own data without having to migrate it to a central site [11].

Some papers are exclusively focused on BIM data collection via platforms with, for instance, the creation of a data storage platform with user access rights and BIM document management [110], and the setup of a platform tool with two main functions: design work management, collaborative workspace for BIM data by discipline, and project work management, costs, procurement [105]. These tools can be considered as prototypes for collaborative BIM platforms, even though they are limited to BIM model cloud-storage and model web-viewing and commenting. The collaboration has a low granularity and stops at the frontier of files: for a given BIM domain-specific model, data sets or objects inside files are not connected to their related data sets of another trade. For instance, the object "wall" itself would appear in two BIM models, the structural one and the architectural one: they are not connected but duplicated.

Other platforms do not limit their function to BIM data centralization and add other types of data like CAD data and ERP data [79], environmental data and sensors data [70, 78], IoT smart devices data [81], lifecycle assessment (LCA) data [47]. These tools process data with high granularity, beyond the frontier of file, and use methodologies like ETL (extract-transform-load), data partial splitting, merging and combining. One paper went beyond the question of BIM collaborative platform, and raised the question of several BIM platform connection [108].

Workflow Orchestration

A third approach addresses the interoperability and collaboration challenge through the BIM workflow aspect, which was raised in 10 papers. This approach goes beyond BIM data technical centralization and connection and considers the human organizational workflows that enable BIM collaboration; in other words, it truly considers both business and technological interoperability. In [50], emphasis is placed on the need to strongly connect BIM workflows to a human work organization, with the prescription of interoperability specification, cross-organizational workflow and human processes based on IDM BuildingSmart recommendations.

From a more technical angle, the human-computer collaborative interface question is addressed in [74] with the proposal of multi-users and multi-trade interface for BIM model viewing, to facilitate model reviews in the context of concurrent engineering. The cross-discipline and cross-model collaboration is addressed through a multi-model approach in 4 papers. In [101] a conceptual multi-model framework based on a central IFC model is developed, to which simulation models and other information repositories are linked. In [102] a linked-data multi model framework is developed, to homogenize data access rather than format in order to create consistent links between various original domain-specific models. A similar approach is used in [103], which describes the setup of a flexible and distributed framework integrating multiple domain-specific languages (DSL), to orchestrate different models and software via extensibility features. In [73] five BIM authoring tools are surveyed, and the demonstration is made that they do not fit the requirement to be part of a process integration design optimization (PIDO) for a multi-disciplinary design optimization (MDO). Such integration would require an improvement of their component interoperability, system automation and model parameterization.

Few papers address the challenge of parametric links between models. As stated in [62], "more efforts are needed to provide parametric descriptions of products in a more standardized way". Similarly, the lack of component interoperability, automation and model parameterization are identified as technical brakes to data exchanges in [73]. The importance of defining the semantics for the links between models is highlighted in [102]. Links clearly need to be made between objects throughout different BIM models to ensure data continuity, data control, and processes like change automatic propagation, which prevents rework. BIM models are typically already parametric, but this parameterization should be extended to the project level, between different teams and trades, outside the borders of files. A given BIM object for trade no. 1 should be connected to its related BIM object or a group of objects in trade no. 2. A step back is made in [85], with a focus on the original purpose of a BIM model. Three conflicting BIM quality indicators are presented: semantic representation, conceptual completeness, and ease of implementation. According to the authors, the tension between these three tendencies is to be considered in a BIM data-exchange process. In other words, a BIM workflow should address the three original conflicting aims of any BIM model.

Semantic Web and Ontologies

A fourth approach is the use of ontologies and semantic web technologies, raised in 9 articles. This technology is used in the IDM/MVD formalization effort, which proposes a new hierarchical structure based on ontological definition of IFC [67], or the development of an ontology-based automated knowledge framework for MVD automatic generation [68]. Semantic web technologies are also used in [89] to represent HBIM data, in [62] to develop a product component catalogue accessible through BIM authoring tool, in [97] where semantic web query results, performed with SPARQL query language, are syntactically translated into an IFC through an ifcXML, and loaded into a BIM authoring tool. A more generic approach is suggested in [104], to allow the AEC various domain-specific to interoperate: the creation of BIM Shared Ontology (BIMSO) to represent the common concepts shared by different AEC domains. The BIMSO acts as a semantic mediator to align the various domain-specific ontologies. The use of semantic web technologies is very recent: the article's publication's dates are distributed between 2015 and 2018. Their use seems to be promising, as stated in a dedicated literature review [29].

To conclude this section, different approaches exist to solve interoperability issues. First of all, the three approaches seem to evolve with time (Fig. 4): the a priori and a posteriori approaches are replaced with a more holistic approach that addresses the interoperability and collaborative challenge via the data flow itself, rather than focusing exclusively on the input or output of this flow.

Considering sub-categories while data automatic deduction, data articulation and data centralization on web-based platforms receive a lot of attention, workflow orchestration methodologies are quite a medium trend. Yet this is the category that addresses

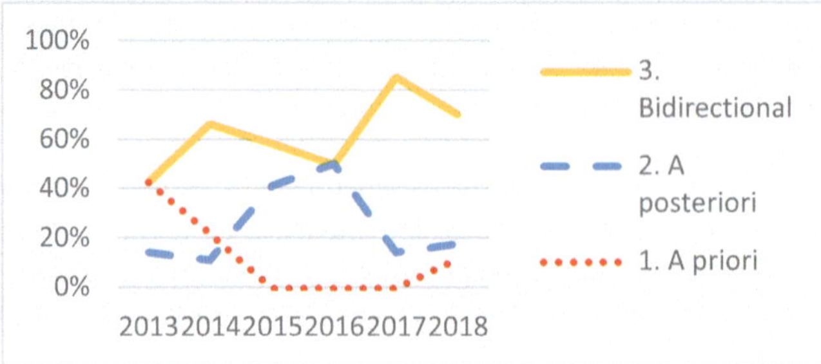

Fig. 4. Evolution of interoperability solutions in time

interoperability with the widest angle, considering both the business and technological aspects in a broad spectrum, from project organization to punctual data exchanges. Finally, while a new, recent trend such as semantic web technologies is promising and rising, a trivial subcategory receives very little attention, although it is the root of any collaborative BIM action: the approach of data exchange via the data retrieval question.

6 Discussion

6.1 Research Trends

As this review has shed light on, issues concerning BIM collaborative and interoperability are addressed by the literature. Many efforts are made, on the one hand, to connect BIM domain-specific models together to infer data between one model and another, to prevent data re-inputting, and to combine automatically data from different sources. On the other hand, cloud-based collaboration for BIM is rising, and BIM data is currently being shared through web-based platforms. However, it seems as though BIM level 3 has still not yet been achieved: in the field, BIM data exchanges are still based on big file submission, which remains the collaborative unit between different trades. BIM maturity levels were defined by the BIM Industry Working Group in 2011 in an official UK government document[2], in which BIM level 3 is characterized by the implementation of a "unique" multidisciplinary model, where every trade would co-model. As mentioned in [85], a "conceptual misunderstanding" persists about BIM models, "being that there ever could be a single set of information or even a single information model that would meet the needs of all project stakeholders." To this light "unique model" in BIM level 3 it cannot be one BIM siloed file; it must be understood as a high granularity cluster of trade-specific interconnected models with a precise user access right management. BIM level 3 can be considered the highest level of collaboration in BIM, and implies that different teams are working on one another's original data, without any data re-inputting or duplication; it requires bidirectional and dynamic interoperability, inter-model parametrization and associativity, the use of federated models via centralized databased. To date, BIM level 3 has not yet been implemented in AEC, with the exception of: (1) BIM models based on product data management and product lifecycle

[2] https://bimportal.scottishfuturestrust.org.uk/page/standards-level-1.

management (PDM/PLM) methodologies using aeronautic or automotive software tools, like CATIA/3D Experience (Dassault Systèmes) or MicroStation (Bentley Systems), on which a high granularity tree structure schema allows such co-modelling systems; (2) the use of collaborative tools like Flux.io[3], which was a platform allowing BIM dynamic interoperability with high granularity, or a similar tool like Speckle[4]. This review shows that the road to BIM level 3 is clearly opened by the convergence of different trends, such as data automatic deduction and transformation, the use of web-based databases, models extensibility, multi-model frameworks, multidisciplinary optimizations, and the rise of semantic web technologies.

6.2 Research Gaps

Three research gaps can be inferred from this literature review. First, collaboration and interoperability are rarely addressed conjointly. It seems that the question of collaboration (the business interoperability level) is less frequently addressed than the interoperability (technological interoperability level); these two levels are often disconnected from each other. Very few articles adopt this double angle by taking into account both the business and technological aspects and how they enable each other. Most of them are in the "workflow orchestration" methodology, which is the concern of only 10 articles. As technical interoperability alone cannot solve business interoperability issues, and reciprocally, the question of the interrelation between interoperability and collaboration could be more addressed.

It can be decomposed into two mirrored sub-questions: (1) what kind of human processes can enhance BIM interoperability? (2) What kind of digital processes can enhance BIM collaboration? The latter would investigate what the typical environment would be that would allow technical interoperability to happen. What kind of business, contractual, physical and human environment could allow this? What kind of project organization would promote robust exchanges? The former would analyze which interoperability uses have an impact on collaboration. For instance, how do structural engineers actually use the IFC MVD "Structural Analysis View"? How does collaboration evolve with the use of semantic web technologies on a given project? How do BIM actors react to a specific interoperability tool, and how does it reconfigure the collaborative practices?

As a lot of new BIM collaborative tools are being brought to the market every day, it would be of interest to measure their actual impact on human collaboration. Few articles address this question, and new studies are needed to provide this kind of insight into the mutual impact between interoperability and collaboration. The second research gap concerns BIM data retrieval. It seems that this very basic need has received little attention in the literature. Very few articles address the challenge of searching for and finding BIM data in a large BIM model. The collaborative aim "data retrieval" in the 2nd analytical framework contains only four articles, and the "BIM querying" solving approach in the 3rd framework contains only two. Nevertheless, to the author's knowledge, this question corresponds to a strong need in the BIM "real-life" field. As stakeholders often receive big BIM files, they need to extract specific information from them, without having to

[3] https://twitter.com/flux_io. This platform closed in April 2018.

[4] https://speckle.works/.

navigate through the entire original BIM model, which is time consuming, human error prone, and requires specific skills (an understanding of how the BIM model is structured, for instance).

The technological answer to this organizational need is: BIM model querying. If a BIM model can be queried efficiently in a robust and reliable way, the result of these queries could become the unit of exchange to the place of the file itself. In a way, this is the original purpose of an MVD: querying and extracting data needed for a specific domain. However, as mentioned, MVD developments are expensive, and unfortunately, not every BIM exchange is based on IFC. What if instead of exchanging files, BIM actors exchange queries results? Additionally, what if they could perform the queries by themselves? That is what is informally done every day in the field, but not in a robust way: actors exchange BIM files, and make partial and manual extractions out of them. If this could be done in a formal and robust way, collaboration could gain fluidity and reliability.

The third research gap is about BIM interoperability for complex geometry buildings, which is addressed by very few articles. During the 1990s and 2000s, the AEC industry investigated a new range of representation, simulation [111], fabrication and construction techniques. The use of 3D free-form and NURBS modelling software for building design has brought a new spectrum of morphologies in architecture, with precursors like Greg Lynn or Zaha Hadid, to name a few, who opened the road for what is often referred to as "computational design". New technologies such as 3D scanning, digital photogrammetry, robotics and additive manufacturing have changed the ways in which buildings can be fabricated and assembled, enabling any type of shape to be constructed easily, with no extra cost. This can be seen as a paradigm shift in the building industry. Unfortunately, the most often used BIM software only supports simple geometry limited to plan-based and single curvature surfaces. Therefore, BIM models whose geometry escape those limits are a mix between a NURBS or free-form modeling software, and a BIM classic software, creating a strong need for an interoperable bridge between these two tools. Given the importance of the "digital turn" [112] in the AEC, this subject could receive more attention in the BIM interoperability research.

7 Conclusion

A lack of BIM technological interoperability reduces the scope of BIM-based collaboration, and this obstacle is acknowledged by industrial and scientific BIM actors. Therefore, the objective of this review was to map the scientific knowledge about BIM interoperability and collaboration, and to identify research trends, gaps and perspectives in order to provide a solid background for future researchers. 80 papers were selected and carefully analysed throughout three distinct analytical frameworks: (1) the industrial context in which the problem is raised (2) the problem characterization (3) the problem resolution.

The strong need for interoperability and collaboration in BIM has been addressed by the literature in various ways and constitutes an active and innovative field. This challenge has been addressed for AEC in general, as well as for specific AEC domains that seem to be more sensitive to the subject, such as the energy-analysis field and the structural domain. A strong effort has been made on the formalization of specific trade knowledge in

BIM data, such as precast concrete. Efforts are increasing in multi-domain optimization, on site operation and transportation infrastructure. The need for BIM connection between different AEC trades is recognized: many applications and prototype software tools address this challenge, and a strong effort is made to reuse data from one trade to another, to link two different domain-specific BIM models together, to combine heterogeneous data sets. These efforts are made on the conception phase, construction phase as well as on the generic AEC industry level. While a priori and a posteriori approaches seem to be in decline, the bidirectional approach is being adopted more and more, and the effort to solve the challenge of interoperability and collaboration is on the data flow level itself, and no longer on the sender or receiver side. The road to strong collaborative levels such as those required by BIM level 3 is opened by the convergence of web-based BIM data hubs, semantic web use and multi-model frameworks, but remains uncertain, as BIM collaboration in the field is still based on data siloes and ad-hoc solutions.

Finally, this review has revealed three research gaps: (1) BIM collaboration and interoperability are rarely conjointly addressed, and the literature needs new studies to analyze the mutual impact between technological and business interoperability; (2) the very basic question of retrieving specific data sets throughout BIM models is not sufficiently addressed, although it is the root of any BIM collaborative action; (3) given the rise of 3D printing, robotics and other complex shape manufacturing technologies, the interoperability for geometrically complex BIM models could receive more attention.

References

1. Eastman, C.M., Eastman, C., Teicholz, P., Sacks, R.: BIM handbook: a guide to building information modeling for owners, managers, designers, engineers and contractors. John Wiley & Sons, Hoboken, NJ (2011)
2. Ma, Z., Ma, J.: Formulating the application functional requirements of a BIM-based collaboration platform to support IPD projects. KSCE J. Civ. Eng. **21**(6), 2011–2026 (2017). https://doi.org/10.1007/s12205-017-0875-4
3. Howard, H.C., Levitt, R.E., Paulson, B.C., Pohl, J.G., Tatum, C.B.: Computer integration: reducing fragmentation in AEC industry. J. Comput. Civ. Eng. **3**, 18–32 (1989). https://doi.org/10.1061/(ASCE)0887-3801(1989)3:1(18)
4. Muller, M.F., Garbers, A., Esmanioto, F., Huber, N., Loures, E.R., Canciglieri, O., Jr.: Data interoperability assessment though IFC for BIM in structural design – a five-year gap analysis. J. Civ. Eng. Manag. **23**, 943–954 (2017). https://doi.org/10.3846/13923730.2017.1341850
5. The Business Value in BIM North-America - Multi-year Trend Analysis and Users Ratings [2007–2012]. McGraw-Hill. https://bimforum.org/wp-content/uploads/2012/12/MHC-Business-Value-of-BIM-in-North-America-2007-2012-SMR.pdf (2012). Accessed 10 Feb. 2019
6. Hu, Z.Z., Zhang, X.Y., Wang, H.W., Kassem, M.: Improving interoperability between architectural and structural design models: an IFC-based approach with web-based tools. Autom. Constr. **66**, 29–42 (2016). https://doi.org/10.1016/j.autcon.2016.02.001
7. Gallaher, M.P., O'Connor, A.C., Dettbarn, J.L., Jr., Gilday, L.T.: Cost Analysis of Inadequate Interoperability in the U.S. Capital Facilities Industry. National Institute of Standards and Technology (2004). https://doi.org/10.6028/NIST.GCR.04-867

8. Grilo, A., Jardim-Goncalves, R.: Value proposition on interoperability of BIM and collaborative working environments. Autom. Constr. **19**, 522–530 (2010). https://doi.org/10.1016/j.autcon.2009.11.003
9. Santos, R., Costa, A.A., Grilo, A.: Bibliometric analysis and review of building information modelling literature published between 2005 and 2015. Autom. Constr. **80**, 118–136 (2017). https://doi.org/10.1016/j.autcon.2017.03.005
10. Zhang, C., Beetz, J., Weise, M.: Interoperable validation for IFC building models using open standards. J. Inf. Technol. Constr. ITcon. 20, 24–39. http://www.itcon.org/2015/2 (2015)
11. Petri, I., Beach, T., Rana, O.F., Rezgui, Y.: Coordinating multi-site construction projects using federated clouds. Autom. Constr. **83**, 273–284 (2017). https://doi.org/10.1016/j.autcon.2017.08.011
12. The Institute of Electrical and Electronics Engineers: IEEE Standard Computer Dictionary – A Compilation of IEEE Standard Computer Glossaries, p. 610 (1990)
13. Paviot, T., Lamouri, S., Cheutet, V.: A generic multiCAD/multiPDM interoperability framework. Int. J. Serv. Oper. Inform. **6**, 124–137 (2011). https://doi.org/10.1504/IJSOI.2011.038322
14. White Paper of the European Interoperability Framework: http://www.urenio.org/e-innovation/stratinc/files/library/ict/15.ICT_standards.pdf (2004)
15. Malsane, S., Matthews, J., Lockley, S., Love, P.E.D., Greenwood, D.: Development of an object model for automated compliance checking. Autom. Constr. **49**, 51–58 (2015). https://doi.org/10.1016/j.autcon.2014.10.004
16. Belsky, M., Eastman, C., Sacks, R., Venugopal, M., Aram, S., Yang, D.: Interoperability for precast concrete building models. PCI J. 59, 144–155 (2014). https://doi.org/10.15554/pcij.03012014.144.155
17. Wang, H., Pan, Y., Luo, X.: Integration of BIM and GIS in sustainable built environment: a review and bibliometric analysis. Autom. Constr. **103**, 41–52 (2019). https://doi.org/10.1016/j.autcon.2019.03.005
18. Chong, H.Y., Lee, C.Y., Wang, X.: A mixed review of the adoption of building information modelling (BIM) for sustainability. J. Clean. Prod. **142**, 4114–4126 (2017). https://doi.org/10.1016/j.jclepro.2016.09.222
19. Sanhudo, L., et al.: Building information modeling for energy retrofitting. A review. Renew. Sustain. Energy Rev. **89**, 249–260 (2018). https://doi.org/10.1016/j.rser.2018.03.064
20. Andriamamonjy, A., Saelens, D., Klein, R.: A combined scientometric and conventional literature review to grasp the entire BIM knowledge and its integration with energy simulation. J. Build. Eng. **22**, 513–527 (2019). https://doi.org/10.1016/j.jobe.2018.12.021
21. Kamel, E., Memari, A.M.: Review of BIM's application in energy simulation: tools, issues and solutions. Autom. Constr. **97**, 164–180 (2019). https://doi.org/10.1016/j.autcon.2018.11.008
22. Ilter, D., Ergen, E.: BIM for building refurbishment and maintenance: current status and research directions. Struct. Surv. **33**, 228–256 (2015). https://doi.org/10.1108/SS-02-2015-0008
23. Pärn, E.A., Edwards, D.J., Sing, M.C.P.: The building information modelling trajectory in facilities management: a review. Autom. Constr. **75**, 45–55 (2017). https://doi.org/10.1016/j.autcon.2016.12.003
24. Wong, J.K.W., Ge, J., He, S.X.: Digitisation in facilities management: a literature review and future research directions. Autom. Constr. **92**, 312–326 (2018). https://doi.org/10.1016/j.autcon.2018.04.006
25. Matarneh, S.T., Danso-Amoako, M., Al-Bizri, S., Gaterell, M., Matarneh, R.: Building information modeling for facilities management: a literature review and future research directions. J. Build. Eng. **24**, 100755 (2019). https://doi.org/10.1016/j.jobe.2019.100755

26. Gao, X., Pishdad-Bozorgi, P.: BIM-enabled facilities operation and maintenance: a review. Adv. Eng. Inform. **39**, 227–247 (2019). https://doi.org/10.1016/j.aei.2019.01.005
27. Miles-Board, T.: The IFC standard – a review of history, development, and standardization. Electron. J. Inform. Technol. Constr. 29 (2012)
28. Laakso, M., Nyman, L.:. Exploring the relationship between research and BIM standardization: a systematic mapping of early studies on the IFC standard (1997–2007). Buildings 6(1), Nr. 7 (2016). https://doi.org/10.3390/buildings6010007.
29. Pauwels, P., Zhang, S., Lee, Y.C.: Semantic web technologies in AEC industry: a literature overview. Autom. Constr. **73**, 145–165 (2017). https://doi.org/10.1016/j.autcon.2016.10.003
30. Tang, S., Shelden, D.R., Eastman, C.M., Pishdad-Bozorgi, P., Gao, X.: A review of building information modeling (BIM) and the Internet of Things (IoT) devices integration: present status and future trends. Autom. Constr. **101**, 127–139 (2016). https://doi.org/10.1016/j.autcon.2019.01.020
31. Abuelmaatti, A., College, B.H., Ahmed, V.: Collaborative technologies for small and medium sized architecture, engineering and construction enterprises: implementation survey. J Inform. Technol. Constr. **19**, 210–224 (2014)
32. Sacks, R., Gurevich, U., Shrestha, P.: A review of building information modeling protocols, guides and standards for large construction clients. ITCON.org ISSN 1874-4753 (2016)
33. Hassan, I.N.: Reviewing the evidence: USE of digital collaboration technologies in major building and infrastructure projects. J. Inform. Technol. Constr. **18**, 40–63 (2013)
34. Li, X., Wu, P., Shen, G.Q., Wang, X., Teng, Y.: Mapping the knowledge domains of building information modeling (BIM): a bibliometric approach. Autom. Constr. **84**, 195–206 (2017). https://doi.org/10.1016/j.autcon.2017.09.011
35. Zhao, X.: A scientometric review of global BIM research: analysis and visualization. Autom. Constr. **80**, 37–47 (2017). https://doi.org/10.1016/j.autcon.2017.04.002
36. Antwi-Afari, M.F., Li, H., Pärn, E.A., Edwards, D.J.: Critical success factors for implementing building information modelling (BIM): a longitudinal review. Autom. Constr. **91**, 100–110 (2018). https://doi.org/10.1016/j.autcon.2018.03.010
37. Abdal, N.B., Yi, S.: Review of BIM literature in construction industry and transportation: meta-analysis. Constr. Innov. **18**, 433–452 (2018). https://doi.org/10.1108/CI-05-2017-0040
38. Cemesova, A., Hopfe, C.J., Mcleod, R.S.: PassivBIM: enhancing interoperability between BIM and low energy design software. Autom. Constr. **57**, 17–32 (2015). https://doi.org/10.1016/j.autcon.2015.04.014
39. Shadram, F., Johansson, T.D., Lu, W., Schade, J., Olofsson, T.: An integrated BIM-based framework for minimizing embodied energy during building design. Energy Build. **128**, 592–604 (2016). https://doi.org/10.1016/j.enbuild.2016.07.007
40. Choi, J., Shin, J., Kim, M., Kim, I.: Development of openBIM-based energy analysis software to improve the interoperability of energy performance assessment. Autom. Constr. **72**, 52–64 (2016). https://doi.org/10.1016/j.autcon.2016.07.004
41. El Asmi, E., Robert, S., Haas, B., Zreik, K.: A standardized approach to BIM and energy simulation connection. Int. J. Des. Sci. Technol. **21**, 59–82 (2015)
42. Kim, J.B., Jeong, W., Clayton, M.J., Haberl, J.S., Yan, W.: Developing a physical BIM library for building thermal energy simulation. Autom. Constr. **50**, 16–28 (2015). https://doi.org/10.1016/j.autcon.2014.10.011
43. Shadram, F., Mukkavaara, J.: An integrated BIM-based framework for the optimization of the trade-off between embodied and operational energy. Energy Build. **158**, 1189–1205 (2018). https://doi.org/10.1016/j.enbuild.2017.11.017
44. Ladenhauf, D., et al.: Geometry simplification according to semantic constraints: enabling energy analysis based on building information models. Comput. Sci. Res. Dev. **31**(3), 119–125 (2014). https://doi.org/10.1007/s00450-014-0283-7

45. Yang, X., Hu, M., Wu, J., Zhao, B.: Building-information-modeling enabled life cycle assessment, a case study on carbon footprint accounting for a residential building in China. J. Clean. Prod. **183**, 729–743 (2018). https://doi.org/10.1016/j.jclepro.2018.02.070
46. Choi, M., Cho, S., Lim, J., Shin, H., Li, Z., Kim, J.J.: Design framework for variable refrigerant flow systems with enhancement of interoperability between BIM and energy simulation. J. Mech. Sci. Technol. **32**(12), 6009–6019 (2018). https://doi.org/10.1007/s12206-018-1151-3
47. Nizam, R.S., Zhang, C., Tian, L.: A BIM based tool for assessing embodied energy for buildings. Energy Build. 170, 1–14 (2018). https://doi.org/10.1016/j.enbuild.2018.03.067.
48. Choi, J., Lee, K., Cho, J.: Suggestion of the core element technology to improve BIM data interoperability based on the energy performance analysis. Int. J. Grid Distrib. Comput. **11**, 157–168 (2018). https://doi.org/10.14257/ijgdc.2018.11.4.14
49. Zhong, B., Gan, C., Luo, H., Xing, X.: Ontology-based framework for building environmental monitoring and compliance checking under BIM environment. Build. Environ. **141**, 127–142 (2018). https://doi.org/10.1016/j.buildenv.2018.05.046
50. Arayici, Y., Fernando, T., Munoz, V., Bassanino, M.: Interoperability specification development for integrated BIM use in performance based design. Autom. Constr. **85**, 167–181 (2018). https://doi.org/10.1016/j.autcon.2017.10.018
51. Alwan, Z., Greenwood, D., Gledson, B.: Rapid LEED evaluation performed with BIM based sustainability analysis on a virtual construction project. Constr. Innov. **15**, 134–150 (2015). https://doi.org/10.1108/CI-01-2014-0002
52. Raffee, S.M., Karim, M.S.A., Hassan, Z.: Building sustainability assessment framework based on building information modelling. ARPN J. Eng. Appl. Sci. 11(8), ISSN 1819-6608 (2016)
53. Guzmán, G.E., Zhu, Z.: Interoperability from building design to building energy modeling. J. Build. Eng. **1**, 33–41 (2015). https://doi.org/10.1016/j.jobe.2015.03.001
54. Kim, Y.J.: BIM to BEM interoperability using web based process model. Int. J. Appl. Eng. Res. **10**(10), 26969–26978 (2015)
55. Solnosky, R., Hill, J.: Formulation of systems and information architecture hierarchies for building structures. J. Inform. Technol. Constr. 18 (2013)
56. Ramaji, I.J., Memari, A.M.: Interpreted information exchange: systematic approach for BIM to engineering analysis information transformations. J. Comput. Civ. Eng. **30**, 04016028 (2016). https://doi.org/10.1061/(ASCE)CP.1943-5487.0000591
57. Ramaji, I.J., Memari, A.M.: Interpretation of structural analytical models from the coordination view in building information models. Autom. Constr. **90**, 117–133 (2018). https://doi.org/10.1016/j.autcon.2018.02.025
58. Liu, Z.Q., Zhang, F., Zhang, J.: The building information modeling and its use for data transformation in the structural design stage. J. Appl. Sci. Eng. 12 (2016). https://doi.org/10.6180/jase.2016.19.3.05
59. Shin, T.-S.: Building information modeling (BIM) collaboration from the structural engineering perspective. Int. J. Steel Struct. **17**(1), 205–214 (2017). https://doi.org/10.1007/s13296-016-0190-9
60. Rio, J., Ferreira, B., Poças-Martins, J.: Expansion of IFC model with structural sensors. Inform. Constr. **65**, 219–228 (2013). https://doi.org/10.3989/ic.12.043
61. Theiler, M., Smarsly, K.: IFC Monitor – an IFC schema extension for modeling structural health monitoring systems. Adv. Eng. Inform. **37**, 54–65 (2018). https://doi.org/10.1016/j.aei.2018.04.011
62. Costa, G., Madrazo, L.: Connecting building component catalogues with BIM models using semantic technologies: an application for precast concrete components. Autom. Constr. **57**, 239–248 (2015). https://doi.org/10.1016/j.autcon.2015.05.007

63. Kim, M.K., Cheng, J.C.P., Sohn, H., Chang, C.C.: A framework for dimensional and surface quality assessment of precast concrete elements using BIM and 3D laser scanning. Autom. Constr. **49**, 225–238 (2015). https://doi.org/10.1016/j.autcon.2014.07.010
64. Lee, Y.C., Eastman, C.M., Solihin, W., See, R.: Modularized rule-based validation of a BIM model pertaining to model views. Autom. Constr. **63**, 1–11 (2016). https://doi.org/10.1016/j.autcon.2015.11.006
65. Lee, Y.C., Eastman, C.M., Solihin, W.: Logic for ensuring the data exchange integrity of building information models. Autom. Constr. **85**, 249–262 (2018). https://doi.org/10.1016/j.autcon.2017.08.010
66. Belsky, M., Sacks, R., Brilakis, I.: Semantic enrichment for building information modeling. Comput.-Aided Civ. Infrastruct. Eng. **31**, 261–274 (2016). https://doi.org/10.1111/mice.12128
67. Venugopal, M., Eastman, C.M., Teizer, J.: An ontology-based analysis of the industry foundation class schema for building information model exchanges. Adv. Eng. Inform. **29**, 940–957 (2015). https://doi.org/10.1016/j.aei.2015.09.006
68. Lee, Y.C., Eastman, C.M., Solihin, W.: An ontology-based approach for developing data exchange requirements and model views of building information modeling. Adv. Eng. Inform. **30**, 354–367 (2016). https://doi.org/10.1016/j.aei.2016.04.008
69. Afsari, K., Eastman, C.M., Castro-Lacouture, D.: JavaScript Object Notation (JSON) data serialization for IFC schema in web-based BIM data exchange. Autom. Constr. **77**, 24–51 (2017). https://doi.org/10.1016/j.autcon.2017.01.011
70. Woo, J.H., Menassa, C.: Virtual Retrofit Model for aging commercial buildings in a smart grid environment. Energy Build. **80**, 424–435 (2014). https://doi.org/10.1016/j.enbuild.2014.05.004
71. Kim, K.P., Park, K.S.: Delivering value for money with BIM-embedded housing refurbishment. Facilities **36**(13/14), 657–675 (2018). https://doi.org/10.1108/F-05-2017-0048
72. Chardon, S., Brangeon, B., Bozonnet, E., Inard, C.: Construction cost and energy performance of single family houses: from integrated design to automated optimization. Autom. Constr. **70**, 1–13 (2016). https://doi.org/10.1016/j.autcon.2016.06.011
73. Díaz, H., Alarcón, L.F., Mourgues, C., García, S.: Multidisciplinary design optimization through process integration in the AEC industry: strategies and challenges. Autom. Constr. **73**, 102–119 (2017). https://doi.org/10.1016/j.autcon.2016.09.007
74. Li, B., Lou, R., Segonds, F., Merienne, F.: Multi-user interface for co-located real-time work with digital mock-up: a way to foster collaboration? Int. J. Interact. Des. Manuf. **11**(3), 609–621 (2016). https://doi.org/10.1007/s12008-016-0335-2
75. Ciribini, A.L.C., Mastrolembo, V.S., Paneroni, M.: Implementation of an interoperable process to optimise design and construction phases of a residential building: a BIM pilot project. Autom. Constr. **71**, 62–73 (2016). https://doi.org/10.1016/j.autcon.2016.03.005
76. Zanchetta, C., Borin, P., Cecchini, C., Xausa, G.: Computational design and classification systems to support predictive checking of performance of building systems. TECHNE – J. Technol. Archit. Environ. 13 (2017). https://doi.org/10.13128/techne-19759
77. Xu, Z., Huang, T., Li, B., Li, H., Li, Q.: Developing an IFC-based database for construction quality evaluation. Adv. Civ. Eng. **2018**, 1–22 (2018). https://doi.org/10.1155/2018/3946051
78. Hernández, J., Martín, L.P., Bonsma, P., Van Delft, A., Deighton, R., Braun, J.D.: An IFC interoperability framework for self-inspection process in buildings. Buildings **8**(2), 32 (2018). https://doi.org/10.3390/buildings8020032
79. Čuš-Babič, N., Rebolj, D., Nekrep-Perc, M., Podbreznik, P.: Supply-chain transparency within industrialized construction projects. Comput. Ind. **65**, 345–353 (2014). https://doi.org/10.1016/j.compind.2013.12.003
80. Zekavat, P.R., Moon, S., Bernold, L.E.: Securing a wireless site network to create a BIM-allied work-front. Int. J. Adv. Robot. Syst. **11**(1), 1 (2014). https://doi.org/10.5772/58441

81. Dave, B., Kubler, S., Främling, K., Koskela, L.: Opportunities for enhanced lean construction management using Internet of Things standards. Autom. Constr. **61**, 86–97 (2016). https://doi.org/10.1016/j.autcon.2015.10.009
82. Patti, E., et al.: Information modeling for virtual and augmented reality. IT Prof. **19**, 52–60 (2017). https://doi.org/10.1109/MITP.2017.43
83. Karaman, S.G., Chen, S.S., Ratnagaran, B.J.: Three-dimensional parametric data exchange for curved steel bridges. Transp. Res. Record in J. Transp. Res. Board. Nr 2331, pp. 27–34 (2013). https://doi.org/10.3141/2331-03
84. Du, J., He, R., Sugumaran, V.: Clustering and ontology-based information integration framework for surface subsidence risk mitigation in underground tunnels. Clust. Comput. **19**(4), 2001–2014 (2016). https://doi.org/10.1007/s10586-016-0631-4
85. Hartmann, T., Amor, R., East, E.W.: Information model purposes in building and facility design. J. Comput. Civ. Eng. **31**, 04017054 (2017). https://doi.org/10.1061/(ASCE)CP.1943-5487.0000706
86. Zhang, J., Liu, Q., Hu, Z., Lin, J., Yu, F.: A multi-server information-sharing environment for cross-party collaboration on a private cloud. Autom. Constr. **81**, 180–195 (2017). https://doi.org/10.1016/j.autcon.2017.06.021
87. Costin, A., Adibfar, A., Hu, H., Chen, S.S.: Building information modeling (BIM) for transportation infrastructure – literature review, applications, challenges, and recommendations. Autom. Constr. **94**, 257–281 (2018). https://doi.org/10.1016/j.autcon.2018.07.001
88. Brumana, R., et al.: Generative HBIM modelling to embody complexity (LOD, LOG, LOA, LOI): surveying, preservation, site intervention—the Basilica di Collemaggio (L'Aquila). Appl. Geomat. **10**(4), 545–567 (2018). https://doi.org/10.1007/s12518-018-0233-3
89. Quattrini, R., Pierdicca, R., Morbidoni, C.: Knowledge-based data enrichment for HBIM: exploring high-quality models using the semantic-web. J. Cult. Herit. **28**, 129–139 (2017). https://doi.org/10.1016/j.culher.2017.05.004
90. Ramaji, I.J., Memari, A.M.: Extending the current model view definition standards to support multi-storey modular building projects. Archit. Eng. Des. Manag. **14**, 158–176 (2018). https://doi.org/10.1080/17452007.2017.1386083
91. Sun, J., Liu, Y.S., Gao, G., Han, X.G.: IFCCompressor: a content-based compression algorithm for optimizing Industry Foundation Classes files. Autom. Constr. **50**, 1–15 (2015). https://doi.org/10.1016/j.autcon.2014.10.015
92. Ryu, J., Lee, J., Choi, J.: Development of process for interoperability improvement of BIM data for free-form buildings design using the IFC Standard. Int. J. Softw. Eng. Appl. 10, 127–138. ISSN 1738-9984 (2016)
93. Koo, B., Shin, B.: Applying novelty detection to identify model element to IFC class misclassifications on architectural and infrastructure building information models. J. Comput. Des. Eng. **5**, 391–400 (2018). https://doi.org/10.1016/j.jcde.2018.03.002
94. Lee, Y.C., Eastman, C.M., Lee, J.K.: Validations for ensuring the interoperability of data exchange of a building information model. Autom. Constr. **58**, 176–195 (2015). https://doi.org/10.1016/j.autcon.2015.07.010
95. Jiang, Y., et al.: Automatic building information model query generation. J. Inf. Technol. Constr. 18 (2015)
96. Gómez-Romero, J., Bobillo, F., Ros, M., Molina-Solana, M., Ruiz, M.D., Martín-Bautista, M.J.: A fuzzy extension of the semantic building information model. Autom. Constr. **57**, 202–212 (2016). https://doi.org/10.1016/j.autcon.2015.04.007
97. Karan, E., Irizarry, J., Haymaker, J.: Generating IFC models from heterogeneous data using semantic web. Constr. Innov. **15**, 219–235 (2015). https://doi.org/10.1108/CI-05-2014-0030
98. Lee, J.K., Kim, M.J.: BIM-enabled conceptual modelling and representation of building circulation. Int. J. Adv. Robot. Syst. **11**, 127 (2014). https://doi.org/10.5772/58440

99. Ahn, K.U., Kim, Y.J., Park, C.S., Kim, I., Lee, K.: BIM interface for full vs. semi-automated building energy simulation. Energy Build. **68**, 671–678 (2014). https://doi.org/10.1016/j.enbuild.2013.08.063
100. Kim, H., Anderson, K.: Energy modeling system using building information modeling open standards. J. Comput. Civ. Eng. **27**, 203–211 (2013). https://doi.org/10.1061/(ASCE)CP.1943-5487.0000215
101. Gupta, A., Cemesova, A., Hopfe, C.J., Rezgui, Y., Sweet, T.: A conceptual framework to support solar PV simulation using an open-BIM data exchange standard. Autom. Constr. **37**, 166–181 (2014). https://doi.org/10.1016/j.autcon.2013.10.005
102. Fuchs, S., Scherer, R.J.: Multimodels — instant nD-modeling using original data. Autom. Constr. **75**, 22–32 (2017). https://doi.org/10.1016/j.autcon.2016.11.013
103. Perisic, A., Lazic, M., Perisic, B.: The extensible orchestration framework approach to collaborative design in architectural, urban and construction engineering. Autom. Constr. **71**, 210–225 (2016). https://doi.org/10.1016/j.autcon.2016.08.005
104. Niknam, M., Karshenas, S.: A shared ontology approach to semantic representation of BIM data. Autom. Constr. **80**, 22–36 (2017). https://doi.org/10.1016/j.autcon.2017.03.013
105. Qing, L., Tao, G., Ping, W.J.: Study on building lifecycle information management platform based on BIM. Res. J. Appl. Sci. Eng. Technol. **7**, 1–8 (2014). https://doi.org/10.19026/rjaset.7.212
106. Das, M., Cheng, J.C.P., Kumar, S.S.: Social BIMCloud: a distributed cloud-based BIM platform for object-based lifecycle information exchange. Visual. Eng. **3**(1), 1–20 (2015). https://doi.org/10.1186/s40327-015-0022-6
107. Grilo, A., Zutshi, A., Jardim-Goncalves, R., Steiger-Garcao, A.: Construction collaborative networks: the case study of a building information modelling-based office building project. Int. J. Comput. Integr. Manuf. **26**, 152–165 (2013). https://doi.org/10.1080/0951192X.2012.681918
108. Afsari, K., Eastman, C., Shelden, D.: Building information modeling data interoperability for cloud-based collaboration: limitations and opportunities. Int. J. Archit. Comput. **15**, 187–202 (2017). https://doi.org/10.1177/1478077117731174
109. Gökçe, K.U., Gökçe, H.U., Katranuschkov, P.: IFC-based product catalog formalization for software interoperability in the construction management domain. J. Comput. Civ. Eng. **27**, 36–50 (2013). https://doi.org/10.1061/(ASCE)CP.1943-5487.0000194
110. Ding, L., Xu, X.: Application of cloud storage on BIM life-cycle management. Int. J. Adv. Robot. Syst. 11(8) (2014). https://doi.org/10.5772/58443
111. Girard, C. : L'architecture, une dissimulation. La fin de l'architecture fictionnelle à l'ère de la simulation intégrale (chapitre 8). In: Modéliser Simuler – Tome 2. Editions Matériologiques, Paris, pp. 245–292 (2014). https://doi.org/10.3917/edmat.varen.2014.01.0245.
112. Carpo, M.: The Second Digital Turn, Design Beyond Intelligence. MIT Press. ISBN: 9780262534024 (2017)
113. Grilo, A., Jardim-Goncalves, R.: Cloud-marketplaces: distributed e-procurement for the AEC sector. Adv. Eng. Inform. **27**, 160–172 (2013). https://doi.org/10.1016/j.aei.2012.10.004
114. Veloso, P., Celani, G., Scheeren, R.: From the generation of layouts to the production of construction documents: an application in the customization of apartment plans. Autom. Constr. **96**, 224–235 (2018). https://doi.org/10.1016/j.autcon.2018.09.013

Implementation of a Holonic Product-Based Platform for Increased Flexibility in Production Planning

Patricio Sáez Bustos(✉) and Carlos Herrera López

Department of Industrial Engineering, University of Concepción, Concepción, Chile
{patricsaez,cherreral}@udec.cl

Abstract. In the Industry 4.0 era, production planning problems are very relevant to production systems and are essential parts of the supply chain. Broadly speaking, production planning problems are tackled using models and methodologies, aiming for optimal solutions. This work introduces realism and stability to optimal production planning strategies using a holonic, product-driven manufacturing platform with increased flexibility. A model based on an anarchic holonic architecture and embedded intelligence logic provides decision-making capacity in a "production lot" in the face of disturbances. The proposed model is validated by comparing the results obtained with a lot-streaming mathematical programming model. Results show that significant changes in lot processing times (disturbances) generate significant changes in completion times. The proposed platform reduces up to 10.95% completion times in face of disturbances, generating significant benefits by increasing flexibility.

Keywords: Industry 4.0 · Holonic manufacturing system · Multi-agent system · Anarchic manufacturing · Lot-streaming · Smart product · Flexibility

1 Introduction

Production planning and control (PPC) is recognized as a complex problem in the industry, that requires achieving customer satisfaction and optimizing available resources. This complexity is explained by a large number of interrelated elements and variables. In specific cases, the problems are theoretical, without real application. Among the most studied production systems are manufacturing production systems organized as flow shops and job shops or task workshops [1–4]. In particular, both types of production systems are present in an industry (independently or jointly), becoming also a central part of the literature's works.

Conventional production planning systems work by developing hierarchies between different product aggregation levels [5], resulting in monotonous and static production scheduling [6]. New perspectives, such as multi-agent systems (MASs) and holonic manufacturing systems (HMSs), have attracted increasing interest in the industry by introducing agility, adaptability, autonomy, and, above all, flexibility in production systems [7, 8]. HMSs manage production in decentralized and distributed decision-making

D. Trentesaux et al. (Eds.): SOHOMA 2021, SCI 987, pp. 180–194, 2021.
https://doi.org/10.1007/978-3-030-80906-5_12

architectures, while MASs provide a greater degree of flexibility and reconfiguration to production systems. In this case, the agents provide a physical representation of the system components, including machines, equipment, products, lots, etc., thus providing different perspectives and control scenarios [9].

An HMS platform with an *anarchic* decision-making architecture was developed to study the need for more flexible production plans [5]. This architecture promotes the need for the decentralization of decisions, delivering this action to the lowest links in the production chain, in this case to production lots. Therefore, production lots exhibit behaviour similar to an intelligent product with features such as those set by [7, 10, 11].

The validation of our platform was performed by comparing the results of a practical example to minimize makespan. A lot-streaming mathematical programming model optimally solved the production planning problems considered in this work. Subsequently, the system was disturbed by changing machines' processing times (typical disturbances in production systems due to failures, machine lock, starving, etc.). In this way, the deterioration impact of planning and the contribution of a distributed decision-making system such as the one proposed can be assessed. Our goal is to show the need for systems that provide more flexibility in decision-making processes.

The article is organized as follows: Sect. 2 describes a bibliographic review associated with flexibility issues in PPC; measures and tools used for the analysis and incorporation of flexibility into production processes are analysed along with the participation of MAS and HMS models for this purpose. Section 3 presents the materials and methods. The construction of the holonic model and the mathematical programming model developed to obtain optimal solutions are also presented. Section 4 describes the experimentation developed for each of the test instances. Section 5 shows our application's main results by considering a standard case and one with disruptions in process times. Finally, Sect. 6 summarizes our work.

2 Bibliographic Review

Flexibility is an attribute that gives manufacturing systems the possibility that at a certain level of variation in the quantities to be produced and/or at interruptions in the production line there will be no significant changes in planning [12]. In 1983, [13] examined the concept of flexibility in manufacturing, defining a framework of attributes that influence the different aspects of flexibility. However, the idea of flexible manufacturing systems was proposed in 1960 by David Williamson, who devised the so-called "24 system" of machines capable of producing 24 h a day without human intervention.

Scheduling problems have been extensively studied in the literature and are a fundamental part of production systems theory. Thus, many articles on flexible manufacturing systems can be found in the literature with techniques that provide the necessary flexibility [14–16]. The techniques analysed are in the fields of simulation, artificial intelligence (AI), and Petri networks, among others. However, there are also works based on mathematical programming, such as that developed by Stecke [17], who used mixed nonlinear programming for machine grouping to minimize part movement. While known to deliver optimal results, such methods can be affected during actual operation by the high execution times of the implemented algorithms [18].

One of the most widely used performance indicators to be minimized in production planning is the completion time (makespan) [18–20]. This concept must be accompanied by other indicators, such as costs or failure rates in more complex models to achieve a closer version of reality. In [21], flexibility was incorporated into a sequential production environment using a new decomposition method combined with sequencing rules (shorter process times are processed first) and a genetic algorithm, which minimize makespan by delivering flexibility to the process.

Distributed processes such as MASs and HMSs provide an excellent opportunity to respond to changes in production environment conditions effectively. J. Yingfeng *et al.* [22] provided real-time production planning supported by a new architecture based on a MAS model in conjunction with the Internet of Things (IoT). This model proposed an optimal machine allocation strategy depending on component status. However, this approach does not incorporate a complete decentralization of the decision-making process. On the other hand, in the work of S. Răileanu *et al.* [1], production planning was implemented by a HMS generating control at different levels. Communication with these levels occurs in the upper layers, with recommendations sent to lower layers which communicate with each other to solve and optimize tasks.

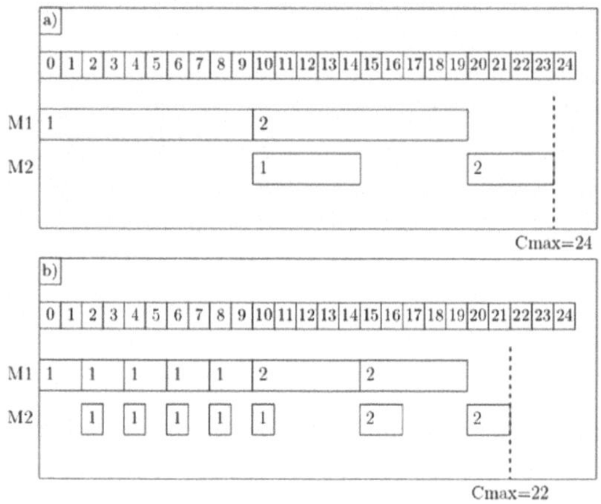

Fig. 1. (a) Scheduling two jobs into two sequential machines (M1 and M2); (b) Scheduling jobs using the lot-streaming technique, dividing work into sublots [23].

3 Materials, Methods and Proposal

The use of scheduling and lot-streaming techniques to solve programming problems has been widely used to reduce completion times optimally. Lot-streaming is a technique that considers "n" jobs on "m" machines, where jobs are divided into sublots to minimize the delay and completion time of tasks (see Fig. 1). In works such as those described in [23–26], lot-streaming is used in various programming problems, the most studied being those about sequential process (flow shop and job shop) organization.

In today's industrial practices, many of the problems include objectives that conflict with each other. These problems are solved by various optimization models, which work with subproblems simultaneously or individually [22, 27]. In this work, an HMS-based approach is proposed for smart lots that make production process decisions for the products contained therein.

The proposed architecture has been simulated on the NETLOGO platform [28]. The architecture is composed of two serial stations in which production processes are developed for batches. At the first station is the M1 machine, where the production process starts sequentially. At the second station, the machines M2, M3, and M4 finish the production process (each lot can be processed on only one second-stage machine), as shown in Fig. 2.

Product lots planning calculates the target global function to minimize completion time or makespan. This objective refers to finishing production as quickly as possible and is calculated by lots at the end of their production processes. Decisions are reused to perform a learning process for future generations of solutions.

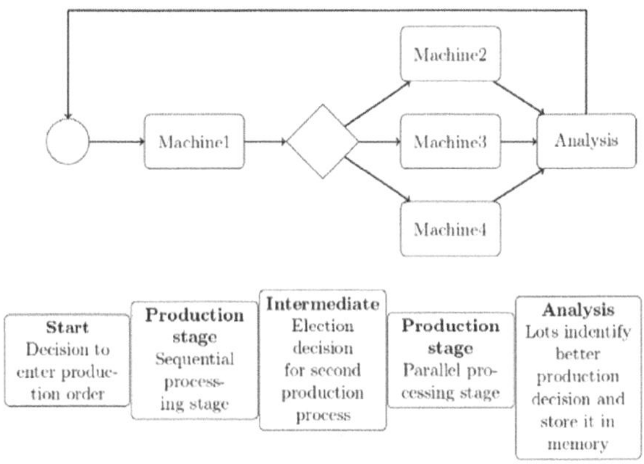

Fig. 2. Stages of the production process

3.1 Interactions Between Agents

A MAS simulation is used to solve this problem for a part lot in an anarchic structure perspective [5]. The type of HMS system is given because the agents of the MAS model represent physical system components. In the work of Yu *et al.* [29], three categories of agents were developed: work agents, machine agents and control agents; however, our system, having an anarchic architecture, delegates authority and autonomy in decision-making to the lowest level of system entities without centralized control or supervision, so there will be only two categories: lot agents and machine agents. Machine agents are static and only receive nonparticipating lot agents; lot agents are dynamic entities that control the realization of the products they represent.

The parameters and steps used in our algorithm are presented in the following pseudocode:

```
Home:
Create lot variable and lot family;
Create variables for lots: process times, layout position, and
memory;
Setting function ();
Create differentiated machines between type I and II (series and
parallel, respectively);
Create undifferentiated lots and positioned in the initial ma-
chine;
Differentiation Function ()
Identify existing lot types and categorizes lots;
Normal production times are allocated per lot;
Initial Position Function ();
Assign the position variable in each lot a random i value be-
tween 1 and n with n s number of lots;
Execute function ();
If the lots have not finished processing then;
   If the lots are in the initial machine then;
      Compare their positions, the one with priority initiates
      processing (Priority: lower position in the layout);
      If lots finish processing then;
         Ask machines in parallel if they are processing
         If machines are not processing then;
            Lot moves to unemployed machine
         Else
            Lot asks lots in process who finishes before
            Starts processing
      Else
         Continue processing
   Else
      Continue processing
Else
   Makespan Function ();
   Calculate completion time
   Lots keep in memory the position in which the best result was
   obtained
   Lots move to starting position
```

According to the pseudocode above, each lot corresponds to an agent defined by the minimum accepted size lot. The generated agents follow the concept of intelligent products proposed by C. Wong *et al.* [27] and C. Herrera [11], where each entity (product lot) consists of the following characteristics:

1. Agent has a unique identity.
2. Agent can communicate effectively with its environment.
3. Agent can retain or store data about itself.

4. Agent implements a language to display its features, production requirements, and so on.
5. Agent can participate or make decisions relevant to its results.

In particular, an agent has the following characteristic profile:

1. *Unique identity*: Each agent has a specific and unrepeatable ID (although there are shared features when talking about lots of the same product).
2. *Effective communication*: Communication between agents (lots) is active and indifferent to location. This is based on the decision-making process and choosing the best sequence for the common goal.
3. *Retention and storage of information about itself*: An agent has a memory of its processing time on each machine. Besides, it saves the position that improves each of the objective functions.
4. *Language*: Based on group consultations with all agents or with agents in sectors of the production process.
5. *Relevant decision making*: Decision to process or not, in addition to the choice of the machine on which to process.

Communication between agents is classified according to the different possibilities of interactions present in the model. These interactions are defined as visual, auditory, and verbal.

Visual interactions refer to whether the product agent displays its entity (level 1), its nearby environment (level 2), or the entire medium (level 3). Auditory interactions correspond to the ability to capture media information from level 1 that identifies the agent as independent of the environment. It develops at a level 2 where it captures local and selective information, to reach a level 3 in which it captures information from the entire environment. Finally, verbal interactions indicate the level of delivery of agent-product information, from disconnection with the medium (level 1), delivery of timely information (level 2), or delivery of information to the entire medium (level 3).

Figure 3 shows the characterization of the agent-product for testing.

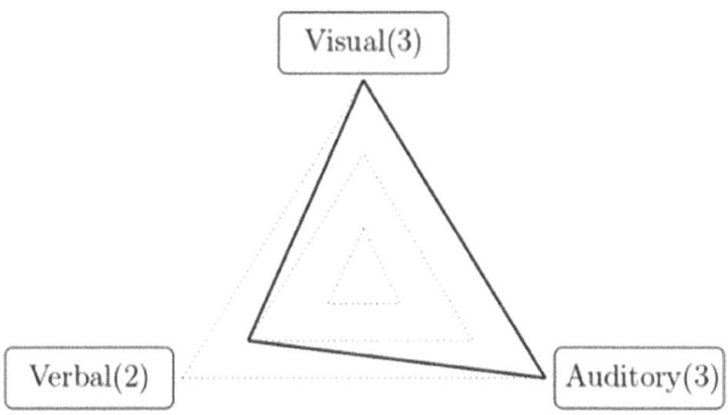

Fig. 3. Characterization of communication for agents

The interaction between agents has a fixed query-based structure. The structure of the interaction between system components is shown in Fig. 4. This representation follows the UML sequence structure and shows the conceptual architecture of an intelligent system.

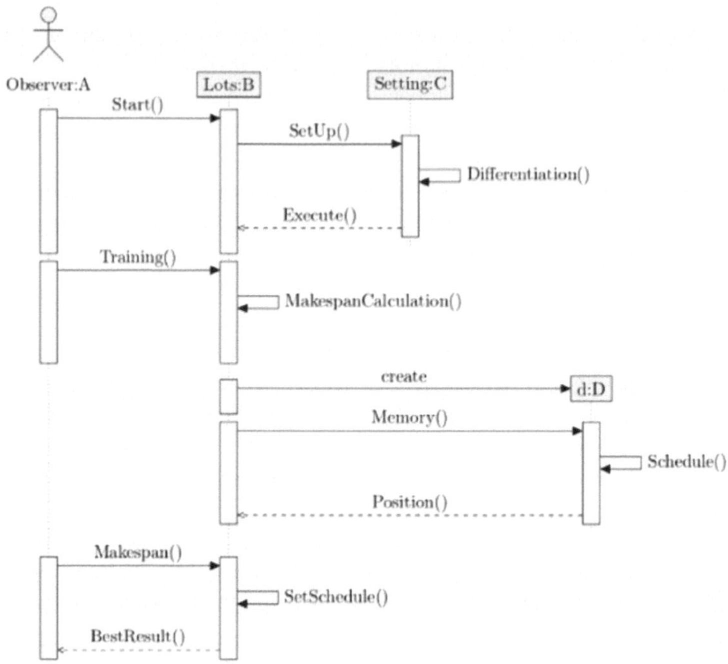

Fig. 4. Sequence UML diagram

The first object in the diagram presented in Fig. 4 is the "observer". It describes user interaction with the model and the receipt of the best results by agents. The second multi-object component (comprises all lots present in the simulation) displays the sequential structure of lot operations, taking the indications entered by the observer. Additionally, it explains the sequence of actions executed by agents (lots), from the configurations of their attributes to indicating the best result to the observer, taking the best position stored in their memory. The third and fourth objects are created from the lots by configuring each agent and saving the best position for each of the objective functions: makespan and utilization.

The structure of our work is exemplified in the UML diagram of classes presented in Fig. 5. Model interactions focus on the observer and lots through a configuration layer that creates machines and lots, allowing lot differentiation depending on the products they contain. The diagram continues with an accurate class, attending to decision-making at the lowest levels, which reads and writes attributes autonomously. Finally, the scheduling object corresponds to the result of the production configurations that the agents run.

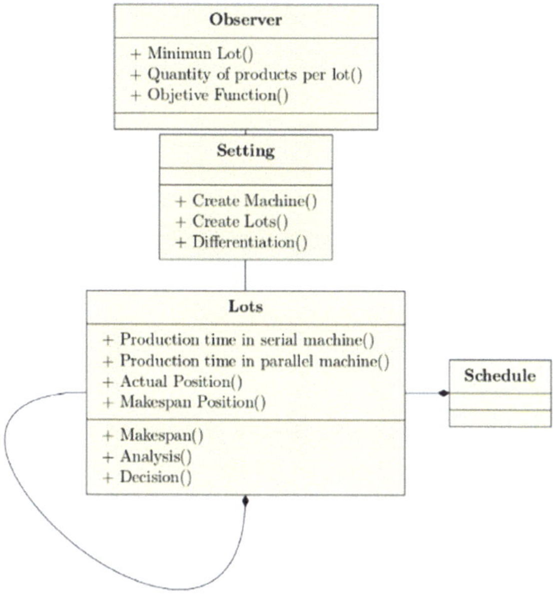

Fig. 5. UML diagram of classes for the proposed model

The lot class corresponds to all the agents present in the model and aims to complete its production process through the sequence of machines established for the system (Fig. 5). This model consists of a sequential machine where the production process begins and continues with the decision-making process of agents, and a second production process involving three identical and unrelated parallel machines. No process has a setup time; however, each machine could have a different rhythm (due to breakdowns or some other cause external to the process). Each machine can serve one lot at a time (with a number of products defined by the minimum lot accepted) and ends in an analysis stage. Machines learn from the generated sequence to position themselves back in the production queue and restart the process.

3.2 Mathematical Programming Model

For comparison, a mathematical lot-streaming model is formulated as well. This mathematical programming model will allow us to develop a generic configuration with optimal results to validate the actual results. The configuration that was considered for the mathematical programming model is shown in Fig. 2 (2-stage hybrid system), where the first stage corresponds to mass production and the second to parallel production. This model was developed and validated in [11].

The problem with the lot-streaming that we address in our work is to divide the quantities to be produced of each product into sub lots to reduce the total sequencing duration (makespan). Sub lots are constrained by minimum amounts defined by the decision-making process, and the model parameters are included in Table 1.

The complete model is developed below:

Table 1. Parameters and variables for the mathematical model

Parameters	
P_i	: *Number of products per lot*
Q_{min}	: *Minimun sub lot size*
TPA_p	: *Unit production time in machine A for product p*
TPB_p	: *Unit production time in machine B for product p*
J	: *Number of lots*
K	: *Number of machines in stage two (parallel)*
M	: *Big number*
Variables	
x_{ijk}	: *Number of product i in lot j assigned to machine k*
w_{ij}	: *1 if product i is assigned to lot j in stage 1 and 0 otherwise*
w_{ijk}	: *1 if product i is assigned to position j in machine k in stage 2 and 0 otherwise*
STA_j	: *Start time of stage 1 for lot j*
STB_{jk}	: *Start time of stage 2 for lot j in machine k*

Objective Function:

$$min\ C_{max}$$

subject to:

$$\sum\nolimits_{j=1}^{J}\sum\nolimits_{k=1}^{K} x_{ijk} = P_i,\ i = 1, 2 \tag{1}$$

$$x_{ijk} \geq Q_{min}{}^{*}w_{ij},\ i = 1,\ 2,\ \forall j,\ \forall k \tag{2}$$

$$x_{ijk} \leq M{}^{*}w_{ijk},\ i = 1, 2,\ \forall j,\ \forall k \tag{3}$$

$$w_{1j} + w_{2j} = 1,\ \forall j \tag{4}$$

$$\sum\nolimits_{k=1}^{K} w_{ijk} = 1,\ i = 1, 2,\ \forall j \tag{5}$$

$$STA_1 = 0 \tag{6}$$

$$STA_j = STA_{j-1} + TPA_1{}^{*}\sum_{k=1}^{K} x_{1jk} + TPA_2{}^{*}\sum_{k=1}^{K} x_{2jk},\ \forall j \tag{7}$$

$$STB_{(j-1)k} \geq STA_j,\ \forall j, \forall k \tag{8}$$

$$STB_{jk} \geq STB_{(j-1)k} + TPB_1{}^{*}x_{1jk} + TPB_2{}^{*}x_{2jk}, \forall j, \ \forall k \tag{9}$$

$$C_{max} \geq STB_{jk} + TPB_1{}^{*}x_{1jk} + TPB_2{}^{*}x_{2jk}, \ \forall k \tag{10}$$

Restrictions (1), (2) and (3) refer to the number of products assigned to each lot and their lower (Q_{min}) and upper limits M. Constraints (4) and (5) are assignments for each lot *j* and machine *k*, respectively. Restrictions (6), (7), (8) and (9) indicate the start times of each lot in both stage one and stage two. Finally, constraint (10) sets the maximum end time value.

4 Experiments

Next, we analyze the consequences of lack of flexibility in production. A practical example was generated with 2 products distributed in minimum lots of 100 units, with a maximum production of 1200 units (total units considering the two products). This example follows the structure presented in Fig. 2, with two production stages to manufacture a product. Machines can work these products with different production times. The first stage starts working semi-finished products and the second stage finishes products.

The programming results generated by the lot-streaming model, as well as those obtained by the MAS platform, will be affected by disturbances in the processing time of the machines when identifying the planning consequences and analysing their behavior.

4.1 Description of Experimentation

Our goal is to verify the need for flexible production plans capable of adapting to disruptions during production processes. Lot production will be simulated in a planning horizon that allows existing lots to be completed. The production will be generated respecting its maximum volume (established at 1200 units) and the demand originated by combinations of minimum lots of both products (e.g., DP1 + DP2 = 1200, with DP1 as the demand for product 1 in lots of 100 units and DP2 as the demand for product 2 in lots of 100 units).

The simulation parameters are as follows (Table 2):

Table 2. Simulation parameters

Parameters	Values
Types of products	2
Lots	Demand dependent (100 products per lot)

(continued)

Table 2. (*continued*)

Parameters	Values
Production	Maximum of 1200 units in total (product 1 + product 2)
Normal production time	Production time: product 1 in stage 1 $\times$ 1 Production time: product 1 in stage 2 $\times$ 3 Production time: product 2 in stage 1 $\times$ 2 Production time: product 2 in stage 2 $\times$ 8
Internal disturbances	
Increased production time for products 1 and 2 in stage 1	100, 200 and 400 (%)
Increased production time for products 1 and 2 in stage 2	100, 200 and 400 (%)

5 Results

Table 3 lists the production data associated with the analysed practical example. Nine volumes of production were assessed. Additionally, the number of lots per product and completion times are set for the mathematical programming (PM) algorithm and multiagent platform (PMAS).

Table 3. Production overview and volume of lot (T.F.: End time in hours)

Demand	Prod. 1	Prod. 2	Lots (Prod. 1)	Lots (Prod. 2)	T.F. PM	T.F. PMAS
Demand1	1000	200	10	2	19	19
Demand2	900	300	9	3	20	20
Demand3	800	400	8	4	22	22
Demand4	700	500	7	5	24	24
Demand5	600	600	6	6	26	26
Demand6	500	700	5	7	28	28
Demand7	400	800	4	8	29	29
Demand8	300	900	3	9	32	32
Demand9	200	1000	2	10	34	34

It is clear from Table 3 that both the linear programming and the solutions found in the adaptive platform are identical, meaning that at any production rate under normal conditions, both solutions are optimal. The increase in production time for lots 1 and 2 in stage 1 is shown in the data expressed in Table 4.

The multiagent simulation results for different production times are identical in both cases (PM and PMAS). However, the production planning obtained under normal

Table 4. Increased production time in stage 1 (times in hours)

Demand	TP stage increase 1 100%		TP stage increase 1 200%		TP stage increase 1 400%	
	Pm	PMAS	Pm	PMAS	Pm	PMAS
Demand1	31	31	45	45	59	59
Demand2	33	33	48	48	63	63
Demand3	35	35	51	51	67	67
Demand4	37	37	54	54	71	71
Demand5	39	39	57	57	75	75
Demand6	41	41	60	60	79	79
Demand7	43	43	63	63	83	83
Demand8	45	45	66	66	87	87
Demand9	48	48	69	69	91	91

conditions is affected by increasing its value by approximately 67% for the 400% increase in processing time. Since this stage is sequential, all processes are delayed; therefore, the agents' adaptation does not influence the completion time.

The results for increasing production time for lots 1 and 2 in stage 2 are shown in Table 5.

Table 5. Increased production time in stage 2 (times in hours)

Demand	TP stage increase 2 100%		TP stage increase 2 200%		TP stage increase 2 400%	
	Pm	PMAS	Pm	PMAS	Pm	PMAS
Demand1	35	34	69	66	137	130
Demand2	39	37	77	71	153	139
Demand3	41	40	81	78	161	154
Demand4	47	45	93	89	185	177
Demand5	51	47	101	91	201	179
Demand6	53	51	103	100	203	196
Demand7	56	55	110	109	218	217
Demand8	59	57	113	111	221	219
Demand9	66	66	130	130	258	258

The results of this simulation demonstrate the importance of flexibility in production scheduling. This is because the completion times obtained from the programming given by the mathematical model are greater than those obtained by the simulation platform. Production planning given by exact models such as PM only guarantees optimality in typical or theoretical situations, while the model generated in the simulation platform can adapt to environmental conditions and modify the initial planning. When this modification is made online as designed in an IoT architecture, that is, as disturbances occur, agents can react and modify their sequencing.

Table 6. Increased production time in stage 2 (times in hours)

Demand	TP increase in stage 2 100%	TP stage increase 2 200%	TP stage increase 2 400%
	Reduction (%)	Reduction (%)	Reduction (%)
Demand1	−2,86	−4,35	−5,11
Demand2	−5,13	−7,79	−9,15
Demand3	−2,44	−3,70	−4,35
Demand4	−4,26	−4,30	−4,32
Demand5	−7,84	−9,90	−10,95
Demand6	−3,77	−2,91	−3,45
Demand7	−1,79	−0,91	−0,46
Demand8	−3,39	−1,77	−0,90
Demand9	0,00	0,00	0,00

While completion times increase considerably compared to the data found under normal conditions, this model improves completion times by adapting to production conditions. Table 6 shows the percentage of damping results for end time in each instance. There is an average damping of 3.92%, with a max. of 10.95% and a min. of 0.

6 Conclusion

This article's objective was to analyze the need of flexibility in production systems and deliver alternatives to existing static models. The proposed platform incorporates two distributed anarchic structures, in an intelligent product approach. This provides a tremendous advantage for production planning systems because human intervention is not needed to change the planning. More interestingly, the production lots reorganize and communicate themselves to achieve the common goal.

Compared to the lot-streaming technique, the platform achieved the same results for both initial conditions and disturbances within the sequential machines. This shows us that the developed platform can identify the optimal solution for all instances of the batch defined. We note that, for sustained increases in production times of 100%, 200%, and 400%, long completion times up to 88% are obtained, where alternatives such as the presented platform would deliver a reduction of up to 10.95% in completion times.

In future research, we will consider larger problems on a real scale. Additionally, the incorporation of new disturbances will be considered when analysing their impacts on production and the effect that the architecture, as the one presented here, would have.

References

1. Răileanu, S.: Production scheduling in a holonic manufacturing system using the open-control concept. UPB Sci. Bull. Series C 72, 1–4. ISSN 1454-234x (2010)
2. Shahzad, A., Mebarki, N.: Learning dispatching rules for scheduling: a synergistic view comprising decision trees. Tabu Search Simul. Comput. **5**, 16 (2016). https://doi.org/10.3390/computers5010003
3. Fan, B.Q., Cheng, T.C.E.: Two-agent scheduling in a flowshop. Eur. J. Oper. Res. **252**(2), 376–384 (2016). https://doi.org/10.1016/j.ejor.2016.01.009
4. Wu, C.-C., et al.: A two-stage three-machine assembly scheduling flowshop problem with both two-agent and learning phenomenon. Comput. Ind. Eng. **130**, 485–499 (2019). https://doi.org/10.1016/j.cie.2019.02.047
5. Ma, A., Nassehi, A., Snider, C.: Anarchic manufacturing. Int. J. Prod. Res. **57**, 2514–2530 (2019). https://doi.org/10.1080/00207543.2018.1521534
6. Role, M., Martinez, E.: Agent-based modeling and simulation of an autonomic manufacturing execution system. Comput. Ind. **63**, 53–78 (2012). https://doi.org/10.1016/j.compind.2011.10.005
7. McFarlane, D., Sarma, S., Chirn, J.: The intelligent product in manufacturing control and management. IFAC **35**, 49–54 (2002). https://doi.org/10.3182/20020721-6-ES-1901.00011
8. Kruger, K., Basson, A.H.: Evaluation criteria for holonic control implementations in manufacturing systems. Manuf. Syst. Mnt. J. Comp. Int. Man. **32**, 148–158 (2019). https://doi.org/10.1080/0951192X.2018.1550674
9. Leitão, P., Rodrigues, N., Barbosa, J., et al.: Intelligent products: the grace experience. Control Eng. Pract. **42**, 95–105 (2015). https://doi.org/10.1016/j.conengprac.2015.05.001
10. Wong, C.Y., McFarlane, D., Zaharudin, A.A., Agarwal, V.: The intelligent product driven supply chain. https://doi.org/10.1109/ICSMC.2002.1173319 (2014)
11. Herrera, C.: Cadre générique de planification logistique dans un contexte de décisions centralisées et distribuées. Université Henry Poincaré - Nancy. NNT: 2011NAN10046 (2011)
12. Yadav, A., Jayswal, S.C.: Modelling of flexible manufacturing system: a review. Int. J. Prod. Res. **56**, 2464–2487 (2018). https://doi.org/10.1080/00207543.2017.1387302
13. Slack, N.: Flexibility as a manufacturing objective. Int. J. Oper. Manag. **3**, 4–13 (1983). https://doi.org/10.1108/eb054696
14. Read, K.K.: Fuzzy rule generation for adaptive scheduling in a dynamic manufacturing environment. Appl. Soft. Comput. J. **8**, 1295–1304 (2008). https://doi.org/10.1016/j.asoc.2007.11.005
15. Zhou, Y., Yang, J.J., Zheng, L.Y.: Multi-agent based hyper-heuristics for multi-objective flexible job shop scheduling: a case study in an aero-engine blade manufacturing plant. IEEE Access **7**, 21147–21176 (2019). https://doi.org/10.1109/ACCESS.2019.2897603
16. Demirel, E., Azelkan, E.C., Lim, C.: Aggregate planning with flexibility requirements profile. Int. J. Prod. Econ. **202**, 45–58 (2018). https://doi.org/10.1016/j.ijpe.2018.05.001
17. Stecke, K.E., Solberg, J.: Loading and control problem for a flexible manufacturing system. Int. J. Prod. Res. **19**, 481–490 (1981)
18. Topaloglu, S., Kilincli, G.: A modified shifting bottleneck heuristic for the reentrant job shop scheduling problem with makespan minimization. Int. J. Adv. Manuf. Technol. **44**, 781–794 (2009). https://doi.org/10.1007/s00170-008-1881-y
19. Ahmadi-Darani, M.H., Moslehi, G., Reisi-Nafchi, M.: A two-agent scheduling problem in a two-machine flowshop. Int. J. Ind. Eng. Comput. **9**, 289–306 (2018). https://doi.org/10.5267/j.ijiec.2017.8.005
20. Gu, J., Gu, M., Cao, C., Gu, X.: A novel competitive co-evolutionary quantum genetic algorithm for stochastic job shop scheduling problem. Comput. Oper. Res. **37**, 927–937 (2010). https://doi.org/10.1016/j.cor.2009.07.002

21. Choi, S.H., Wang, K.: Flexible flow shop scheduling with stochastic processing times: a decomposition-based approach. Comput. Ind. Eng. **63**, 362–373 (2012). https://doi.org/10.1016/j.cie.2012.04.001
22. Wang, J., Zhang, Y., Liu, Y., Naiqi, W.: Multiagent and bargaining-game-based real-time scheduling for internet of things-enabled flexible job shop. IEEE Internet Things J. **6**(2), 2518–2531 (2019). https://doi.org/10.1109/JIOT.2018.2871346
23. Tseng, C., Liao, C.J.: A discrete particle swarm optimization for lot-streaming flow-shop scheduling problem. Eur. J. Oper. Res. **191**, 360–373 (2008). https://doi.org/10.1016/j.ejor.2007.08.030
24. Kumar, S., Bagchi, T.P., Sriskandarajah, C.: Lot streaming and scheduling heuristics for m-machine no-wait flowshops. Comput. Ind. Eng. **38**, 149–172 (2000). https://doi.org/10.1016/S0360-8352(00)00035-8
25. Potts, C.N., Baker, K.R.: Flow shop scheduling with lot streaming. Oper. Res. Lett. **8**, 297–303 (1989). https://doi.org/10.1016/0167-6377(89)90013-8
26. Trietsch, D., Baker, K.: Basic techniques for lot streaming. Oper. Res. **41**(6), 1065–1076 (1993). https://doi.org/10.1287/opre.41.6.1065
27. Gharaei, A., Jolai, F.: A multi-agent approach to the integrated production scheduling and distribution problem in multi-factory supply chain. Appl. Soft. Comput. J. **65**, 577–589 (2018). https://doi.org/10.1016/j.asoc.2018.02.002
28. Wilensky, U.: NetLogo. Center for Connected Learning and Computer-Based Modeling Northwestern University Evanston. http://ccl.northwestern.edu/netlogo/ (1999)
29. Yu, F., Wen, P., Yi, S.: A multi-agent scheduling problem for two identical parallel machines to minimize total tardiness time and makespan. Adv. Mech. Eng. **10**, 1–14 (2018). https://doi.org/10.1177/1687814018756103

People and Industry in the Era of Society 5.0

Location-Routing for a UAV-Based Recognition System in Humanitarian Logistics: Case Study of Rapid Mapping

Paula Saavedra, Alejandro Pérez Franco, and William J. Guerrero(✉)

Facultad de Ingeniería, Universidad de La Sabana, Bogota, Colombia
william.guerrero1@unisabana.edu.co

Abstract. Unmanned Aerial Vehicles (UAV) have the potential to improve the first recognition activities that are necessary for humanitarian relief operations in the aftermath of a sudden disaster. Rapid mapping of the affected area is one example of these activities. UAVs provide solutions to the challenge of obtaining reliable information about the status and location of an affected zone and the coordination of the different rescue teams involved in relief operations. This paper proposes a UAV-based rapid mapping system for the first stage of recognition in a post-disaster situation, helping to recognize the status or level of damage at a specific zone. This recognition system is modelled based on an adapted integrated capacitated location-routing (CLRP) model corresponding to the operating specifications of UAVs that represent challenging constraints. This model seeks to determine the optimal location of the UAV's hubs that should be placed before the disaster happens, and the optimal routing of UAVs after the disaster has happened. The adapted model is tested with data generated trying to emulate a small city. The results show a near-optimal location, number of UAV needed, and the routes for each one. From these results, organizational insights are provided to put in place the system. Also, relevant research directions in UAV location routing for rapid mapping are proposed.

Keywords: Humanitarian logistics · Drone · Routing · Optimization · Disaster response · Rapid mapping

1 Introduction

In recent years, significant attention has been brought to studying challenges in humanitarian logistics due to the increase of damages resulting from natural and man-made disasters. Examples include the Sumatra tsunami in 2004 [27], the hurricane Katrina in 2005 [7], the Haiti earthquake in 2010 [3], the Pakistan flood in 2010 [18], the Mexico City earthquake in 2017 [1], and the hurricane Maria in Puerto Rico in 2017 [29]. Further, the COVID-19 global outbreak has motivated the use of new technologies to improve the humanitarian response in several regions of the world [26]. Within this context, post-disaster relief poses major

D. Trentesaux et al. (Eds.): SOHOMA 2021, SCI 987, pp. 197–207, 2021.
https://doi.org/10.1007/978-3-030-80906-5_13

logistical challenges such as the coordination of agencies to collect, transport, and storage humanitarian supplies.

Humanitarian logistics is known as "the process of planning, implementing and controlling the efficient, cost-effective flow and storage of goods and materials as well as related information, from point of origin to point of consumption for the purpose of alleviating the suffering of vulnerable people" [25]. Briefly, "for humanitarians, logistics represents the processes and systems involved in mobilizing people, resources, skills, and knowledge to help vulnerable people affected by disaster" [28]. Humanitarian logistics has a lead role in preparing for, responding to, and recovering from sudden disasters. In addition, humanitarian logistics involves the efficient management of resources, such as the flow of human capital, material resources, or economic aids with the purpose of relieving the suffering of an affected population.

Humanitarian logistics has been studied by the operations research community to develop different mathematical models with the aim to optimize the use of resources and save lives. For example, a Discrete Choice Model is proposed by [8] to model the decision that a person makes after a disaster in order to reduce its deprivation cost; the results are useful to estimate the social cost of humanitarian relief operations. Further, a Markov Decision Process is proposed in [13] to manage the inventory of perishable items properly assuming the demand and supply are stochastic variables allowing, the authors to determine the optimal ordering policy. Also, inventory-routing decisions are optimized by [11] to deliver humanitarian aid. Thus, current research on humanitarian logistics has proven the important role of operations management in improving the disaster response, both in the preparedness and response stages [28].

After a disaster has happened, there are limitations to reach the affected disaster zones. These limitations are mostly due to the lack of resources (vehicles or rescue teams) or disruption of roads, causing significant groups of unattended population. A considerable amount of deaths could be avoided if the rescue team arrives on time to the affected zone. However, rescue operations are often delayed by road disruptions or the procurement of the vehicles in charge of transporting rescue teams. Furthermore, recognition is one of the most important and challenging activities to perform during a rescue operation. Recognition activities reveal the level of damage of the streets and buildings and help to obtain information about the affected population in a post-disaster scenario. If this information is not obtained accurately, it could lead to unfair distribution of relief resources where some areas in need of assistance and aid are not considered in the response plan. If recognition is carried out properly and quickly, efficient logistics may be performed and lives may be saved.

In order to make an efficient humanitarian operation, the use of technology is essential. The first challenge in obtaining such efficiency is to obtain reliable information about the situation in real-time, to reach difficult zones, and to coordinate the different rescue teams. Furthermore, in most cases, existing maps are outdated, given the changes in infrastructure produced by the disaster. Thus, technology has developed tools that reduce the time required to obtain information and response times. One useful method to achieve this goal is the use of Unmanned Aerial Vehicle (UAV) [12]. UAVs have been used in different areas of logistics in the past and are currently deployed for Humanitarian Logistics by the International Emergency Drone Organization (IEDO). A UAV can be used in the first stage of recognition in post-disaster relief in order to identify the most affected zones. After the recognition stage, it is possible to build maps, assess the situation, and manage resources properly considering the status or level of damage of each zone. Since the use of UAVs is remote, no individuals are put at risk during the operations. Also, UAVs work with greater speed, resist strong pressure and temperature changes. Their small size allows them to move around quickly through small spaces which makes them great autonomous devices with the ability to obtain information about risky zones [6].

Also, the use of UAVs has been proposed for a variety of projects before. For example, a project focused on creating a UAV capable of providing humanitarian aid by transporting supplies to critical zones is proposed by [5]. Moreover, [14] propose the use of UAV helicopters for relief distribution operation for a population that is located on collapsed or inaccessible zones. Furthermore, [10] studied rapid humanitarian deliveries with a location-routing problem (LRP) as well as using UAVs. Therefore, we conclude that the use of UAV for humanitarian logistics is a feasible project. As opposed to previous literature, our focus is mainly on the use of UAV for recognition activities, whereas related papers are focused on distribution and supply chain activities. Thus, significant changes in the model are proposed to address the challenge of covering a zone to recognize damages and identify potential victims rapidly. In the model proposed by [2], the problem of routing drones in the aftermath of a disaster is studied and solved by a learnheuristic approach (a combination of heuristics with machine learning). In their setting, the location of hubs from where drones depart is fixed. We consider this decision in order to design the system in our context. A recent literature review on the use of UAVs for logistics is presented by [4,22], and [24].

The LRP is a mathematical model that integrates decisions for the location of facilities (plants, depots, warehouses, hubs, etc.) which are linked with routing decisions. [20] and [9] are the most recent literature reviews on the topic. [20] has made a survey on the classical problem, including the computation of lower bounds and exact methods, heuristics, and metaheuristics. Furthermore, [9] studied some variants to classical problems such as the Inventory Location-Routing Problem (ILRPs) or the Pickup-and-delivery Location-Routing Problem. This literature review also shows that there is no concrete study that uses a Location-Routing Problem to solve recognition or rapid mapping problems in humanitarian logistics.

This paper proposes the use of a UAV-based recognition system for the first stage in a post-disaster situation, helping to identify the status or level of damage of a specific zone. To design the operation of the system, we focuse on optimizing the decisions of locating the UAV prior to the disaster and the routing of UAV after the disaster has happened. Our proposal is to adapt the Mixed Integer Programming (MIP) model for the Capacitated Location-Routing

Problem (CLRP) bearing in mind the constraints associated to the operating UAV. These constraints include the maximum radius signal and a maximum battery autonomy of the devices.

Thus, the contributions of this paper are the following: 1) we present a discussion of the use of UAV for the recognition activities in the aftermath of a disaster; 2) we adapt a mathematical model based on the LRP to design the operation of UAV for recognition activities in humanitarian logistics; 3) we analyze a scenario in which the near-optimal location and routing decisions are made to efficiently cover a zone after a disaster has happened.

This paper is structured as follows: Sect. 2 illustrates the mathematical model. Section 3 presents the results obtained and the insights found. Finally, Sect. 4 presents the conclusions and future research.

2 Problem Statement

The proposed model is a modified version of LRP proposed by [19]. Consider a set of candidate hubs (I) in which UAVs may be located before a disaster happens. These hubs are secure buildings, usually public infrastructure belonging to the military, fire departments, etc., with reinforced structures capable of resisting earthquakes, hurricanes, and other risks. Also, consider a set of nodes (J) which constitute points of recognition. These points are in the air and are the locations in which UAVs may take pictures and videos of the areas that could have been damaged by the disaster. The objective of the model is to find the best location for UAVs prior to the disaster, and once it happens, how should the UAVs be programmed to visit all the nodes in order to collect the information about damages and victims. Thus, the LRP studied in this paper for UAV operation can be defined as:

Sets

- V Nodes in the graph $V = I \cup J$
- I set of potential UAV hubs $I = \{1, \cdots, m\}$
- J set of nodes to be served $J = \{1, \cdots, n\}$
- K set of available UAV $K = \{1, \cdots, o\}$

Parameters

- O_i opening cost for hub $i \in I$
- C_{ij} traveling cost between any node i and $j \in V$
- F fixed cost of using any UAV
- D_{ij} traveling time between any node i and $j \in V$
- R_{ij} distance between hub $i \in I$ and node $j \in J$
- A maximum distance between UAVs and hub
- Q autonomy time of each UAV
- M big number

Decision Variables

- X_{ijk} = 1 if UAV k travels from node $i \in V$ immediately to node $j \in V$; 0 otherwise.
- Y_i = 1 if hub $i \in I$ is opened; 0 otherwise.
- W_{ij} = 1 if node $j \in J$ is allocated to hub $i \in I$
- T_{jk} = Arrival time of UAV $k \in K$ to node $j \in V$

Mathematical Formulations

$$\min Z = \sum_{i \in I} O_i Y_i + \sum_{i \in V} \sum_{j \in V} \sum_{k \in K} C_{ij} X_{ijk} + + \sum_{i \in I} \sum_{j \in J} \sum_{k \in K} F X_{ijk} \tag{1}$$

subject to:

$$\sum_{i \in V} \sum_{k \in K} X_{ijk} = 1 \, \forall j \in J \tag{2}$$

$$\sum_{i \in V} \sum_{j \in V} D_{ij} X_{ijk} \leq Q \, \forall k \in K \tag{3}$$

$$\sum_{i \in V} X_{ijk} - \sum_{i \in V} X_{jik} = 0 \, \forall j \in J, \forall k \in K \tag{4}$$

$$\sum_{j \in J} X_{ijk} \leq 1 \, \forall i \in I, \forall k \in K \tag{5}$$

$$T_{jk} \geq T_{ik} + D_{ij} X_{ijk} - M(1 - X_{ijk}) \, \forall i \in J, \forall j \in V, \forall k \in K \tag{6}$$

$$\sum_{u \in J} X_{iuk} + \sum_{u \in V} X_{ujk} \leq 1 + W_{ij} \, \forall i \in I, \forall j \in J, \forall k \in K \tag{7}$$

$$W_{ij} \leq Y_i \, \forall i \in I, \forall j \in J \tag{8}$$

$$R_{ij} W_{ij} \leq A \, \forall i \in I, \forall j \in J \tag{9}$$

$$X_{ijk} = 0 \, \forall i \in I, \forall j \in I, \forall k \in K \tag{10}$$

$$X_{iik} = 0 \, \forall i \in V, \forall k \in K \tag{11}$$

$$\sum_{i \in I} W_{ij} = 1 \, \forall j \in J \tag{12}$$

$$Y_i, W_{ij} \in \{0, 1\} \, \forall i \in I, \forall j \in I, \forall k \in K \tag{13}$$

$$X_{ijk} \in \{0, 1\} \, \forall i \in V, \forall j \in V, \forall k \in K \tag{14}$$

$$T_{jk} \geq 0 \, \forall j \in V, \forall k \in K \tag{15}$$

The objective function 1 computes the sum of all costs of the system including the opening cost, travelling cost, and fixed costs associated with running each UAV hub. Constraint 2 forces that each zone is served by one single UAV. Constraint 3 defines a constraint related to the battery capacity. Constraint 4 ensures the flow conservation i.e. each UAV must return to the origin depot. Constraint 5 ensures that each UAV leaves maximum one depot. Constraint 6 forces subtour elimination. Constraint 7 forces that a UAV can be assigned to a depot only if a route linking them is opened. Constraint 8 ensures that customers can only be assigned to an open depot. Constraint 9 is associated with the maximum radius that the signal from a hub can cover; that is, a drone will never be further from the hub than distance A. Equation 10 ensures that no hubs are connected. Equation 11 ensures that there is no arc between the same nodes. Constraint 12 ensures that each customer is assigned to one hub. Constraints 13–15 are associated with the domain of each decision variable.

3 Case Study

To analyze and validate the proposed model, we build an instance inspired from the city of Chía, Colombia using UAV parameters taken from Phantom Pro-4 [19], about maximum distance for radio signal and maximum speed. The city is interesting from a humanitarian point of view since [21] has identified several risks that make a significant low-income population vulnerable in Colombia. Earthquakes and floods are significant risks faced by the city. The city has an area of 80 km^2 and a population density of 1700 people per Km^2. The city was founded in 1537, and several buildings are old, which makes some of them highly vulnerable to seismic activity [16]. Further, the city is located near Bogotá river, which has caused floods in the past.

For the case study, we designed a system that aims to evaluate the damages in the entire city, for example in the case of an earthquake. Furthermore, in the case of a flood, the affected area of the city would be limited to areas near the river. Thus, a system that covers the entire city would also be able to respond to both risks.

The coordinates of each node are generated with Geogebra online tool [20] that emulates a small city as shown in Fig. 1. We estimated the parameters D_{ij} and R_{ij} using `Google Maps`. The distances are computed using the Euclidean distance formula since UAVs can flight over buildings. The fixed cost F and the opening cost O_i are set to US$500 and US$100 respectively. These costs are assumed based on average market prices for drones at the time of the research. Additionally, we assumed that for each minute of travelling the cost is US$1. This cost considers the operational cost for the drone, plus a penalty for each minute required to obtain the information.

The proposed mathematical model is coded in GAMS IDE active with CPLEX 12.3 license, and has been tested on a HP Laptop that has 2 GHz AMD 6 - 5200 and 4 GB of RAM. A time limit of 9000 s (2.5 h) is used.

In the test instance, 36 nodes and four UAVs are considered. Tests were made in larger instances but the computational resources were insufficient. This instance is solved in 2.5 h and it is not the guaranteed optimal solution. The solution is depicted in Fig. 2 with an objective function value of $1895. With 36 nodes, 9 of them are potential hubs and 27 zones to be recognized. It is possible to observe that the solution uses 3 UAVs and each one has a different route shown in Fig. 2. The chosen hubs are nodes 2, 3, and 8. The UAV departing

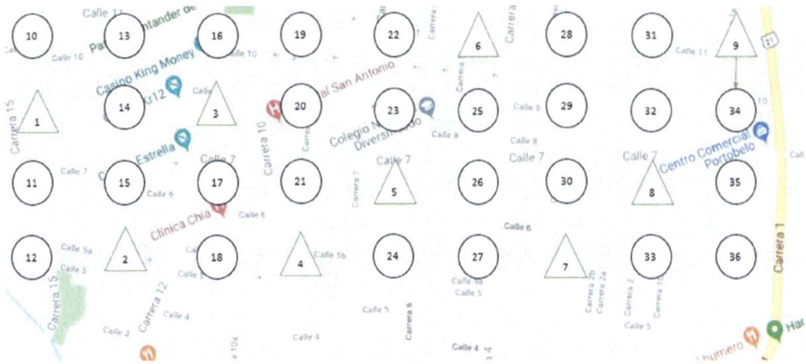

Fig. 1. Initial Graph. Triangles are candidate hubs and circles are damaged zones. Source: Google Maps

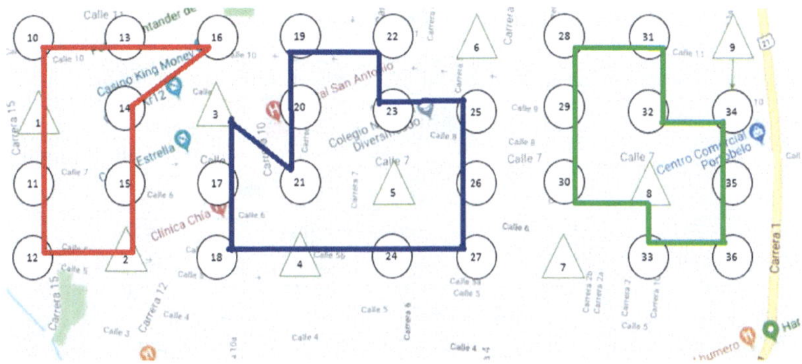

Fig. 2. Proposed solution. Triangles are candidate hubs and circles are damaged zones. Lines represent the routing of UAVs.

from hub 2 needs 23 min to capture the information of its assigned nodes. The UAV departing from the hub 3 needs 36 min to capture the information, and the UAV departing from hub 8 needs 26 min to capture the information. Since the UAVs will be activated simultaneously to perform the screening and routing at the same time, the total time required for the system to collect the information about the state of the city is the time of the longest UAV (36 min in this case). As a matter of fact, this satisfies the operational constraint of the crisis room that is expected to start working up to 1 h after a disaster has happened.

Another important feature of this solution is that routes do not overlap at any time. In this case, it is important to minimize the risk of UAV collision. Consider that wind conditions may affect the real trajectory of the UAV. Thus, our future research agenda includes assessing the capability of the UAV to follow automatically the determined path and to compute the minimum distance that must be kept between two UAVs to avoid a collision and to include it as a constraint in the model. Literature on the topic is presented by [17].

Further, random instances are created, using data similar to the real case to evaluate other scenarios and the capability of the model to react to different input data. The location of candidate hubs is modified in each case. Our goal is to evaluate different scenarios in which one of the candidate facilities will not be able to install a hub. In total, 5 instances have been tested using this methodology; instance 1 corresponds to the real case. Again, a time limit of9000 s is imposed. Table 1 presents a summary of the results for these instances. Column 2 presents the total objective value, column 3 lists the routing cost, column 4 gives the cost of using UAVs, and column 5 presents the cost of locating hubs. Column 6 presents the CPU time to obtain the solution. Out of the 5 instances, none could be solved to proven optimality within 2.5 h. Instance 2 could not find an integer solution without exceeding our computational resources. These instances are available upon request at `william.guerrero1@unisabana.edu.co`.

Table 1. Summary of results for test instances

Instance	Total cost	Routing cost	UAV cost	Hub fix cost	CPU s
1	1895	95	1500	300	9000
2	–	–	–	–	OutOfMemory
3	2517	117	2000	400	9000
4	2511	111	2000	400	9000
5	2520	120	2000	400	9000

Important insights are deducted from this study. First, between 3 and 4 UAVs are required to map the city in approximately 30 min. This is a significant reduction when compared to making these operations with terrestrial vehicles, which is estimated at 2 h when no road disruptions exist. Furthermore, reasonably good quality solutions are computed by commercial solvers within 2.5 h. These initial theoretical solutions are useful to understand how the UAVs should be programmed to follow a path, and compare their real trajectories with different environmental conditions against the theoretically optimal solutions. Further, pilot tests for the UAVs can be made in order to analyze the optimal tours in both directions to evaluate if there is a difference in terms of travel times under different climate conditions as well. These are future research works.

4 Conclusions

The activities carried out in humanitarian logistics have limitations and challenges that preclude the effective result of humanitarian operations. The restricted number of resources (vehicles or rescue teams) or disruptions of roads are some of the limitations existing in humanitarian operations. Furthermore, inappropriate management of resources can cause a lack of assistance to a specific unattended population. In order to improve the performance of the operations, a better recognition process should be put in place. Recognition and early assessment of the damages is one of the most important and difficult activities to carry out in a post-disaster situation. The UAV-based recognition system proposed in this paper was developed for the first stage in a post-disaster situation with a LRP adapted model considering the constraints associated with operating a UAV.

The presented model can be applied in problems of real-life scenarios to plan the location of UAV hubs and the routes for the devices. The studied problem is challenging for large instances because it is an NP-hard problem [20]. Our results are based on near-optimal solutions to the problem. However, it may be possible to get higher quality solutions using metaheuristics.

The proposed system can be useful in other situations than disaster relief. For example, the UAVs can be used for population surveillance and security, such as the quick response to robberies or riot events. In these situations, the UAV can help the authorities to understand the urban dynamics more frequently, without the limitations of surveillance cameras that have fixed locations. For further reading, see [23].

Finally, we suggest that in future research more sophisticated solution methods are developed such as metaheuristics for this problem, bearing in mind constraints associated with UAVs operation systems such as the battery autonomy and connectivity, avoiding route overlapping, in order to avoid risks associated to UAV collision, and selecting the best location in which UAV can assess damages for a given building or block. Solving larger instances is a challenge as well, due to the fact this is an NP-hard problem [25]. Additionally, designing UAVs exclusively for this purpose allows better recognition processes for other scenarios. Also, a deep sensitivity analysis of the solution when costs vary is proposed together with a comparison of the results using different other devices and UAV models.

Further, redundant information may be collected by the UAV since the cameras they carry may be able to capture wide-frame photos and even 360 degrees videos that provide significant information. In these cases, not every node should be visited, but only one node within a cluster of nodes as proposed by [15]. This is also future research.

References

1. Atienza, V.M.C., Singh, S.K., Schroeder, M.O.: Qué ocurrió el 19 de septiembre de 2017 en méxico? Revista Digital Universitaria **18**(7), 1 (2017)
2. Bayliss, C., Juan, A.A., Currie, C.S.M., Panadero, J.: A learnheuristic approach for the team orienteering problem with aerial drone motion constraints. Appl. Soft Comput. **92**, 106280 (2020)
3. Bilham, R.: Lessons from the haiti earthquake. Nature **463**(7283), 878–879 (2010)
4. Bist, B.: Literature survey on unmanned aerial vehicle. Int. J. Pure Appl. Math. **119**(12), 4381–4387 (2018)
5. Boehm, D., Chen, A., Chung, N., Malik, R., Model, B., Kantesaria, P.: Designing an unmanned aerial vehicle (UAV) for humanitarian aid. Technical report, Rutgers School of Engineering (2017)

6. Bravo, R.Z.B., Leiras, A., Oliveira, F.L.C.: The use of UAVs in humanitarian relief: an application of POMDP-based methodology for finding victims. Prod. Oper. Manag. **28**(2), 421–440 (2019)
7. Brunkard, J., Namulanda, G., Ratard, R.: Hurricane katrina deaths, louisiana, 2005. Disaster Med. Public Health Prep. **2**(4), 215–223 (2008)
8. Cantillo, V., Serrano, I., Macea, L.F., Holguín-Veras, J.: Discrete choice approach for assessing deprivation cost in humanitarian relief operations. Socio-Econ. Plann. Sci. **63**, 33–46 (2018)
9. Drexl, M., Schneider, M.: A survey of variants and extensions of the location-routing problem. Eur. J. Oper. Res. **241**(2), 283–308 (2015)
10. Escribano Macias, J.J., Angeloudis, P., Ochieng, W.: Integrated trajectory-location-routing for rapid humanitarian deliveries using unmanned aerial vehicles. In: 2018 Aviation Technology, Integration, and Operations Conference, p. 3045 (2018)
11. Espejo-Díaz, J.A., Guerrero, W.J.: A bi-objective model for the humanitarian aid distribution problem: analyzing the trade-off between shortage and inventory at risk. In: Figueroa-García, J.C., Duarte-González, M., Jaramillo-Isaza, S., Orjuela-Cañon, A.D., Díaz-Gutierrez, Y. (eds.) WEA 2019. CCIS, vol. 1052, pp. 752–763. Springer, Cham (2019). https://doi.org/10.1007/978-3-030-31019-6_63
12. Estrada, M.A.R., Ndoma, A.: The uses of unmanned aerial vehicles-UAV's-(or drones) in social logistic: natural disasters response and humanitarian relief aid. Procedia Comput. Sci. **149**, 375–383 (2019)
13. Ferreira, G.O., Arruda, E.F., Marujo, L.G.: Inventory management of perishable items in long-term humanitarian operations using Markov decision processes. Int. J. Disaster Risk Reduction **31**, 460–469 (2018)
14. Golabi, M., Shavarani, S.M., Izbirak, G.: An edge-based stochastic facility location problem in UAV-supported humanitarian relief logistics: a case study of tehran earthquake. Nat. Hazards **87**(3), 1545–1565 (2017)
15. Guerrero, W.J., Velasco, N., Prodhon, C., Amaya, C.-A.: On the generalized elementary shortest path problem: a heuristic approach. Electron. Notes Discrete Math. **41**, 503–510 (2013)
16. Hube, M.A., María, H.S., Arroyo, O., Vargas, A., Almeida, J., López, M.: Seismic performance of squat thin reinforced concrete walls for low-rise constructions. Earthq. Spectra **36**(3), 1074–1095 (2020)
17. Lin, Y., Saripalli, S.: Sampling-based path planning for UAV collision avoidance. IEEE Trans. Intell. Transp. Syst. **18**(11), 3179–3192 (2017)
18. Martius, O., et al.: The role of upper-level dynamics and surface processes for the Pakistan flood of July 2010. Q. J. R. Meteorol. Soc. **139**(676), 1780–1797 (2013)
19. Prins, C., Prodhon, C., Ruiz, A., Soriano, P., Calvo, R.W.: Solving the capacitated location-routing problem by a cooperative lagrangean relaxation-granular tabu search heuristic. Transp. Sci. **41**(4), 470–483 (2007)
20. Prodhon, C., Prins, C.: A survey of recent research on location-routing problems. Eur. J. Oper. Res. **238**(1), 1–17 (2014)
21. Cuervo, S.X.R., Galindo, A.L.B., et al.: Análisis de los escenarios de riesgo por fenómenos amenazantes para el municipio de chia cundinamarca como herramienta de planificación territorial. Technical report, Universidad Distrital Francisco Jose De Caldas (2015)
22. Rojas Viloria, D., Solano-Charris, E.L., Muñoz-Villamizar, A., Montoya-Torres, J.R.: Unmanned aerial vehicles/drones in vehicle routing problems: a literature review. Int. Trans. Oper. Res. (2020, in press)

23. Solano-Pinzón, N., Pinzón-Marroquín, D., Guerrero, W.J.: Surveillance camera location models on a public transportation network. Ingeniería y Ciencia **13**(25), 71–93 (2017)
24. Thibbotuwawa, A., Bocewicz, G., Nielsen, P., Banaszak, Z.: Unmanned aerial vehicle routing problems: a literature review. Appl. Sci. **10**(13), 4504–4524 (2020)
25. Thomas, A.: Leveraging private expertise for humanitarian supply chains. Forced Migr. Rev. **21**, 64–65 (2004)
26. Ting, D.S.W., Carin, L., Dzau, V., Wong, T.Y.: Digital technology and COVID-19. Nat. Med. 1–3 (2020)
27. Titov, V., Rabinovich, A.B., Mofjeld, H.O., Thomson, R.E., González, F.I.: The global reach of the 26 December 2004 Sumatra Tsunami. Science **309**(5743), 2045–2048 (2005)
28. Tomasini, R.M., Van Wassenhove, L.N.: From preparedness to partnerships: case study research on humanitarian logistics. Int. Trans. Oper. Res. **16**(5), 549–559 (2009)
29. Zorrilla, C.D.: The view from puerto rico-hurricane maria and its aftermath. N. Engl. J. Med. **377**(19), 1801–1803 (2017)

A Local Search Algorithm for the Assignment and Work Balance of a Health Unit

Néstor Díaz-Escobar[1], Pamela Rodríguez[1], Verónica Semblantes[1], Robert Taylor[1], Daniel Morillo-Torres[2(✉)], and Gustavo Gatica[1]

[1] Universidad Andres Bello, Santiago, Chile
ggatica@unab.cl
[2] Pontificia Universidad Javeriana - Cali, Cali, Colombia
daniel.morillo@javerianacali.edu.co

Abstract. In any healthcare service, guidelines regarding the number of staff and how to respond to patient demand must be followed. In Chile, to ensure there is 24/7 care, coordinators use a manual allocation model system called "The Fourth Shift" (TFS) to assign staff. The model has a four-day shift pattern which allocates 48 h of work and 48 h of rest. However, scheduling healthcare workers is always a challenge, as there are administrative, legal and individual constraints. A balanced shift assignment, meaning one that considers work hours and specific staff requests, has a significant impact on an overall work environment. To find a fair balance, this paper proposes a two-phase heuristic. The first is a constructive phase and the second is a local search phase. This paper simultaneously incorporates six Key Performance Indicators (KPIs) and chance events aiming at leveling the workload for healthcare workers. The heuristics are validated with one-month shifts for a healthcare service with 12 nurses. The results validate the effectiveness of the proposed approach by disrupting the solution with five cumulative scenarios.

Keywords: Healthcare service · Shift assignment · Heuristics · Nurse rostering

1 Introduction

One of the main problems that healthcare services face is meeting the 24/7 demand for care and therefore, staff [10]. However, some healthcare services do not have the technological or systematic tools available to support an improved shift allocation process. The process is usually done manually, whereby a coordinator assigns shifts to the staff on a weekly basis [22].

The proper execution of healthcare operations is dependent on the medical tasks carried out by nurses [8]. In fact, nurse labour accounts for approximately 25% of a hospital's total operative budget and 44% of its direct care costs [11]. Consequently, offering flexible schedules for nurses is a primary goal in assuring

D. Trentesaux et al. (Eds.): SOHOMA 2021, SCI 987, pp. 208–222, 2021.
https://doi.org/10.1007/978-3-030-80906-5_14

workforce stability because it can make the profession seem more appealing, especially in a context where there is high turnover and a lack of personnel [23]. However, scheduling hospital staff always presents a challenge in situations where there are special requests or contingencies [15]. The manual process is often tedious as well as highly complex, even when creating feasible assignments. In relevant academic and industrial areas, literature refers to this as the Nurse Rostering Problem (NRP) [2,6,12,14,33].

The base information and primary sources for this study were gathered from interviews with nurses at a Chilean public healthcare service [20,21,30]. In this paper, an analysis of the current shift allocation process for a Health Unit is carried out, by using the Business Process Model Notation (BPMN) [35]. Six relevant Key Performance Indicators (KPIs) are taken into account and a constructive heuristic of allocation is proposed to find a scheduling solution.

The document is structured as follows: Sect. 2 presents the problem description; Sect. 3 conducts a bibliographic review of the problems with assigning nurses; Sect. 4 details the proposed model; Sect. 5 shows the results of the computational experiment; Finally, Sect. 6 presents the conclusions and future work.

2 Problem Description

To ensure 24/7 care is implemented, a shift system (or rostering system) for nurses called "The Fourth Shift" (TFS) was designed in Chile. While it is not officially regulated under government laws, it is in fact law-based in order to avoid incurring improper practices [21]. This shift system is widely used in most healthcare centres throughout the country. TFS generally distributes nurses' work into four days. This distribution guarantees care over a Long shift (12 daytime hours), a Night shift (12 nighttime hours) and two full days of rest. Although the Chilean government does not specify an exact timeframe for each shift, hospitals typically define a Long shift as being from 8:00 am–8:00 pm, and a Night shift as being from 8:00 pm–8:00 am of the next day [24].

Without the loss of generality, this study focuses on a group of 12 nurses working 12-h hospital shifts. One limitation is that in any given day, and during each shift, three nurses need to be working. In compliance with TFS, it is recommended that four different shifts be assigned when designing the nurse roster as shown in Table 1.

Table 1. Defining shifts for assignment based on TFS.

Shift	Day 1		Day 2		Day 3		Day 4	
	8:00 am	8:00 pm	8:00 am	8:00 pm	8:00 am	8:00 pm	8:00 am	8:00 pm
A	Long	–	–	Night	Off		Off	
B	Off		Long	–	–	Night	Off	
C	Off		Off		Long	–	–	Night
D	–	Night	Off		Off		Long	–

Table 2. Constraints considered in the investigation.

Constraints	Hard	Soft
Demand by shift	X	–
Compliance by Fourth Shift system	X	–
Average weekly working hours	X	X
Days off	X	X
Night shifts	X	–
Free full weekends	X	–
Free Sundays	X	X
Times with two consecutive days off	X	X

Nurse rostering problems are completely dependent on the business or hospital that requires said allocation; the needs, contracts, types of services, and laws, among many other elements, all have their own particularities. However, the general limitations can be classified as Hard Constraints, and Soft Constraints [4]. The former are conditions that cannot be breached under any circumstances; for example, the maximum number of work hours defined by law cannot be exceeded. The latter are conditions that are expected to be met, but that if breached, would still make the solution feasible. Usually for such constraints, a penalty function or a maximum violation tolerance is defined. It is also possible for one constraint to be simultaneously Hard and Soft; for example, the number of contracted hours cannot exceed a limit, yet has an expected value. Table 2 shows the main constraints that we will consider in this study.

3 Related Works

The impact of the Work-Life Balance (WLB) to date is not established on the health workforce in the long term, given the frequent job rotation [13,29]. Therefore, efforts are needed to understand the health consequences of these professionals and their needs. In addition to the human component, the WLB research has benefits for economic development, where changes in the nature of work occur in a globalised context [18]. On the other hand, due to the existence of long shifts, and given the lack of health care personnel, overtime is used to cover the offer [26]. The short-term negative effects of these rotations of healthcare personnel can be reduced by minimizing the number of shifts. However, it impacts on the economic well-being of staff [25]. In this paper, an algorithm is proposed to respond to health worker requests while being equitable to the workload.

There are different areas where the NRP can be applied: bus scheduling, university exams, courses, wards, and hospital beds, among others [7]. These applications can be modelled as Timetabling Problems, where certain criteria must be considered to generate feasible timetables at the operational level. These criteria include: the duration of the schedule, use of resources, customer preferences and adherence to regulations [19].

NRP is considered an NP-Hard class problem [9], which makes finding optimal solutions for mid to large-sized problems non-viable. Generally, the NRP seeks to schedule nurses while considering requests (staff, nurses, administrators or physical) in a way that meets the hospitals' demand [5,16]. In addition to the staff constraints there are others that fall within the legal scope of each country [16]. Relevant literature confirms that nurses' satisfaction, as well as their work performance, is a result of the combination of their workload and working conditions [3,34]. The importance of handling staff constraints lies in that nurses manage records and documents that contribute to healthcare coordination, which is fundamental to the proper functioning of a hospital [6].

The methods for solving this problem can be classified into two groups: the exact method and the heuristic method [32]. The former delivers an optimal solution, however, this method tends to be computationally expensive when it comes to solving real-life problems. The latter can find a reasonable solution in a significantly short amount of time, although it is sometimes unclear how close it is to the optimal solution. There are a variety of methods proposed in the literature that demonstrate how to handle this problem: Constraint Programing (CP) [24], Particle Swarm Optimization (PSO) [36], Mixed Integer Programming (MIP) [27], Tabu Search (TS) [1] and Simulated Annealing (SA) [32]. There are even hybrid meta-heuristics approaches that allow finding NRP solutions [2,14,28]. However, most of them do not present informed decision-making using KPIs. On the other hand, we can apply for the heuristics, which are mainly used to generate initial solutions or improve previously found solutions. Even though heuristics are simpler algorithms than metaheuristics, they can allow resources to be used more efficiently to meet the demands of hospitals, patients, and particular staff requirements.

In recent years, realistic models have grown in importance within the research surrounding the NRP. These models deal with the unexpected occurrence of random events in the NRP, such as unprecedented demand, accidents, etc. [17]. This paper therefore proposes a reactive heuristic algorithm, which equally makes assignments by defining thresholds for each performance indicator (KPI). In addition, due to its nature and flexibility, it reactively updates KPIs and reschedules staff when given a contingency. Thus, the scheduling is always feasibly maintained within the established thresholds. This creates a roster which allows for shift rescheduling for a team of 12 nurses in a public healthcare service in Chile.

4 Algorithm Proposed

To solve the NRP, a local search heuristic is proposed, which is divided into two phases: a constructive phase and a search phase. The first creates a TFS-aligned allocation base month, and the second phase performs a local search and exchange algorithm. A set of Nurses from clinical units were involved in this project. The nurse coordinator performs weekly assignation and defines the priority of the KPIs. For this reason, the algorithm proposed in the local search phase follows this guideline.

4.1 Phase 1. Construction

The construction phase creates a base assignment for an ideal situation: a complete staffing of nurses, which is repeated weekly, every month. However, it does not take into account special requests from hospital personnel (vacations, days off, etc.). Of course, since this base assignment does not consider individual requests it is not applicable in a real-world environment. However, it is still used as an initial solution. The challenge is creating more flexible schedules that respect new requests from nurses, adhere to current institutional regulations (compliant with TFS system), and meet the minimum nurse requirement for each shift.

In general, this phase can be described in the following four steps:

1. Determine the four types of shifts in accordance with TFS: A (Long, Night, Free, Free), B (Free, Long, Night, Free, C (Free, Free, Long, Night), and D (Night, Free, Free, Long), (for more details see Table 1).
2. Create an empty matrix with the number of nurses as row numbers and the days of the months as columns.
3. Repeat for each nurse until the last day of the month. Take shift A as candidate shift, and verify whether the current shift type (the 12-h time slot) already has the maximum number of nurses. If they do not have them, shift A is assigned. Otherwise, it must be changed to the next type of shift (B, C, or D) until the restriction is met.
4. The final result is a base shift matrix.

Figure 1 shows the ideal roster for a one-week period (the base assignment), built at this phase. It consists of assignments for each 12-h time slots for shifts A, B, C and D without any special requests. The days each nurse should work are displayed in yellow.

4.2 Phase 2. Local Search

In this phase the initial solution built in Phase 1 is incorporated and each KPI or requirement is incorporated one by one. For each addition, the local search exchanges shifts until all requests and requirements are considered. This phase can be summarized in the following steps:

Shift	Nurse	Day 1		Day 2		Day 3		Day 4		Day 5		Day 6		Day 7	
		Monday		Tuesday		Wednesday		Thursday		Friday		Saturday		Sunday	
		Long	Night	Long	Night	Long	Night	Long	Night	Long	Night	Long	Night	Long	Night
A	1	Worked	Off	Off	Worked	Off	Off	Off	Off	Worked	Off	Off	Worked	Off	Off
	2	Worked	Off	Off	Worked	Off	Off	Off	Off	Worked	Off	Off	Worked	Off	Off
	3	Worked	Off	Off	Worked	Off	Off	Off	Off	Worked	Off	Off	Worked	Off	Off
B	4	Off	Off	Worked	Off	Off	Worked	Off	Off	Off	Off	Worked	Off	Off	Worked
	5	Off	Off	Worked	Off	Off	Worked	Off	Off	Off	Off	Worked	Off	Off	Worked
	6	Off	Off	Worked	Off	Off	Worked	Off	Off	Off	Off	Worked	Off	Off	Worked
C	7	Off	Off	Off	Off	Worked	Off	Off	Worked	Off	Off	Off	Off	Worked	Off
	8	Off	Off	Off	Off	Worked	Off	Off	Worked	Off	Off	Off	Off	Worked	Off
	9	Off	Off	Off	Off	Worked	Off	Off	Worked	Off	Off	Off	Off	Worked	Off
D	10	Off	Worked	Off	Off	Off	Off	Worked	Off	Off	Worked	Off	Off	Off	Off
	11	Off	Worked	Off	Off	Off	Off	Worked	Off	Off	Worked	Off	Off	Off	Off
	12	Off	Worked	Off	Off	Off	Off	Worked	Off	Off	Worked	Off	Off	Off	Off

Fig. 1. Fourth Shift ideal assignment.

1. Read the requests made by the medical or managerial staff from the CSV file.
2. Repeat according to the number of requests and days of the month.
3. Calculate the numbers of hours worked per nurse and organize them in descending order.
4. Make feasible shift exchanges between nurses until the selected requirement is cumulatively met (satisfying the requirements of past iterations). The exchanges are made with nurses that have the least number of hours worked.
5. Update the performance indicators to measure workload and constraints.
6. Validate the range for the exchange to be considered approved. If the range is met, save the assignment. If it is not met, return to step 4.
7. Return the month assignment as the final result.

Algorithm 1 shows the pseudo-code for the proposed algorithm. For its execution, the set of nurses and their requests for the month (to be scheduled) are received as input. The thresholds in the algorithm are defined by the competent unit. Thus, in this case, an allocation that allows each KPI to be met by 65% is considered a feasible solution. It should be noted that in the event of chance events, the personnel responds to the need for a service, even if it entails overworking. Typically, mathematical models do not consider these contingency reallocations, but the proposed heuristic does allow it.

Finally, the following are the definitions of the performance indicators established by the case study:

- Average weekly working hours (KPI1): the average number of hours worked by each nurse is calculated during the first four weeks of the month, according to Eq. (1), where Tt_{ij} is the number of 12-h shifts worked per nurse j (Long or Night) in month i. J is defined as the set of nurses. For the evaluation, between 42 and 45 (on average) hours a week is considered an acceptable range.

$$KPI1_i = \frac{Tt_{ij}}{4} \quad \forall j \in J \tag{1}$$

Algorithm 1. Proposed Heuristics (Req, DayMonth, Dem)

```
Input Req, DayMonth, Dem // Requirements, days per month to assign, demand
for nurses
Output MatAsig // Mapping matrix
Begin Algorithm
  Shifts = A, B, C, D
  Procedure Create Month Base // Phase 1
  Procedure Local Search // Phase 2
    For i = 1 to KPIs
      While requirement is not fulfilled
        //Calculation list of hours worked by nurse in descending order
        timeMatrix = shiftExchange(modifiedMatrix, timeMatrix, Nurses)
        If is requirement fulfilled? == true
          modifiedMatrix = timeMatrix
          break
        End If
      End While
    End For
  Return modifiedMatrix
End Algorithm
```

- Days off (KPI2): it is considered a day off when a nurse does not work long shifts or night shifts. They are free from 8:00 am to 8:00 pm the next day. The indicator is calculated using Eq. (2) where Dm_i is the number of days in a month i. It is considered that on average, each nurse should have 15 days off per month, with a range of 12 to 18 days.

$$KPI2_i = Dm_i - Tt_{ij} \quad \forall j \in J \tag{2}$$

- Night shifts (KPI3): this indicator calculates each nurse's night shifts (Night) using Eq. (3), where Tn_{ij} is the number of night shifts that nurse j has worked in a month i. It is considered that on average, each nurse should work eight night shifts, therefore between six to ten shifts per month.

$$KPI3_i = Tn_{ij} \quad \forall j \in J \tag{3}$$

- Full weekends off (KPI4): this indicator calculates how many consecutive Saturdays and Sundays off a nurse has, using Eq. (4), where SDC_{ij} is the number of consecutive Saturdays and Sundays that nurse j has in the month i. It is considered that each nurse should have at least one weekend off a month.

$$KPI4_i = SDC_{ij} \quad \forall j \in J \tag{4}$$

- Sundays off (KPI5): this indicator calculates the Sundays off in a month, using Eq. (5), where D_{ij} is the number of Sundays off that nurse j has in the month i. It is ideal to have two Sundays off a month with a normal range of one to three.

$$KPI5_i = D_{ij} \quad \forall j \in J \tag{5}$$

- Two consecutive days off (KPI6): This indicator calculates the number of times each nurse has two consecutive days off in the month, using Eq. 6, where LL_{ij} is the number of times that nurse j has two consecutive days off in the month i. Between six to eight occurrences is considered an acceptable range.

$$KPI6_i = LL_{ij} \quad \forall j \in J \tag{6}$$

5 Computational Experiments

This section presents the proposed algorithm's results for shift allocation, which considers both the base assignment of 12 nurses and four special events. These events reflect nurse requests made at the beginning of the month, with the exception of the last (unexpected) event that appears in the middle of the month. The following shows the variation of the KPIs defined in Sect. 4 for each event cumulatively, i.e., assignments are made considering all previous events. The proposed algorithm was developed in Python. It was carried out on a PC with an Intel $i5$ 7th generation processor running at a speed of 2.50 GHz and with 8 GB of RAM memory. Given the size of the test instances, the average computation times, are close to zero seconds. Finally, the allocation found which meets all requests and requirements will be illustrated. The events were based on special requests, and without losing generality the following were selected:

- Event No. 1: Nurse 1 requests all Thursdays off.
- Event No. 2: Nurse 2 requests three full weekends off, accumulating with the previous events.
- Event No. 3: Nurse 12 is on medical leave for 10 days, the first day of the month, accumulating with the previous events.
- Event No. 4: Nurse 4 requests three Fridays off, accumulating with the previous events.
- Event No. 5: month with fortuitous event, nurse 5 is on leave for seven days in the middle of the month, so shifts must be reassigned, accumulating with the previous events.

The nurses work on average 42 h a week in the base month. Figure 2 depicted KPI1 variation in relation to the hours worked. This number does not change significantly with scheduled events, but a fortuitous event drops this indicator to a 75% compliance rate. This means that three nurses are not within the established measurement range (42 to 45 h). However, this remains within the acceptable range as it is greater than 65%.

For KPI2, it is acceptable for each nurse to take off between 12 and 16 days a month. Figure 3 shows that KPI2 is within the set ranges, as the number of days off mostly varies between 12 and 16. While the nurse number 5 and 12 are outside of the established range, this indicator surpasses the minimum compliance (65%) for all events.

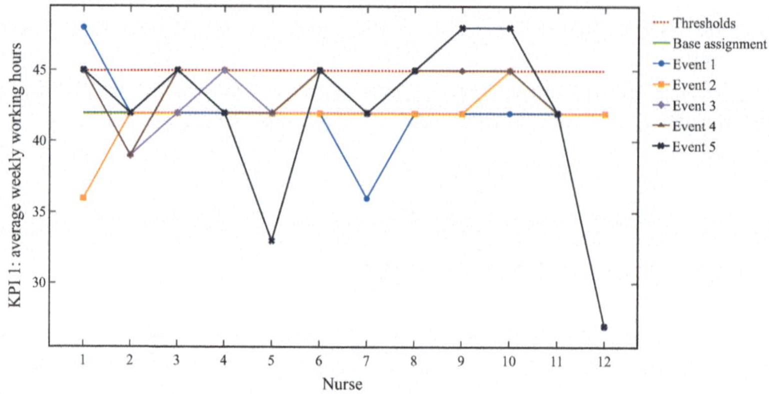

Fig. 2. Variation of KPI based on the months for each event and base assignment

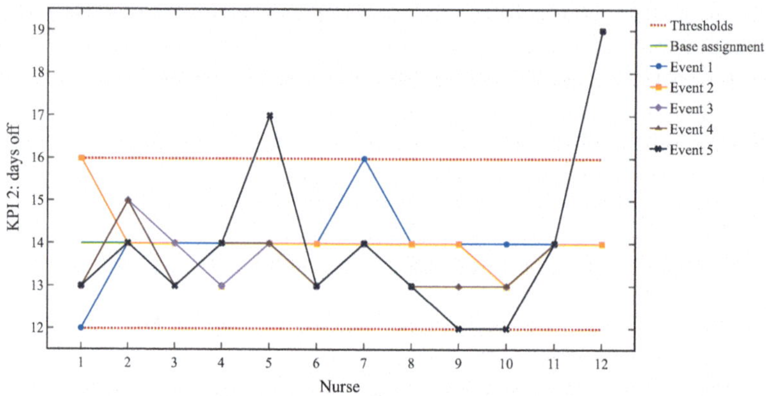

Fig. 3. Variation of KP12 based on the months for each event and base assignment

KPI3 remains 100% compliant for events 1 and 2, since nurses work between six and eight night shifts. For events 3 and 4 the indicator meets 91.67%, given that a nurse works four night shifts. Finally, when a chance event occurs, two nurses work an out of range amount, so the acceptance percentage drops to 83.34%. Even with night shift variation, this indicator remains within the established range for each event, with a value greater than 65%. Figure 4 shows the variation of KPI3.

In KPI4 the nurses should take at least a full weekend off in the month. In events 1 and 2 there is not a significant KPI fluctuation with compliance values between 91.6% to 83.33%, respectively. For the following events 3 nurses are not within the expected range so compliance decreases to 75% as seen in Fig. 5. Still, the compliance value is bigger than the minimum required (65%).

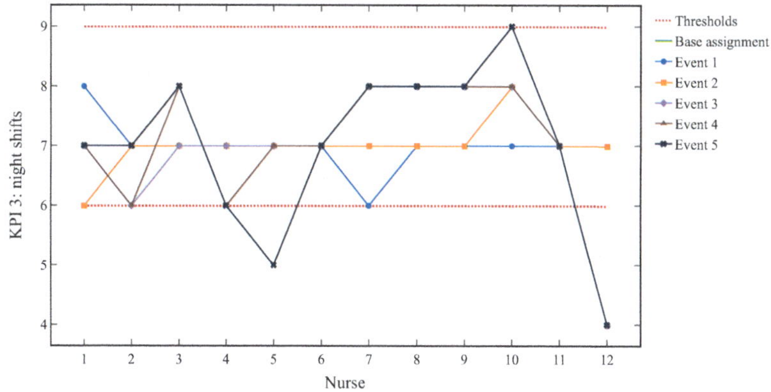

Fig. 4. Variation of KP12 based on the months for each event and base assignment

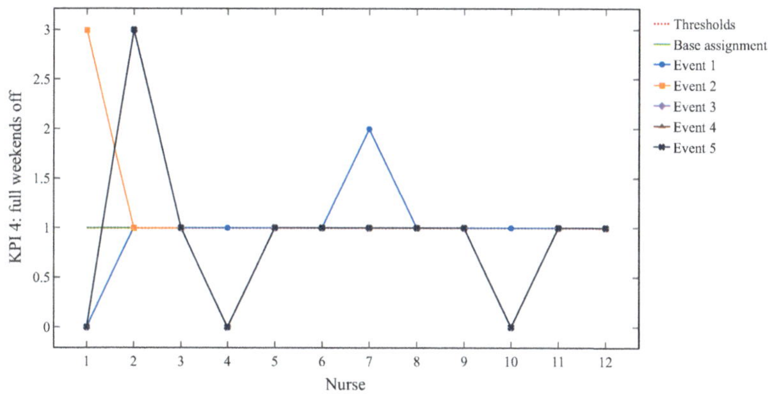

Fig. 5. KPI4 variation based on the months for each event and assignment base

Regarding KPI5 (Sundays off in the month), Fig. 6 shows that when all events occur, nurses have between one to three Sundays off. Therefore, this indicator is not altered in regards to the base assignment, which demonstrates a 100% compliance.

KPI6 determines an acceptable range of 6 to 8 times with two consecutive days off per nurse. It can be seen in Fig. 7 that for events 1, 2, 3 and 4 compliance is 91.67%, 83.34%, 75% and 83.34%, respectively. In a chance event, it shows the indicator varies to 66.67%. All events meet the established minimum.

Table 3 summarizes the acceptance levels for each KPI and corresponding event. It is important to note that these should not be less than 65%. As a result, all indicators are found to be within expectations. Thus, the allocation is accepted. It is notable that 33.3% of the final indicators are at 100%. The Appendix shows the nurse roster found for the month studied, highlighting the shifts worked, and applying the already defined 5 events.

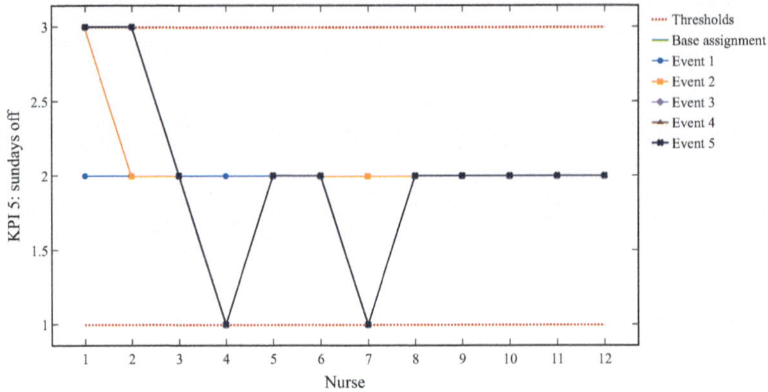

Fig. 6. KPI5 variation based on the months of each event and base assignment

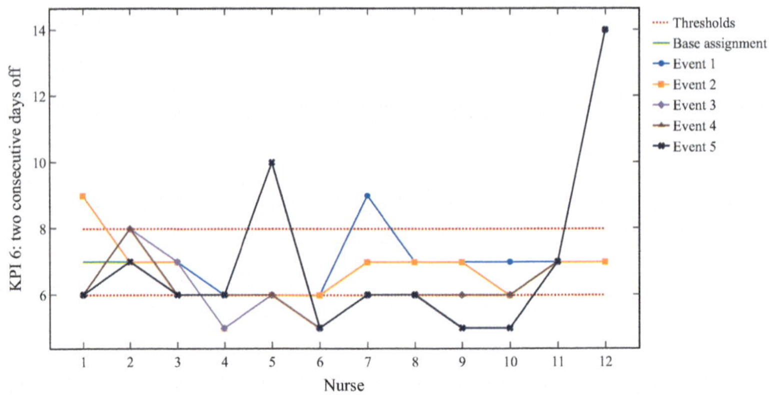

Fig. 7. KPI6 variation based on the months for each event and base assignment.

Table 3. Percentage of KPI compliance based on the events implemented of the proposed algorithm.

	Base assignment	Event 1	Event 2	Event 3	Event 4	Event 5
KPI1	100%	83.34%	91.67%	83.33%	83.34%	75.00%
KPI2	100%	100%	100%	100%	83.34%	83.34%
KPI3	100%	100%	100%	91.67%	91.67%	83.34%
KPI4	100%	91.67%	83.33%	75.00%	75.00%	75.00%
KPI5	100%	100%	100%	100%	100%	100%
KPI6	100%	91.67%	83.33%	75.00%	83.34%	66.67%

6 Conclusions and Future Work

This research addressed the Nurse Rostering Problem for a Chilean healthcare provider, focusing on how to comply with the TFS system and how an ideal shift allocation is affected by nurses' specific requests. For the solution, a two-phase reactive heuristic was proposed: a constructive one, responsible for the ideal allocation; and a local search, which considers each request and makes shift exchanges in order to achieve a feasible solution. The proposed algorithm focuses on feasibility. The priority defined by nurses based on KPIs is used in the local search, rather than a multi-objective approach. Six KPIs were determined for the assignments, and based on the applied case's criteria, 65% was the minimum acceptance rate for each indicator. The algorithm was subsequently tested with five events (specific requests), four of them planned and the last being fortuitous. The results showed that the algorithm is computationally efficient, capable of adapting to changes and can find feasible high-quality solutions in all analyzed cases.

The number of consecutive hours worked by each nurse proved very relevant in this study. Due to a general lack of health-care workers in Chile, improving nurses' schedules contributes to bettering overall working environments for hospital staff [31]. It is clear that a well allocated shift assignment produces good working conditions for nurses, improves their quality of life, and makes a public sector career (one marked by long shifts and staff shortage) more appealing. For these reasons, many efforts have been made in the last 40 years in the development of algorithms. However, nurse rostering is still a recurring problem in operational research. Therefore, allocation studies, such as this one, are helpful in opening a new line of research for determining how to work towards better scheduling solutions for the healthcare sector. Actually, two generalizations of the problem are currently being worked on, one with stochastic aspects and a multi-target version.

Appendix

Figures 8 and 9 show the nurse rostering found for the month studied taking into account the 5 events described in Sect. 5, highlighting the shifts worked (yellow).

Shift	Nurse	Day 1		Day 2		Day 3		Day 4		Day 5		Day 6		Day 7	
		M		T		W		R		F		S		U	
		L	N	L	N	L	N	L	N	L	N	L	N	L	N
A	1	1	0	0	1	0	0	0	0	1	0	0	1	0	0
	2	1	0	0	1	0	0	0	0	1	0	0	0	0	0
	3	1	0	0	1	0	0	0	0	1	0	0	1	0	0
B	4	0	0	1	0	0	1	0	0	0	0	1	0	0	1
	5	0	0	1	0	0	1	0	0	0	0	1	0	0	1
	6	0	0	1	0	0	1	1	0	0	0	1	0	0	1
C	7	0	0	0	0	1	0	0	1	0	0	0	0	1	0
	8	0	0	0	0	1	0	0	1	0	1	0	0	1	0
	9	0	1	0	0	1	0	0	1	0	0	0	0	1	0
D	10	0	1	0	0	0	0	1	0	0	1	0	1	0	0
	11	0	1	0	0	0	0	1	0	0	1	0	0	0	0
	12	0	0	0	0	0	0	0	0	0	0	0	0	0	0

Shift	Nurse	Day 8		Day 9		Day 10		Day 11		Day 12		Day 13		Day 14	
		M		T		W		R		F		S		U	
		L	N	L	N	L	N	L	N	L	N	L	N	L	N
A	1	0	0	1	0	0	1	0	0	0	0	1	0	0	0
	2	1	0	1	0	0	1	0	0	0	0	1	0	0	1
	3	0	0	1	0	0	1	0	0	0	0	1	0	0	1
B	4	0	0	0	0	1	0	0	1	0	0	0	0	1	0
	5	0	0	0	0	1	0	0	1	0	0	0	0	1	0
	6	0	0	0	0	1	0	0	1	0	0	0	0	1	0
C	7	0	1	0	1	0	0	1	0	0	1	0	0	0	1
	8	0	1	0	0	0	0	1	0	0	1	0	0	0	0
	9	0	1	0	0	0	0	1	0	0	1	0	0	0	0
D	10	1	0	0	1	0	0	0	0	1	0	0	1	0	0
	11	1	0	0	1	0	0	0	0	1	0	0	1	0	0
	12	0	0	0	0	0	0	0	0	1	0	0	1	0	0

Fig. 8. Shift assignment for the first and second week.

Shift	Nurse	Day 15		Day 16		Day 17		Day 18		Day 19		Day 20		Day 21	
		M		T		W		R		F		S		U	
		L	N	L	N	L	N	L	N	L	N	L	N	L	N
A	1	1	0	0	1	0	0	0	0	1	0	0	1	0	0
	2	0	0	0	0	1	0	0	1	0	1	0	0	0	0
	3	0	0	0	0	1	0	0	1	0	1	0	0	1	0
B	4	0	1	0	0	0	0	1	0	0	0	0	0	1	0
	5	0	0	0	0	0	0	0	0	0	0	0	0	0	0
	6	0	1	0	0	0	0	1	0	0	1	0	0	0	0
C	7	0	0	0	0	1	0	0	1	0	0	0	0	1	0
	8	1	0	0	1	0	0	0	0	1	0	0	1	0	0
	9	1	0	0	1	0	0	1	0	1	0	0	1	0	0
D	10	0	1	1	0	0	1	0	0	0	0	1	0	0	1
	11	0	0	1	0	0	1	0	0	0	0	1	0	0	1
	12	0	0	1	0	0	1	0	0	0	0	1	0	0	1

Shift	Nurse	Day 22		Day 23		Day 24		Day 25		Day 26		Day 27		Day 28	
		M		T		W		R		F		S		U	
		L	N	L	N	L	N	L	N	L	N	L	N	L	N
A	1	0	0	1	0	0	1	0	0	0	0	1	0	0	1
	2	0	1	0	0	0	0	1	0	0	1	0	0	0	0
	3	0	1	0	0	0	0	1	0	0	1	0	0	0	0
B	4	1	0	0	1	0	0	0	0	1	0	0	1	0	0
	5	1	0	0	1	0	0	0	0	1	0	0	1	0	0
	6	1	0	0	1	0	0	0	0	1	0	0	1	0	0
C	7	0	1	0	0	0	0	1	0	0	1	0	0	0	0
	8	0	0	1	0	0	1	0	0	0	0	1	0	0	1
	9	0	0	1	0	0	1	0	0	0	0	1	0	0	1
D	10	0	0	0	0	1	0	0	1	0	0	0	0	1	0
	11	0	0	0	0	1	0	0	1	0	0	0	0	1	0
	12	0	0	0	0	1	0	0	1	0	0	0	0	1	0

Fig. 9. Shift assignment for the third and fourth week.

References

1. Adamuthe, A.C., Bichkar, R.S.: Tabu search for solving personnel scheduling problem. In: 2012 International Conference on Communication, Information Computing Technology (ICCICT), pp. 1–6 (2012). https://doi.org/10.1109/ICCICT.2012.6398097
2. Awadallah, M.A., Bolaji, A.L.A., Al-Betar, M.A.: A hybrid artificial bee colony for a nurse rostering problem. Appl. Soft Comput. J. **35**, 726–739 (2015). https://doi.org/10.1016/j.asoc.2015.07.004
3. Azaiez, M.N., Al Sharif, S.S.: A 0-1 goal programming model for nurse scheduling. Comput. Oper. Res. **32**(3), 491–507 (2005). https://doi.org/10.1016/S0305-0548(03)00249-1
4. Blöchliger, I.: Modeling staff scheduling problems. A tutorial. Eur. J. Oper. Res. **158**(3), 533–542 (2004). https://doi.org/10.1016/S0377-2217(03)00387-4
5. Burke, E.K., De Causmaecker, P., Berghe, G.V., Van Landeghem, H.: The state of the art of nurse rostering. J. Sched. **7**(6), 441–499 (2004). https://doi.org/10.1023/B:JOSH.0000046076.75950.0b
6. Cheang, B., Li, H., Lim, A., Rodrigues, B.: Nurse rostering problems - a bibliographic survey. Eur. J. Oper. Res. **151**(3), 447–460 (2003). https://doi.org/10.1016/S0377-2217(03)00021-3
7. Chen, P.S., Zeng, Z.Y.: Developing two heuristic algorithms with metaheuristic algorithms to improve solutions of optimization problems with soft and hard constraints: an application to nurse rostering problems. Appl. Soft Comput. **93**, 106336 (2020). https://doi.org/10.1016/j.asoc.2020.106336
8. Cornell, P., et al.: Transforming nursing workflow, Part 1: the chaotic nature of nurse activities. J. Nurs. Adm. **40**(9), 366–373 (2010). https://doi.org/10.1097/NNA.0b013e3181ee4261
9. De Causmaecker, P., Vanden Berghe, G.: A categorisation of nurse rostering problems. J. Sched. **14**(1), 3–16 (2011). https://doi.org/10.1007/s10951-010-0211-z
10. Del-Fierro-Gonzáles, V., Mix-Vidal, A.: Orientaciones Técnicas para el Rediseño al Proceso de Atención de Urgencia de Adulto, en las Unidades de Emergencia Hospitalaria. Technical report, Ministerio de Salud, Chile (2018). https://iopscience.iop.org/article/10.1088/1757-899X/844/1/012044
11. El Adoly, A.A., Gheith, M., Nashat Fors, M.: A new formulation and solution for the nurse scheduling problem: a case study in Egypt. Alex. Eng. J. **57**(4), 2289–2298 (2018). https://doi.org/10.1016/j.aej.2017.09.007

12. Hadwan, M., Ayob, M.: A constructive shift patterns approach with simulated annealing for nurse rostering problem. In: Proceedings 2010 International Symposium on Information Technology - Visual Informatics, ITSim 2010, vol. 1 (2010). https://doi.org/10.1109/ITSIM.2010.5561304
13. Jamieson, I., Kirk, R., Andrew, C.: Work-life balance: what generation Y nurses want. Nurse Leader **11**(3), 36–39 (2013). https://doi.org/10.1016/j.mnl.2013.01.010. http://www.nurseleader.com/article/S154146121300030X/fulltext. http://www.nurseleader.com/article/S154146121300030X/abstract. https://www.nurseleader.com/article/S1541-4612(13)00030-X/abstract
14. Jaradat, G.M., et al.: Hybrid elitist-ant system for nurse-rostering problem. J. King Saud Univ. Comput. Inf. Sci. **31**(3), 378–384 (2019). https://doi.org/10.1016/j.jksuci.2018.02.009
15. Jaumard, B., Semet, F., Vovor, T.: A generalized linear programming model for nurse scheduling. Eur. J. Oper. Res. **107**(1), 1–18 (1998). https://doi.org/10.1016/S0377-2217(97)00330-5
16. Legrain, A., Bouarab, H., Lahrichi, N.: The nurse scheduling problem in real-life. J. Med. Syst. **39**(1), 1–11 (2014). https://doi.org/10.1007/s10916-014-0160-8
17. Legrain, A., Omer, J., Rosat, S.: An online stochastic algorithm for a dynamic nurse scheduling problem. Eur. J. Oper. Res. **285**(1), 196–210 (2020). https://doi.org/10.1016/j.ejor.2018.09.027
18. Lewis, S., Gambles, R., Rapoport, R.: The constraints of a "work-life balance" approach: an international perspective. Int. J. Hum. Resour. Manag. **18**(3), 360–373 (2007). https://doi.org/10.1080/09585190601165577
19. Liu, Z., Liu, Z., Zhu, Z., Shen, Y., Dong, J.: Simulated annealing for a multi-level nurse rostering problem in hemodialysis service. Appl. Soft Comput. **64**, 148–160 (2018). https://doi.org/10.1016/j.asoc.2017.12.005
20. Melita Rodríguez, A., Cruz Pedreros, M., Merino, J.M.: Burnout in nursing professionals working in health centers at the eighth region, of Chile. Ciencia y Enfermeria **14**(2), 75–85 (2008). https://doi.org/10.4067/S0717-95532008000200010
21. de Hacienda, M.: Fija Texto Refundido, Coordinado Y Sistematizado De La Ley No 18.834, Sobre Estatuto Administrativo (2005). https://www.leychile.cl/N?i=236392&f=2018-06-05&p=
22. Osores, F., Cabrera, G., Linfati, R., Umaña-Ibañez, S., Coronado-Henández, J., Gatica, G.: Design of an information system for optimizing the programming of nursing work shifts. IOP Conf. Ser. Mater. Sci. Eng. **844**, 012044 (2020). https://doi.org/10.1088/1757-899x/844/1/012044
23. Ozkarahan, I.: A flexible nurse scheduling support system. Comput. Methods Programs Biomed. **30**(2–3), 145–153 (1989). https://doi.org/10.1016/0169-2607(89)90066-7
24. Pizarro, R., Rivera, G., Soto, R., Crawford, B., Castro, C., Monfroy, E.: Constraint-based nurse rostering for the Valparaíso Clinic Center in Chile. In: Communications in Computer and Information Science, CCIS, vol. 174, pp. 448–452 (2011). https://doi.org/10.1007/978-3-642-22095-1_90
25. Pryce, J., Albertsen, K., Nielsen, K.: Evaluation of an open-rota system in a Danish psychiatric hospital: a mechanism for improving job satisfaction and work-life balance. J. Nurs. Manag. **14**(4), 282–288 (2006). https://doi.org/10.1111/j.1365-2934.2006.00617.x
26. Rogers, A.E., Hwang, W.T., Scott, L.D., Aiken, L.H., Dinges, D.F.: The working hours of hospital staff nurses and patient safety. Health Aff. **23**(4), 202–212 (2004). https://doi.org/10.1377/hlthaff.23.4.202

27. Santos, H.G., Toffolo, T.A.M., Gomes, R.A.M., Ribas, S.: Integer programming techniques for the nurse rostering problem. Ann. Oper. Res. **239**(1), 225–251 (2014). https://doi.org/10.1007/s10479-014-1594-6
28. Stølevik, M., Nordlander, T.E., Riise, A., Frøyseth, H.: A hybrid approach for solving real-world nurse rostering problems. In: Lee, J. (ed.) CP 2011. LNCS, vol. 6876, pp. 85–99. Springer, Heidelberg (2011). https://doi.org/10.1007/978-3-642-23786-7_9
29. Tanaka, S., Maruyama, Y., Ooshima, S., Ito, H.: Working condition of nurses in Japan: awareness of work-life balance among nursing personnel at a university hospital. J. Clin. Nurs. **20**(1–2), 12–22 (2011). https://doi.org/10.1111/j.1365-2702.2010.03354.x
30. Trapp, U.A., Larrain, S.A.I., Santis, E.M.J., Olbrich, G.S.: Causas de abandono de la práctica clínica hospitalaria de enfermería. Ciencia y enfermería **22**, 39–50 (2016). https://doi.org/10.4067/S0717-95532016000200004
31. Turchi, V., et al.: Night work and quality of life. A study on the health of nurses. Ann. Ist. Super. Sanita. **55**(2), 161–169 (2019)
32. Turhan, A.M., Bilgen, B.: A hybrid fix-and-optimize and simulated annealing approaches for nurse rostering problem. Comput. Ind. Eng. **145**, 106531 (2020). https://doi.org/10.1016/j.cie.2020.106531
33. Valouxis, C., Gogos, C., Goulas, G., Alefragis, P., Housos, E.: A systematic two phase approach for the nurse rostering problem. Eur. J. Oper. Res. **219**(2), 425–433 (2012). https://doi.org/10.1016/j.ejor.2011.12.042
34. Van der Heijden, B.I.J.M., Houkes, I., Van den Broeck, A., Czabanowska, K.: "I just can't take it anymore": how specific work characteristics impact younger versus older nurses' health, satisfaction, and commitment. Front. Psychol. **11** (2020). https://doi.org/10.3389/fpsyg.2020.00762
35. White, S.A., Miers, D.: BPMN Guía de Referencia y Modelado. Future Strategies Inc (2010)
36. Wu, T.H., Yeh, J.Y., Lee, Y.M.: A particle swarm optimization approach with refinement procedure for nurse rostering problem. Comput. Oper. Res. **54**, 52–63 (2015). https://doi.org/10.1016/j.cor.2014.08.016

Ensuring Ethics of Cyber-Physical and Human Systems: A Guideline

Damien Trentesaux(✉)

LAMIH UMR CNRS 8201, Université Polytechnique Hauts de France, 59313 Valenciennes, France
damien.trentesaux@uphf.fr

Abstract. Designing and using cyber-physical systems integrating humans put ethics at stakes. Meanwhile, even if works are done to assess the life-cycle of these systems and the technologies used to develop them, ensuring the ethics of cyber-physical and human systems remains insufficiently studied. The objective of this paper is to suggest a guideline to ensure ethics of such systems. A case study is proposed to illustrate the guideline. This guideline is at its first step of development and needs several improvements to gain maturity and applicability.

Keywords: Cyber-physical system · Human-machine system · Ethics · Guideline

1 Introduction

A Cyber-Physical System is the integration through networks of computing elements (cyber part) and physical systems (physical part) [1]. Cyber-Physical and Human Systems (CPHS) are "human-in-the-loop" CPS [2].

The advances in Information and Communication Technologies (e.g., Artificial Intelligence) and their application to various sectors (e.g., Industry 4.0 or transportation sector with autonomous cars) enable the development of new abilities such as learning, adaptation and evolution. Their integration into CPHS is often proposed, see [3]. Meanwhile, the integration of these new abilities coupled with the complexity of the design and control of a CPHS as well as its nature to evolve in more open, uncontrolled environments engage risks related to the human. Such risks are typically relevant to safety and security levels, and more recently ethics [4].

Ethics is not a new field of research, but engineering or operating ethics in the scientific world is new [5]. The definition of Ethics remains unstable and evolves with time; it is also multidisciplinary [6]. Ethics often refers to morality [7]. In our work, we consider ethics as "*the strive for the good life, with oneself and others, in just/fair institutions*" (translated from [8]). In this paper we apply this concept to a CPHS whose ethical behaviour must be studied. From our point of view, "ethical" is not equivalent to "safe". Safety is "*the absence of undesirable events and accidents*" [9]. For sure, an unsafe system does not behave ethically. Meanwhile, a safe system may not behave ethically, since the functions it operates safely may not be ethical (a clear example is the

D. Trentesaux et al. (Eds.): SOHOMA 2021, SCI 987, pp. 223–233, 2021.
https://doi.org/10.1007/978-3-030-80906-5_15

use of personal data gathered by a fully functional system). Consequently, ethics contains other dimensions such as trustworthiness, privacy, objectivity, altruism, kindness, caring and benignity [10].

Designing and using a CPHS while paying attention to ethical aspects, and in a more general view paying attention to ethical aspects during the whole lifecycle of a CPHS is getting importance in the scientific community [10, 11]. The objective of this paper is to suggest a guideline to ensure that a CPHS is ethical, considering its whole lifecycle. For that purpose, Sect. 2 contains a survey of related works and fields of research, and Sect. 3 details our guideline. Section 4 contains an application of our guideline in the context of a manufacturing CPHS. This paper concludes with a list of improvements to operate.

2 Related Work

From our perspective, several research fields present connections with our research topic.

The first one is risk assessment. The link with our topic is clear since we consider ethical risks in the lifecycle of a CPHS. Risk assessment (and management) is an important and intense field of research [12]. Risk is globally seen as "*the potential for realization of unwanted, negative consequences of an event*" and a mean to measure it is to set a function of "*a triplet (C, Q, K), where C is some specified consequences, Q a measure of uncertainty associated with C (typically probability) and K the background knowledge that supports C and Q (which includes a judgement of the strength of this knowledge)*" [12]. Risk assessment is often accompanied with functional and dysfunctional analysis as operated in safety studies. Several methods have been developed, inspired from the concept of risk assessment. In the industrial sector, the Failure mode and effects analysis (FMEA) is a method that instantiates the introduced function as a function of Probability (P), Severity (S) and Detection (D) to evaluate the possible failure of a system and its impact on the system operation.

Life Cycle Assessment (LCA) is a second interesting source of inspiration, mainly because the method tries to evaluate the impact on a specific view (here, the environment) on the entire lifecycle of a system (often, a product), which we intend to do with an ethics view. LCA is "*a comprehensive analysis approach to quantify and assess the consumption of resources and the environmental impacts associated with a product (or service) throughout its life cycle*" [13]. LCA is nowadays a mature approach; some norms have been elaborated relatively, one of the most famous being ISO 14000 series [14]. It is based on the construction of a relation graph of a system (product) for the different phases of its lifecycle (production, transport, use, disposal…). From this graph, a description of the activities led for each phase is done (e.g., assembly) and for each activity, the resources (material, energy…) needed and generated are identified, enabling to assess the environmental impact of the system. In [15], a framework inspired from the concept of LCA was proposed, considering environmental aspects to facilitate the maintenance of a CPS. An interesting aspect relevant to environment is that somehow, one may consider that environment is connected to ethics: an ethical behaviour is also a behaviour that aims to protect the environment, to spare rare natural resources, to limit the generation of wastes, etc.

More recently, researchers have rediscovered the concept of technology assessment (developed in the 70's) and started to evaluate the ethical impact of technologies [16], typically technologies fostered by Industry 4.0 principles. This approach "*will serve as a tool for identifying adverse effects of new technologies at an early stage [...] and can be conducted on the basis of a check-list that refers to nine crucial ethical aspects of technology; (1) Dissemination and use of information, (2) Control, influence and power, (3) Impact on social contact patterns, (4) Privacy, (5) Sustainability, (6) Human reproduction, (7) Gender, minorities and justice, (8) International relations, and (9) Impact on human values*" [16]. Such an approach is interesting in the sense that it tries to evaluate the ethical implication of a technology. Meanwhile, it is designed to focus on technologies, not on systems integrating various technologies, such as CPHS.

The concept of ethical assessment has also been proposed. For example, the EU project SATORI (https://satoriproject.eu/) aims to develop a common European framework for ethical assessment of research and innovation. In the project, ethical principles concern the research integrity, the social responsibility, the respect for human research participant, the protection of animals used in research, etc. The project suggests the creation of ethics committees working on standardized application forms. These forms contain information about the subject research project (including the stakeholders, information on the social impact of the research…), and a self-assessment of ethical considerations. Such an approach suits well the context of medical activities or activities led on animals for example.

Finally, it is worth mentioning initiatives clearly relevant to specific aspects of ethics, the main ones being privacy and data protection. In that context, the General Data Protection Regulation (GDPR) has been designed and voted at EU level. The GDPR applies to every organization handling data. It ensures that these organizations will consider data privacy from the design of their systems. Such initiatives ensure a certain level of ethical behaviour of these organizations, but the focus is set on only one aspect relevant to ethics with regards to information handling, processing and dissemination.

From this short overview, the topic of our research presents connections with existing approaches and fields. Consequently, the proposed guideline, described in the next section tries to integrate as much as possible some principles discussed in this section.

3 The Proposed Guideline

In this section, the proposed guideline is described. Up to now, it has been designed to be mainly used the industrial sector (production, logistics, transportation…). Medical or military domains are for example not concerned.

3.1 Specifications and Global Framework

The proposed guideline has been built on the following specifications. First of all, the studied CPHS is composed of sub-systems, either one purely cyber-physical or with the human in the loop. In this paper, the recursive systemic decomposition is not considered.

Second, for each sub-system, a primary function is defined, supported by a cyber-physical equipment interacting or interoperating with a human. This function is supposed to be safe (that is, from a safety studies point of view) to clearly show that ethics requires functional safety as a prerequisite but concerns notions not covered by safety. For each phase of the lifecycle of the considered subsystem (classified as beginning of life, middle of life and end of life), stakeholders are identified. They may be different according to the phase studied in the lifecycle but all are concerned with ethics. For example, a researcher making experiments with students (early beginning of life of the CPHS) is concerned, as well as a maintainer upgrading the CPHS in its middle of life. The guideline considers each of these phases. It must suggest classical closed loop processes to enable the progress of CPHS design regarding experiments and knowledge generated about its ethical or unethical behaviour when used, maintained or disposed.

Third, since ethics are related to the human, the society, the environment in which the CPHS evolves and the cyber-physical equipment and technology supporting the function, the guideline must consider at least these different dimensions. For this scope, a list of "sieves" relevant to these dimensions is suggested. The term "sieve" is used as a metaphor: it is a mean to filter among the different dimensions the one that is studied to emphasis on specific ethical aspects relevant to these different dimensions. Hence, each sieve will enable the analysis of specifically dependent subjects (some aspects relevant to ethics from human perspective are not consistent with others on technology).

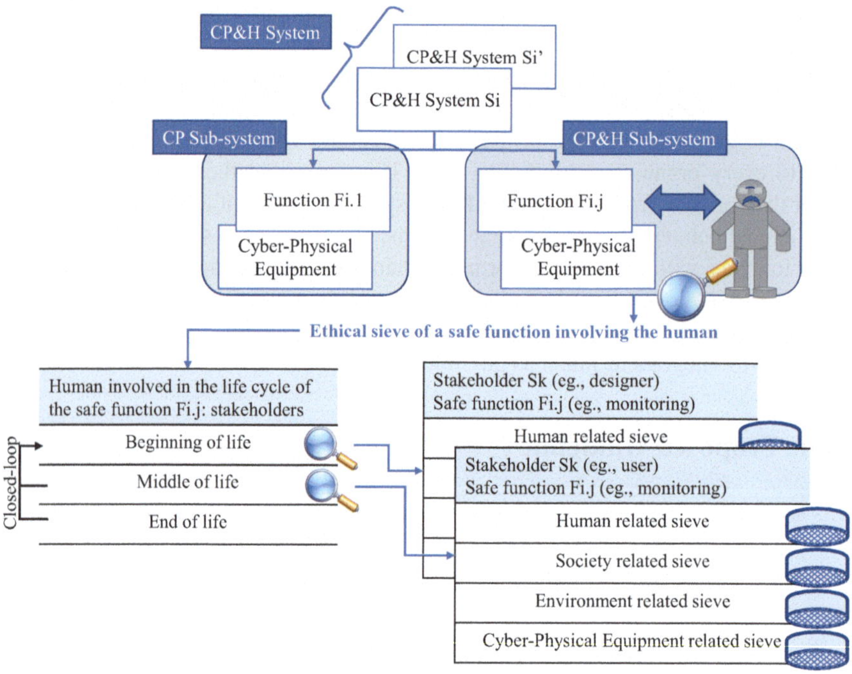

Fig. 1. The global framework of the proposed guideline

Figure 1 depicts the global framework of the guideline proposed. The list of sieves can be particularized according to the specific application fields where the guideline is applied.

3.2 The Sieves

Figure 2 details the four minimal and necessary sieves (whose number may be augmented according to specific contexts) and the relevant main questions asked concerning the sieve. Contributions relevant to the introduced ethical technology assessment can be connected to the fourth sieve.

Sieves	Safe function Fi.j: the question the sieve will try to answer
Human related sieve	Does the safe function Fi.j induce ethical risks for the stakeholders involved in the lifecycle of the CP&HS?
Society related sieve	Does the safe function Fi.j induce ethical risks for the society not directly involved in the lifecycle of the CP&HS?
Environment related sieve	Does the safe function Fi.j induce ethical risks because of the environment in which the CP&HS evolves?
Cyber-Physical equipment related sieve	Does the safe function Fi.j induce ethical risks because of the equipment and the technology that support the function?

Fig. 2. The different sieves and their main question

Answering each main question requires the answering of a series of sub-questions. Figures 3 and 4 contain a list of sub-questions relevant to the first and third sieves (the others are not described in this paper).

Human related sieve	
Stakeholder Sk; Safe function Fi.j	Answer
Does Fi.j require Sk data as input?	Yes/No
Does Fi.j require to apply/to be subjected to a force in relation to Sk?	Yes/No
Does Fi.j exchange data with Sk ?	Yes/No
Does Fi.j generate Sk data as output?	Yes/No
Does Fi.j have to store Sk-related data?	Yes/No
Does Fi,j require physical connection with Sk?	Yes/No

At least one positive answer triggers the answer "YES" to the relevant main question, here: "Does the safe function Fi.j induce ethical risks for the stakeholders involved in the lifecycle of the CP&HS? "

An ethical risk analysis must be triggered on this topic

Fig. 3. The human related sieve

As an output of each of these sieves, at least one positive answer triggers the ethical risk analysis and mitigation for the relevant topic.

3.3 Ethical Risk Analysis and Mitigation

For each positive answer to any sub-question, an ethical risk analysis and mitigation process is triggered, as depicted Fig. 5. It enables the identification of the stakes relevant to ethical aspects, specifically for the sub-question asked. As proposed by the risk assessment approach, the ethical risk is evaluated, and different paradigms relevant to ethics can be used to mitigate the risk.

Environment related sieve	
Stakeholder Sk; Safe function Fi.j	Answer
Does Fi.j apply in an open or uncontrolled environment?	Yes/No
Is it possible than some perturbations have not been identified ?	Yes/No
Is it possible for Fi.j to be used in other environments by Sk ?	Yes/No
Is it possible for the environment to evolve with time ?	Yes/No

At least one positive answer triggers the answer "YES" to the relevant main question, here: "Does the safe function Fi.j induce ethical risks because of the environment in which the CP&HS evolves? "

An ethical risk analysis must be triggered on this topic

Fig. 4. The environment related sieve

In this paper, two paradigms are suggested: *deontology* (one decides according to ethical rules "must/must not") and *utilitarianism* (one decides according to the possible ethical implications). Others are possible, but these two paradigms are often proposed in the literature when one seeks to apply ethics or to implement (operate) ethics [10]. If required, the mitigation of the ethical risk may be used as an iteration to revise the specification, the design and the production of the CPHS (as fostered by the closed loop PLM - Product Lifecycle Management [17]).

3.4 Output of the Guideline

The guideline finishes when mitigations (deontology or utilitarianism-related) are defined for each ethical risk. Each ethical risk is identified for each sub-question, for each main question (one per sieve), for each stakeholder, for each phase of the lifecycle of the CPHS and last, for each couple (safe function, subsystem) of the CPHS. Consequently, the size of the study may increase rapidly with the complexity of the studied CPHS, but

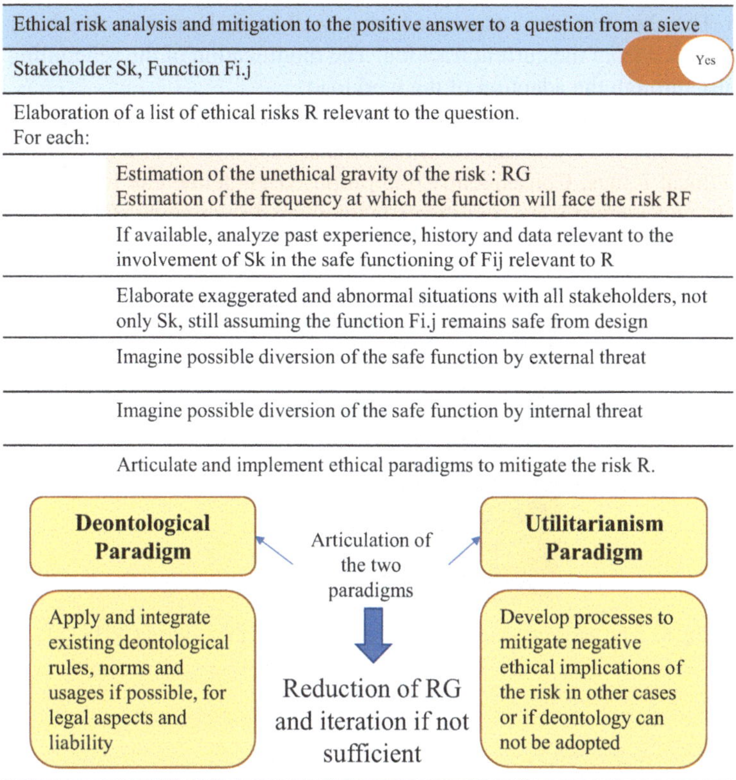

Fig. 5. Ethical risk analysis and mitigation.

from our point of view, ethics engage huge challenges that need to be addressed and studied in depth.

Obviously, when defined, these mitigations remain to be developed, tested, implemented and potentially recursively studied using the same guideline if they lead to the development of new subsystems and safe functions.

4 Case Study: The Operator 4.0

In this section we present a case study to illustrate our guideline. A flagship type of applications concerns the concept of Operator 4.0 [18] based partially on the principle of symbiotic systems [19] and that exploits the advances of Industry 4.0 technologies. Even if a lot of works are currently done on this topic, the concept of Operator 4.0 generates ethical risks that remain still unstudied up to now [20]. In this context, we consider the development of a CPHS where the production is adapted according to the human fatigue [21]. In their work, the authors present an original human-aware CPS able to dynamically change the schedule of tasks to human operators, regarding the cognitive fatigue level during execution. The human operator fatigue is estimated using a galvanic skin response (GSR) signal that tracks electric currents on his skin, see Fig. 6. The objective is to ensure the human well-being by minimizing his cognitive fatigue estimated from the current tracking. The modification of the scheduling is done dynamically through the adaption of the workload.

Even if this work is obviously done according to the will to ease the work of operators and even if results show how the contribution can ease him/her, the contribution does generate ethical issues. Consequently, the proposed guideline can help point out the ethical risks and suggest ways to mitigate them when designing this CPHS.

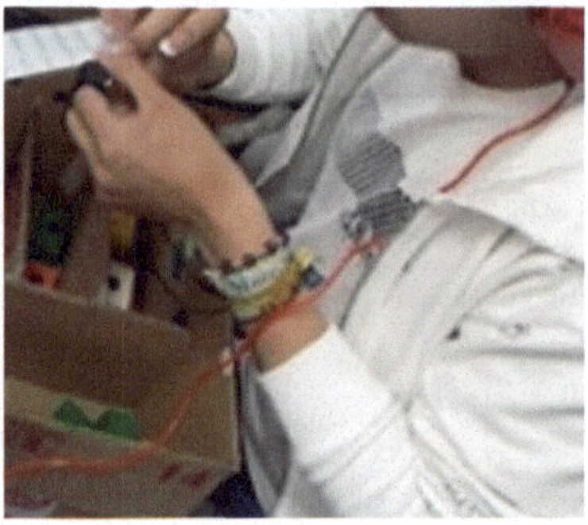

Fig. 6. The galvanic skin response sensor used in the studied CPHS [21]

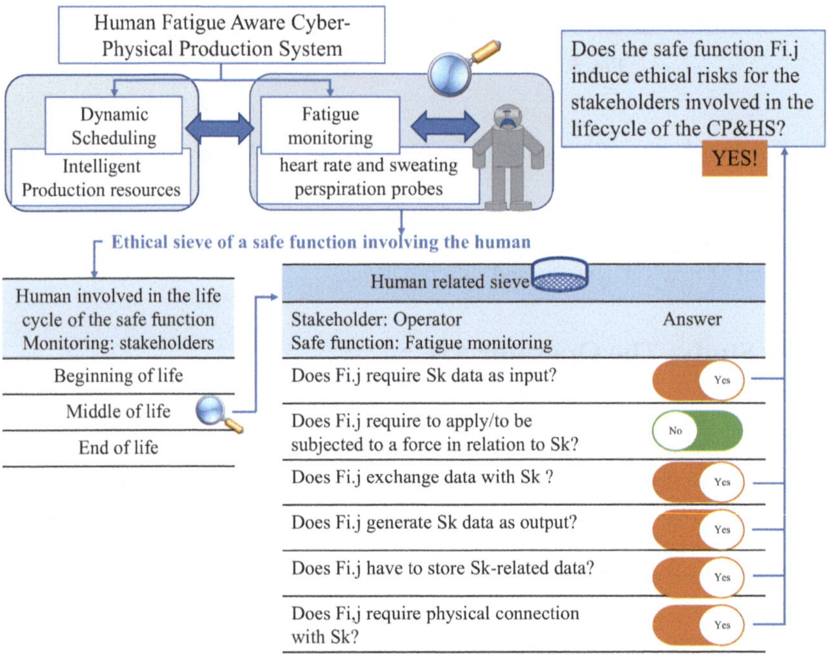

Fig. 7. A focus on the human-related sieve for the case study

The application of the guideline is not fully provided. The focus of the illustration is set on the use phase (CPHS middle of life) and concerns the stakeholder "operator". It is also set on the human sieve for the couple: safe function: "fatigue monitoring", sub-system: probes, as depicted Fig. 7. From this sieve, one can see that five sub-questions get positive answers, which means that each of these positive answers needs an ethical risk analysis and mitigation.

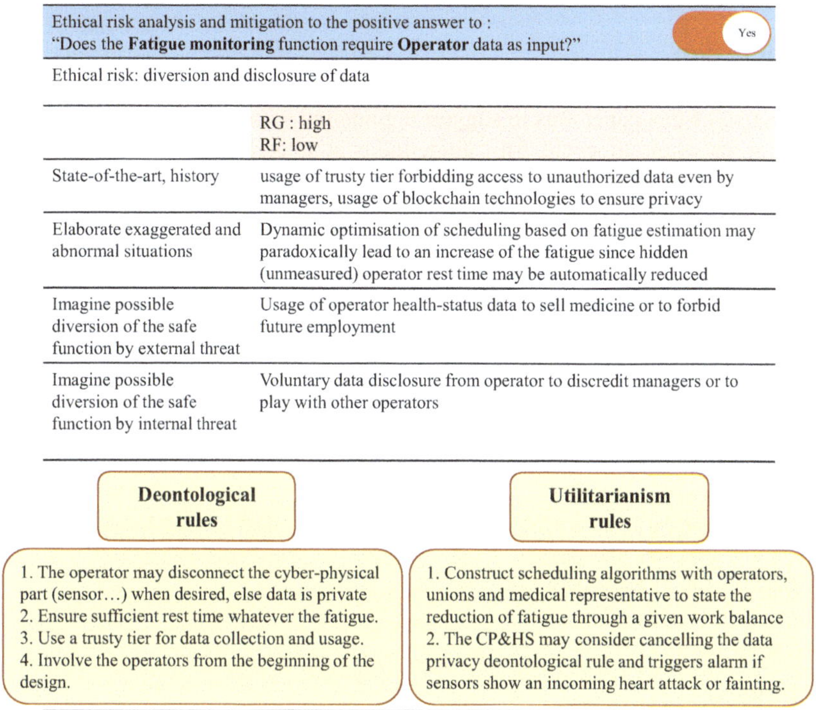

Fig. 8. Focus on an ethical risk to the positive answer to operator data gathering

Figure 8 contains an illustrative study of the ethical risks for one of the sub-questions whose answer was positive: "does the fatigue monitoring function require operator data as input?" (the first sub-question in Fig. 6). From this study, some ethical risks can be identified, among them, the "diversion and disclosure of data". For this risk, some suggestions can be imagined in order to mitigate it. According to the guideline, these suggestions concern:

- *Deontological paradigm*: the definition of static ethical rules such as "the operator may disconnect the probes when desired" (the scheduling system must then be adapted to comply with this rule, generating new interesting research avenues).
- *Utilitarianism paradigm*: the construction of adequate scheduling algorithms (e.g., multicriteria-based, considering weights of criteria and the aggregation function), the

definition of possible decisions to make in abnormal situations, potentially breaking deontological rules (e.g., in case of emergency), etc.

These suggestions are the proposed outputs of the guideline; they help mitigate ethical risks of the studied CPHS for the considered couple (safe function, sub-system). One can see that in return they generate new types of original scientific problems, generated from the need to consider ethics.

5 Conclusion

The objective of this paper was to suggest a guideline to ensure the ethics of a CPHS, considering its whole lifecycle. This guideline is at an early stage of its development. It is intended to develop the awareness of stakeholders involved in the lifecycle of a CPHS about ethical stakes relevant to the CPHS, including researchers, scientists and industrialists in charge of its development.

Future works concern the increase of the maturity of the guideline, both from a methodological point of view (sub-questions, risks analysis and mitigation process, etc.) and from a development point of view (software development, etc.). The author finally hopes that the proposal, even imperfect and subject to numerous improvements, will foster the scientific and technological developments of ethical CPHS.

Acknowledgment. Parts of the work presented in this paper are carried out in the context of: Surferlab, a joint research lab with Bombardier and Prosyst, partially funded by the European Regional Development Fund (ERDF), Hauts-de-France; the HUMANISM No ANR-17-CE10-0009 research program; the project "Droit des robots et autres avatars de l'humain", IDEX "Université et Cité" of Strasbourg University.

References

1. Lee, E.A.: The past, present and future of cyber-physical systems: a focus on models. Sensors **15**, 4837–4869 (2015). https://doi.org/10.3390/s150304837
2. Sowe, S.K., Zettsu, K., Simmon, E., de Vaulx, F., Bojanova, I.: Cyber-physical human systems: putting people in the loop. IT Prof. **18**, 10–13 (2016). https://doi.org/10.1109/MITP.2016.14
3. Cardin, O.: Classification of cyber-physical production systems applications: proposition of an analysis framework. Comput. Ind. **104**, 11–21 (2019). https://doi.org/10.1016/j.compind.2018.10.002
4. Trentesaux, D., Caillaud, E.: Ethical stakes of Industry 4.0. In: IFAC-PapersOnLine. Elsevier, Berlin (2020)
5. Fisher, M., List, C., Slavkovik, M., Winfield, A.: Engineering moral machines, Informatik-Spektrum (2016)
6. Trentesaux, D., Rault, R., Caillaud, E., Huftier, A.: Ethics of autonomous intelligent systems in the human society: cross views from science, law and science-fiction. In: Borangiu, T., Trentesaux, D., Leitão, P., Cardin, O., Lamouri, S. (eds.) SOHOMA 2020. SCI, vol. 952, pp. 246–261. Springer, Cham (2021). https://doi.org/10.1007/978-3-030-69373-2_17

7. Allen, C., Smit, I., Wallach, W.: Artificial morality: top-down, bottom-up, and hybrid approaches. Ethics. Inf. Technol. **7**, 149–155 (2005). https://doi.org/10.1007/s10676-006-0004-4
8. Ricoeur, P.: Soi-même comme un autre, Seuil (1990)
9. Aven, T.: What is safety science? Saf. Sci. **67**, 15–20 (2014). https://doi.org/10.1016/j.ssci.2013.07.026
10. Trentesaux, D., Karnouskos, S.: Ethical behaviour aspects of autonomous intelligent cyber-physical systems. In: Borangiu, T., Trentesaux, D., Leitão, P., Giret Boggino, A., Botti, V. (eds.) SOHOMA 2019. SCI, vol. 853, pp. 55–71. Springer, Cham (2020). https://doi.org/10.1007/978-3-030-27477-1_5
11. van Gorp, A.: Ethical issues in engineering design processes; regulative frameworks for safety and sustainability. Des. Stud. **28**, 117–131 (2007). https://doi.org/10.1016/j.destud.2006.11.002
12. Aven, T.: Risk assessment and risk management: review of recent advances on their foundation. Eur. J. Oper. Res. **253**, 1–13 (2016). https://doi.org/10.1016/j.ejor.2015.12.023
13. Zhang, Y., Luo, X., Buis, J.J., Sutherland, J.W.: LCA-oriented semantic representation for the product life cycle. J. Clean. Prod. **86**, 146–162 (2015). https://doi.org/10.1016/j.jclepro.2014.08.053
14. ISO 14040:2006 - Environmental management -- Life cycle assessment -- Principles and framework. http://www.iso.org/iso/catalogue_detail?csnumber=37456. Accessed 10 June 2016
15. Sénéchal, O., Trentesaux, D.: A framework to help decision makers to be environmentally aware during the maintenance of cyber physical systems. Environ. Impact Assess. Rev. **77**, 11–22 (2019). https://doi.org/10.1016/j.eiar.2019.02.007
16. Palm, E., Hansson, S.O.: The case for ethical technology assessment (eTA). Technol. Forecast. Soc. Chang **73**, 543–558 (2006). https://doi.org/10.1016/j.techfore.2005.06.002
17. Kiritsis, D.: Closed-loop PLM for intelligent products in the era of the Internet of things. Comput. Aided Des. **43**, 479–501 (2011). https://doi.org/10.1016/j.cad.2010.03.002
18. Romero, D., Bernus, P., Noran, O., Stahre, J., Fast-Berglund, Å.: The operator 4.0: human cyber-physical systems & adaptive automation towards human-automation symbiosis work systems. In: Nääs, I., et al. (eds.) APMS 2016. IAICT, vol. 488, pp. 677–686. Springer, Cham (2016). https://doi.org/10.1007/978-3-319-51133-7_80
19. Gill, K.S. (ed.): Human Machine Symbiosis: The Foundations of Human-centred Systems Design. Springer-Verlag, London (1996)
20. Pacaux-Lemoine, M.-P., Trentesaux, D.: Ethical risks of human-machine symbiosis in industry 4.0: insights from the human-machine cooperation approach. IFAC-PapersOnLine **52**, 19–24 (2019). https://doi.org/10.1016/j.ifacol.2019.12.077
21. Paredes-Astudillo, Y.A., et al.: Human fatigue aware cyber-physical Production system. In: Presented at the 1st IEEE International Conference on Human-Machine Systems, Rome (2020)

FarmBot Simulator: Towards a Virtual Environment for Scaled Precision Agriculture

Victor Alexander Murcia, Juan Felipe Palacios, and Giacomo Barbieri(✉)

Mechanical Engineering Department, Universidad de los Andes, Bogotá, Colombia
{va.murcia10,jf.palacios,g.barbieri}@uniandes.edu.co

Abstract. Precision agriculture (PA) is considered the future of agriculture since it allows optimizing returns on inputs while preserving resources. To investigate PA strategies, test benches can be applied to reproduce the use of the technology in scaled farms before implementation. The FarmBot CNC robot farming machine has the premises to become a test bench for PA, but its control software must be adapted for implementing farming strategies based on the feedback of sensors. In line with the Virtual Commissioning technology, a FarmBot Simulator is proposed in this paper to support the development of a control software able to implement different PA strategies, before its deployment into the physical test bench. A case study is developed to demonstrate that the proposed FarmBot Simulator allows the verification of different PA strategies through the implementation of a Software-in-the-Loop simulation. Due to the importance of food security in the 'worldwide landscape', we hope that the FarmBot approach may become a future strategy for the investigation in PA due to its low-cost implementation and its open source software.

Keywords: Society 5.0 · Precision agriculture · FarmBot · Virtual Commissioning · Test bench

1 Introduction

Society 5.0 is defined as "A human-centered society that balances economic advancement with the resolution of social problems by a system that highly integrates cyber and physical space" [1]. Food security is considered as a current and future social problem since global population exceeds 7.2 billion and is continuously increasing. In 2050, the population has been estimated to reach 9.6 billion [2], and productivity needs to be almost 50% higher than that of 2012 [3].

Traditional *soil-based agriculture* allows producing large amounts of food but with a negative impact on the environment [4]. It is estimated that 87% of the yearly consumed freshwater is utilized for agriculture [5], and high percentages of water and fertilizer are lost in the ground due to leaching [6]. Traditional agricultural systems are not sustainable on a long run and farmers are under pressure to reduce or eliminate nutrient-laden water discharges to the environment.

Precision agriculture (PA) is a farming management concept based on observing, measuring, and responding to inter and intra-field variability in crops. The goal of PA

D. Trentesaux et al. (Eds.): SOHOMA 2021, SCI 987, pp. 234–246, 2021.
https://doi.org/10.1007/978-3-030-80906-5_16

research is to define a decision support system for the whole farm management with the objective to optimize returns on inputs while preserving resources [7]. PA has been enabled by the integration of GPS with technologies of the digital transformation such as real-time Sensors, Internet of Things, Big Data and Artificial Intelligence. The farmers' and/or researchers' ability to locate the crop precise position in the field allows for the creation of maps able to represent the spatial variability of as many variables as can be measured. Variable rate technology (e.g. seeders, sprayers, etc.) uses this data in conjunction with satellite imagery to optimally distribute resources.

In the context of PA, test benches can be used for reproducing scaled farms to investigate the use of technology before its implementation. The *AgroLab Uniandes* was founded in 2019 with this objective and contains different production systems spacing from traditional agriculture to hydroponics and aquaponics [8]. Among the available technologies, a FarmBot was implemented for investigating PA strategies in soil-based agriculture.

The *FarmBot project* developed an open source precision agriculture CNC robot farming machine [9]. The *FarmBot Uniandes* was mounted on the top of four pots for concurrently investigating different farming strategies, see Fig. 1. In this way, different strategies can be implemented since the crops are physically separated in four clusters.

Fig. 1. FarmBot installed in the AgroLab Uniandes

The FarmBot developers designed a '*FarmBot Operating System*' that runs on a Raspberry Pi controller. The Raspberry Pi communicates with the '*Farm Design*' web application which allows the user to configure the layout of the scaled farm and to schedule farming operations at a specific day and time. However, in the current version of the 'FarmBot Operating System', farming operations cannot be controlled based on the feedback of sensors. Therefore, a different control software must be developed to convert the FarmBot into a test bench for PA.

In the industrial automation domain, the *Virtual Commissioning* technology is used to virtually validate the code before its deployment [10]. Therefore, the objective of this work is to develop a *FarmBot Simulator* for supporting the development of a control software able to implement different PA strategies, before its deployment into the physical test bench. The paper is structured as follows: Virtual Commissioning is summarized in Sect. 2, while the proposed FarmBot Simulator is described in Sect. 3. Section 4 uses the simulator for the development of a control software able to implement different

PA strategies. The results obtained are discussed in Sect. 5, while Sect. 6 presents the conclusions and sets the directions for future work.

2 Virtual Commissioning

Innovation is defined as the application of better solutions that meet new requirements, unarticulated needs, or existing market needs [11]. Although a universally framework for identifying all the types of innovation is not available, few categories are generally utilized to describe a product/process innovation or a disruptive/incremental/radical/architectural innovation [12].

Process innovations are innovations in the way an organization conducts its business, such as in the techniques of producing or marketing goods or services. Architectural innovations are innovations that result from the application of existing technology to a different market. In this work, we propose to apply a widespread technology of the industrial automation domain for improving the process of designing and verifying control strategies for a scaled PA application. Therefore, the presented solution can be defined as a process and architectural innovation.

Virtual Commissioning (VC) consists in the early development and validation of the control software using a simulation model [13]. VC is performed by connecting a simulation model that reproduces the behaviour of the physical plant to the hardware or emulated controller which contains the software to be validated. VC requires a *virtual plant model* fully described at the level of sensors and actuators.

VC enables the evaluation of the system's functionality, performance and safety before its physical assembly and commissioning [10]. Nowadays, VC is implemented in many automation applications [14], due to its ability to speed up the commissioning process. Without VC, the production system must be stabilized solely by real commissioning with real physical plants and real controllers, which is expensive and time consuming. An industrial study showed that real commissioning time is reduced approximately by 75% when VC is implemented beforehand [15].

As shown in [10], there can be two possible configurations in VC: (i) *Hardware-in-the-Loop* (HiL) commissioning involving a virtual plant and a real controller; (ii) *Software-in-the-Loop* (SiL) commissioning involving a virtual plant and a virtual controller. HiL commissioning is generally utilized for validating hard real-time applications, while SiL commissioning for soft real-time ones. Since the FarmBot does not need to fulfil hard real-time requirements, Software-in-the-Loop Commissioning is proposed within this work.

Finally, the following requirements must be fulfilled to develop a simulator for SiL commissioning in accordance with the Farmbot project mission and able to verify different PA strategies:

- *Open source software*: since the 'Farmbot project' is defined as an open source precision agriculture CNC farming project, the FarmBot Simulator runs with open source software.
- *Interface*: SiL commissioning is valuable only if the verified control software can be deployed into the controller without the need of changes, which may constitute a

source of error. Therefore, the developed simulator must implement the same interface of the 'FarmBot Operating System' to enable the seamless deployment of the verified code.

- *Movement and irrigation*: only these two functionalities will be implemented in the first version of the simulator. The reason is to obtain an early feedback from the scientific community before the development of additional functionalities.
- *Graphical display*: a graphical display representing the system state must be available to facilitate the debug operation.

3 FarmBot Simulator

In this section, the developed FarmBot Simulator for designing PA strategies is presented. The architecture implemented for the SiL commissioning is shown in Sect. 3.1, while Sect. 3.2 describes the software selected for the SiL simulation. The FarmBot Simulator is illustrated in Sect. 3.3.

3.1 Software-in-The-Loop Commissioning

The FarmBot control architecture is illustrated in the upper part of Fig. 2 (physical domain).

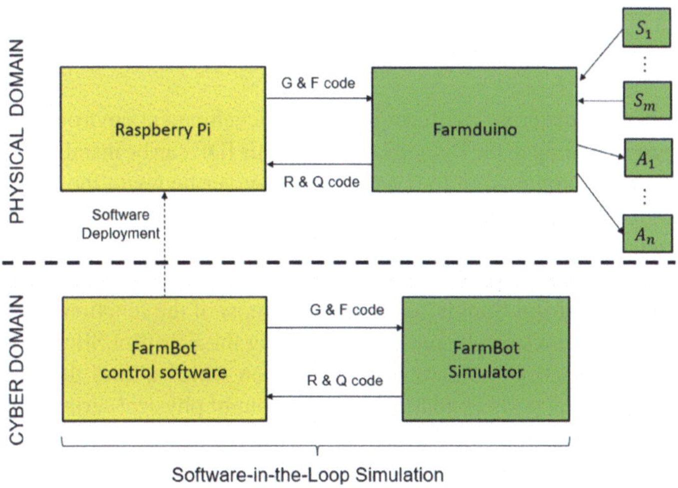

Fig. 2. Architecture implemented for the Software-in-the-Loop commissioning

The FarmBot sensors and actuators are controlled through a Farmduino board[1] programmed with an Arduino MEGA 2560 microcontroller. The Farmduino acts as a middleware interfacing the Farmbot I/Os with the Raspberry Pi microcontroller. The Raspberry Pi constitutes the 'brain' of the system since it contains the control software and

[1] https://farm.bot/products/v1-5-farmduino.

connects the robot to the 'Farm Design' web application through a MQTT Gateway[2]. FarmBot's Raspberry Pi runs a custom operating system named '*FarmBot OS*' to execute scheduled events and upload sensor data. The OS communicates with the Farmduino over a serial connection to send G- and F-code commands, and receives collected data through R- and Q- code. G-code is the code used in numerical control programming, while F-, R- and Q-code contain custom functions created by the FarmBot developers[3]. In the current version of the 'FarmBot OS', farming operations are scheduled through the 'Farm Design' web application and cannot be controlled based on the feedback of sensors. Therefore, a different control software must be developed to convert the FarmBot into a test bench for PA.

Since the Farmduino code will be left unchanged, the FarmBot Simulator must implement the Raspberry-Farmduino interface; see lower part of Fig. 2 (cyber domain). In this way, the control software can be designed and validated using the simulator, and seamlessly deployed into the FarmBot's Raspberry Pi.

3.2 Software-in-the-Loop Simulation

Since the 'Farmbot project' is defined as an open source precision agriculture CNC farming project, the SiL simulation must be implemented with open source software. In order to allow the seamless deployment of the verified code, the 'FarmBot control software' and the FarmBot Simulator must communicate through a serial port and exchange functions written with the G-, F-, R- and Q-code illustrated in Sect. 3.1.

The chain of software selected for the SiL simulation is shown in Fig. 3 and consists of:

- *Spyder*[4]: open source cross-platform integrated development environment (IDE) for scientific programming in the Python language. This IDE can be installed in Raspberry PI allowing the deployment of the control software verified with the SiL simulation.
- *Visual Studio*[5]*:* IDE provided from Microsoft®. Visual Studio supports 36 programming languages and its most basic edition, the Community edition, is available free of charge. Since Farmduino is programmed in C++, this programming language is selected for the FarmBot Simulator. In this way, most of the functions provided from the FarmBot developers can be reused for building the FarmBot Simulator
- *Free Virtual Serial Ports*[6]*:* Windows® application which allows the user to create software virtual serial ports emulating the behaviour of physical serial ports. Through this software, Spyder and Visual Studio are interfaced with a virtual serial port. This interface enables the seamless deployment of the verified control software since the communication functions are the same for both the cyber and the physical domain.

---

[2] https://software.farm.bot/docs/intro.

[3] https://github.com/FarmBot/farmbot-arduino-firmware.

[4] https://www.spyder-ide.org/.

[5] https://visualstudio.microsoft.com/.

[6] https://freevirtualserialports.com/.

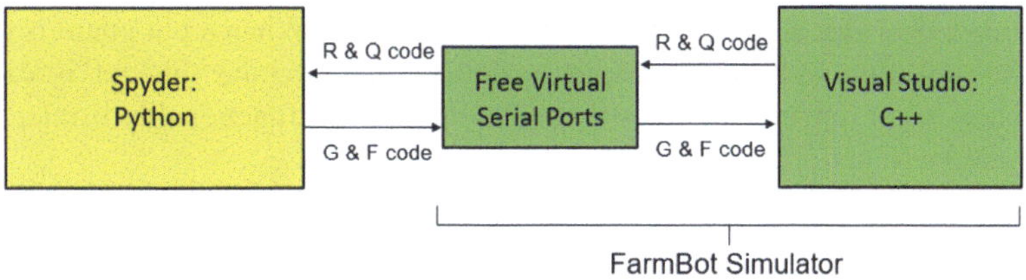

Fig. 3. Software selected for the Software-in-the-Loop simulation

3.3 FarmBot Simulator

The *Farmbot simulator* was built adapting the classes provided from the Farmduino controller. Figure 4 illustrates a hierarchical representation of the *Farmduino* classes. The classes implemented with changes are shown in green, the ones completely changed in blue, while the ones not implemented in red. Changes were performed to adapt the Farmduino classes to run in the virtual environment. However, the FarmBot Simulator presents the same interface of the Farmduino controller, making the seamless deployment of the verified code viable.

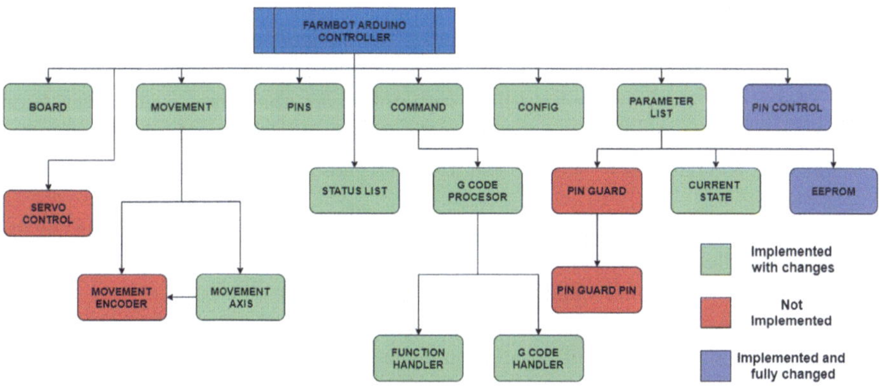

Fig. 4. Software architecture of the Farmduino code

Next, a description of the Farmduino classes shown in Fig. 4 is provided:

- *Farmbot Arduino Controller*: contains the setup and loop functions and constitutes the main class that instantiates all the others.
- *Board*: defines the version of the Farmbot PCB board.
- *Pins*: configures the physical pins of the ATMEGA 2560 controller.
- *PinControl*: controls the pins of the ATMEGA 2560. In our simulator, this class is replaced with a new one called "*Arduino Pins*". The implemented class virtualizes each physical pin storing its commanded value, the 'pin mode' (i.e. Input/Output) and the 'pin type' (i.e. Analog/Digital).

- *PinGuard*: provides an extra layer of safety to the pins. When a pin guard is set, the firmware will automatically set the pin to a defined state after the 'timeout' is reached[7]. This class is not implemented in the simulator since this functionality is relative to a hardware malfunction.
- *PinGuardPin*: assigns a pin to the 'pinGuard' class.
- *Config:* contains the configuration parameters as firmware parameters (e.g. reply timeouts, etc.), parameters for the movement, etc.
- *Parameter List*: sets the configuration parameters. Default parameters are loaded in case they have not been defined from the user.
- *EEProm*: manages the EEPROM memory of the ATMEGA 2560. In the simulator, this class is replaced with a "csv" file used to store the values of the configuration parameters. These parameters are retained even if the Farmbot Simulator is stopped
- *Movement*: responsible for the robot movement and for the update of the robot position.
- *ServoControl*: controls the FarmBot servomotors. This class in not implemented since the FarmBot Uniandes works with stepper motors.
- *MovementEncoder*: creates, configures, and obtains the feedback of the encoders. In the simulation, this class is not implemented since the virtual environment does not model physical no-idealities; e.g. obstacles, axis misalignment, etc. For each motor, the actual position has been calculated from the steps commanded to the motor and not from the feedback of encoders.
- *MovementAxis:* controls the axis movement using the feedback of encoders and limit-switch sensors. Since encoders are not virtualized in the FarmBot Simulator and the FarmBot Uniandes does not have limit switch sensors, the actual position has been assumed as the one computed starting from the steps commanded to the stepper motors.
- *Command*: receives the F- and Q-code functions and converts them into executable commands.
- *GCodeProcessor*: instantiates the command to be executed using the functions of the 'GCode Handler' or 'Function Handler'.
- *FunctionHandler*: contains all the 'F-code' functions.
- *GCodeHandler*: contains all the 'G-code' functions.
- *StatusList*: contains the list of possible states; e.g. moving, idle, emergency stop, etc.
- *CurrentState*: sets the current state.

The Farmduino Simulator was implemented in Visual Studio and a graphical display was generated as shown in Fig. 5. In the left-hand side, three FarmBot actuators are placed: the '*Vacuum Pump*' used for the seeding operation, the '*Led Strip*' used to light up the scaled farm at night, and the '*Valve*' used for the watering operation. When one of the actuators is active, its symbol gets a bright colour.

In the right-hand side of Fig. 5, the serial communication can be configured, initialized and its state is visualized. A '*Command History*' window presents the last commands sent to the simulator.

Finally, the *FarmBot 'farm'* is represented in the centre of Fig. 5. Two orthogonal views are shown allowing the user to locate the FarmBot end-effector in the three spatial dimensions. The movements commanded from the 'FarmBot control software' are scaled

[7] https://software.farm.bot/docs/pin-guard.

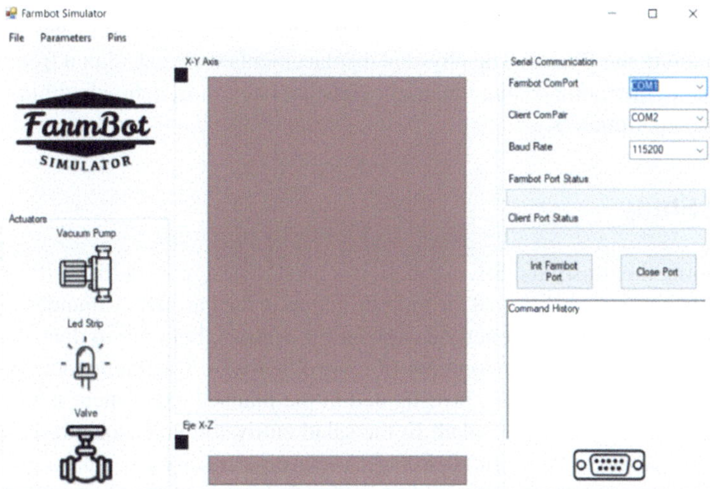

Fig. 5. Graphical display of the Farmbot Simulator

to effectively mimic the displacement that the end-effector would have in the physical system. Next, it is illustrated how the scaling operation was implemented.

Visual Studio is a dots per inch (DPI) aware application, which means that its display is automatically scaled. The scaling operation is performed through a scale factor that can be set in Visual Studio. For instance, a 256×256 pixel window will have a 128× 128 pixel size if the scale factor is set to 200%. In our case study, a scale factor of 100% is considered. The equivalence in between the *'FarmBot farm'* physical dimensions and the display in pixels is shown in Table 1.

Table 1. Equivalence in between the physical dimensions and the display in pixels

Axis	Physical dimensions (mm)	Virtual dimensions (pixels)
X	0–3000	0–360
Y	0–3000	14–366
Z	0–500	17–45

Starting from the table, the equations for converting the physical displacements into the virtual ones were identified:

$$d_{px}^{x} = d_{mm}^{x} * \frac{360}{3000} \tag{1}$$

$$d_{px}^{y} = 14 + d_{mm}^{y} * \frac{366 - 14}{3000} \tag{2}$$

$$d_{px}^{z} = 17 + d_{mm}^{y} * \frac{45 - 17}{500} \tag{3}$$

where d^i_{mm} is the physical displacement along the ith axis, and d^i_{px} is the virtual displacement along the ith axis. The physical displacements were calculated from the steps commanded to the motors using the motor constant [step/mm] defined within the motor configuration parameters.

4 Case Study

In this section, the illustrated SiL simulation is used to design a control software able to concurrently implement different farming strategies in the FarmBot Uniandes. The focus of this case study is not to properly design an experiment, but only to demonstrate that the simulator enables the development of control software for implementing different farming strategies. The variable investigated in the planned experiment is the effect of soil moisture in the growth of salad. In the salad cultivator, soil moisture is suggested to constantly remain in between 60–80%[8]. A hygrometer sensor is placed on each pot, and irrigation is triggered at different thresholds. Irrigation starts when soil moisture is below the lower bound and stops when its value is above the upper bound.

Next, the thresholds planned for each pot are listed:

- Pot 1: $60\% \leq \textit{Relative Humidity}\ (RH) \leq 65\%$
- Pot 2: $65\% \leq RH \leq 70\%$
- Pot 3: $70\% \leq RH \leq 75\%$
- Pot 4: $75\% \leq RH \leq 80\%$

To adapt the *FarmBot Simulator* to the planned experiment, four *pots* are added to the display of the scaled farm as shown in Fig. 6. Furthermore, four *linear 'Gauges'* are inserted to simulate the soil moisture sensed by each hygrometer. The user can change their values to verify the correct functioning of the system.

Then, the Python control software is designed using the Spyder IDE. Production systems are characterized with different operational modes [16, 17]. The following operational modes are implemented for the FarmBot Uniandes:

- *Manual operational mode:* the operator can individually move and activate all the FarmBot actuators. This operational mode is useful in case of malfunctions to identify the source of errors
- *Automatic operational mode:* the system implements the desired functionality autonomously, without the need of an operator.

The *HMI* (Human Machine Interface) designed for the FarmBot Uniandes is shown in Fig. 7. The HMI is characterized by: (i) Push buttons: for controlling the system and switching from one operational mode to the other; (ii) Leds: for indicating the status of the system; (iii) Textboxes: for showing the values sensed by the hygrometer sensors.

---

[8] https://www.infoagro.com/hortalizas/lechuga.htm.

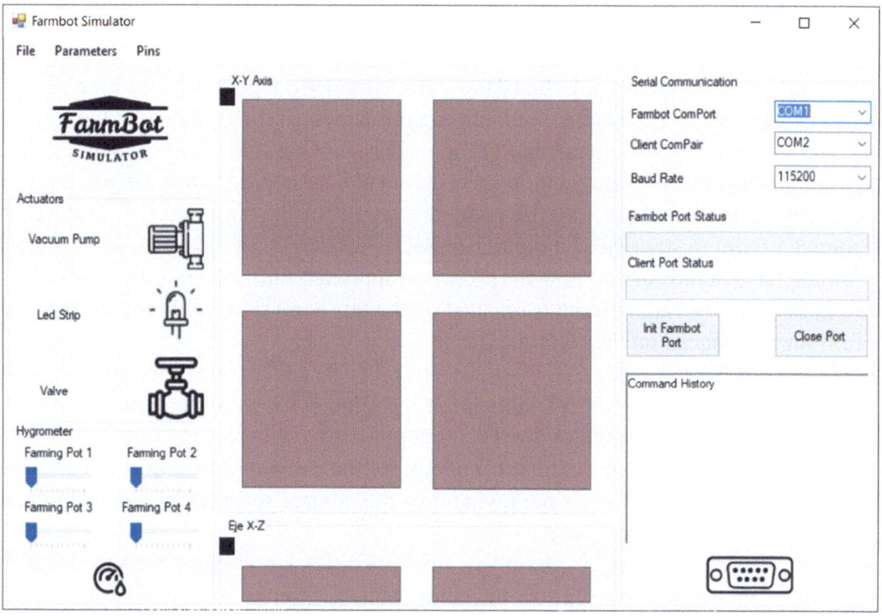

Fig. 6. Farmbot Simulator customized for the FarmBot Uniandes

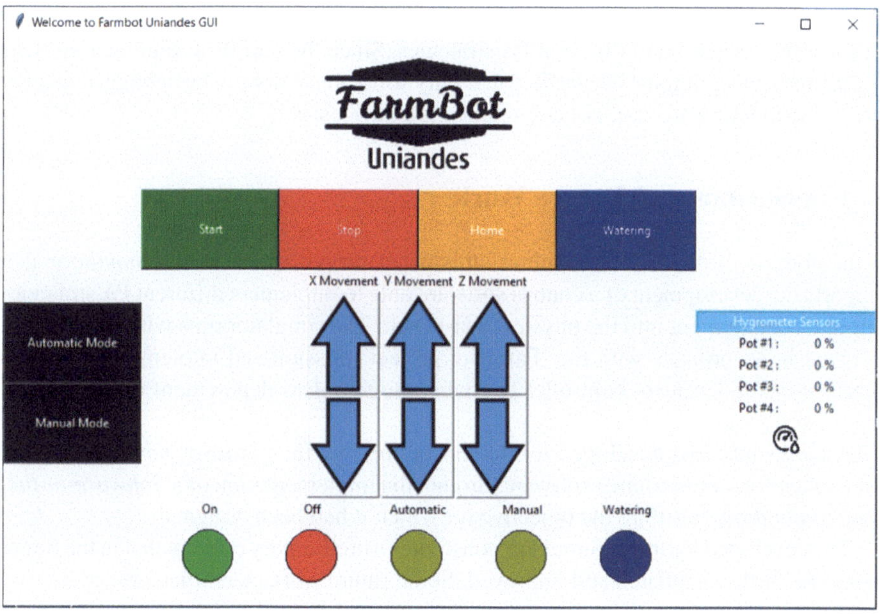

Fig. 7. HMI implemented for the FarmBot Uniandes

5 Results and Discussion

Spyder and Visual Studio are interfaced through a virtual serial port using the Free Virtual Serial Ports software. Then, a SiL simulation is implemented for the debug of the control software. A video[9] is made available to the reader showing that the developed control software enables the implementation of different PA strategies. Furthermore, readers can download the simulator and the code developed for this case study[10] to develop customized farming strategies on their Farmbot.

Finally, it is demonstrated how the proposed approach fulfils the requirements identified in Sect. 2. The requirement is recalled on the left-hand side, while the strategy for its fulfilment is placed on the right-hand side:

- *Open source software* → the SiL simulation is implemented with open source software complying with the mission of the 'Farmbot project'.
- *Interface* → the Farmbot Simulator implements the same interface of the 'FarmBot Operating System'. Seamless deployment of the control software seems viable and will be verified in future work.
- *Movement and irrigation* → these two functionalities are implemented in this first version of the simulator and allow the generation of different irrigation strategies.
- *Graphical display* → a graphical display able to mimic the system configuration is available to facilitate the debug operation.

In summary, the proposed approach fulfils the objective to verify the control software for the implementation of different PA strategies. Since the FarmBot Simulator consists of a virtual model, it must be clarified that only software errors can be debugged and not errors due to hardware, e.g. not communicating devices, etc.

6 Conclusions and Future Work

In the context of precision agriculture, this paper proposes a *FarmBot Simulator* that supports the development of a control software able to implement different PA strategies before its deployment into the physical test-bench. The simulator runs with open source software in accordance with the 'FarmBot project' mission and implements the same interface of the FarmBot controller to enable the seamless deployment of the verified control software.

A case study was developed for demonstrating that the FarmBot Simulator allows the verification of the control software through the implementation of a *Software-in-the-Loop Simulation*, fulfilling the objective for which it had been designed.

The developed FarmBot Simulator constitutes a preliminary concept that in the future should be further validated and improved. Some future works identified are:

---

[9] https://www.youtube.com/watch?v=I_7CE4BaFrs&list=PLXnpmcz3YbS9pPO8y2lxz8V7sNdVn4Gih.

[10] https://github.com/GiacomoBarbieri1/FarmBot-Simulator.

- *Farming functionality*: in this first version of the simulator only the movement and irrigation functionalities are implemented. A future version will include additional functionalities useful for the development of PA strategies such as seeding and crop monitoring through the FarmBot Webcam. Moreover, Spyder will be interfaced with the 'Farm Design' web application for the generation of control software that allows to remotely monitor and control the robot.
- *Seamless deployment*: the Farmbot Simulator implements the same interface of the 'FarmBot Operating System'. The control software will be deployed in the FarmBot Uniandes controller to evaluate the seamless deployment of the verified code
- *PA test-bench*: a farming experiment will be designed and implemented to certify that the FarmBot can be utilized as a test bench for PA.

References

1. Fukuyama, M.: Society 5.0: Aiming for a new human-centered society. Jpn. Spotlight **1**, 47–50 (2018)
2. United Nations: World urbanization prospects: The 2014 revision-highlights. Department of Economic and Social Affairs (2014). https://esa.un.org/unpd/wup/publications/files/wup2014-highlights.pdf
3. World Bank Group: Global Monitoring Report 2015/2016: Development Goals in an Era of Demographic Change (2016). http://pubdocs.worldbank.org/en/503001444058224597/Global-Monitoring-Report-2015.pdf
4. Tilman, D., Cassman, K.G., Matson, P.A., Naylor, R., Polasky, S.: Agricultural sustainability and intensive production practices. Nature **418**(6898), 671–677 (2002)
5. Kumar, R.R., Cho, J.Y.: Reuse of hydroponic waste solution. Environ. Sci. Pollut. Res. **21**(16), 9569–9577 (2014). https://doi.org/10.1007/s11356-014-3024-3
6. Mitsch, W.J., Gosselink, J.G., Zhang, L., Anderson, C.J.: Wetland Ecosystems. John Wiley & Sons, Hoboken (2009)
7. McBratney, A., Whelan, B., Ancev, T., Bouma, J.: Future directions of precision agriculture. Precis. Agric. **6**(1), 7–23 (2005)
8. Zapata, F., Barbieri, G., Ardila, Y., Akle, V., Osma, J.: AgroLab: a living lab in Colombia for research and education in urban agriculture. In: Cumulus Conference Proceedings - The Design After (2019)
9. Aronson, R.: FarmBot, (2019). https://farm.bot/
10. Lee, C.G., Park, S.C.: Survey on the virtual commissioning of manufacturing systems. J. Comput. Des. Eng. **1**(3), 213–222 (2014)
11. Maranville, S.: Entrepreneurship in the business curriculum. J. Educ. Bus. **68**(1), 27–31 (1992)
12. Schilling, M.A., Shankar, R.: Strategic Management of Technological Innovation. McGraw-Hill Education, Chennai (2019)
13. Reinhart, G., Wünsch, G.: Economic application of virtual commissioning to mechatronic production systems. Prod. Eng. Res. Devel. **1**(4), 371–379 (2007)
14. Lechler, T., Fischer, E., Metzner, M., Mayr, A., Franke, J.: Virtual Commissioning–Scientific review and exploratory use cases in advanced production systems. In: 26th CIRP Design Conference, vol. 81, pp. 1125–1130 (2019)
15. Koo, L.J., Park, C.M., Lee, C.H., Park, S., Wang, G.N.: Simulation framework for the verification of PLC programs in automobile industries. Int. J. Prod. Res. **49**(16), 4925–4943 (2011)

16. Barbieri, G., Battilani, N., Fantuzzi, C.: A PackML-based design pattern for modular PLC code. IFAC-PapersOnLine **48**(10), 178–183 (2015)
17. Barbieri, G., Gutierrez, D.: A GEMMA-GRAFCET methodology to enable digital twin based on real-time coupling. Procedia Comput. Sci. **180**, 13–23 (2021, in press)

Logistics and Supply Chains

Connectivity Through Digital Supply Chain Management: A Comprehensive Literature Review

Iván Henao-Hernández, Andrés Muñoz-Villamizar, and Elyn Lizeth Solano-Charris(✉)

Operations and Supply Chain Management Research Group, International School of Economic and Administrative Sciences, Universidad de La Sabana, Chía, Colombia
{ivanhehe,andres.munoz1,erlyn.solano}@unisabana.edu.co

Abstract. The COVID-19 pandemic, the evolution of Industry 4.0, and its implementation in logistics lead to the acceleration of Digital Supply Chains (DSC). These networks adopt technologies such as Sensor Nodes, Radio Frequency Identification or Artificial Intelligence that allow managers to get traceability of the Supply Chain. Traceability of the Supply Chain improves decision-support systems and provides valuable real-time information to increase the effectiveness of the decision-making process. In this article, the academic contributions on connectivity approaches for DSC are analysed and classified according to four main groups, i.e., warehousing management, production systems, transportation networks and reverse logistics. Finally, conclusions and future research lines are presented.

Keywords: Connectivity · Digital supply chains · Literature review

1 Introduction

The COVID-19 pandemic has generated significant changes in all areas of life, including supply chains (Ivanov and Dolgui 2020). As stated in this paper, pandemic became a test for supply chains regarding their robustness, flexibility, recovery, and resilience. Thus, the transformation of traditional supply chains into Digital Supply Chains (DSC) has been accelerated during the pandemic, making connectivity a crucial factor. In this context, connectivity can support DSC to improve visibility across the end-to-end supply chain and support agility and resiliency in the DSC (Deloitte 2020; Kayikci 2018; Ben-Daya *et al.* 2019).

Companies that have highly digitized supply chains can expect efficiency gains of 4.1% annually and revenue increase by 2.9% a year (Strategy 2016). Furthermore, managing real-time visibility of the entire supply chain will increase the system resiliency and generate savings of $210 billion in operating costs for 2025 (World Economic Forum 2016).

These high improvement levels make some authors investigate the impact of DSC in different industries (Zhengxia and Laisheng 2010; Doyle *et al.* 2015). Some literature

D. Trentesaux et al. (Eds.): SOHOMA 2021, SCI 987, pp. 249–259, 2021.
https://doi.org/10.1007/978-3-030-80906-5_17

reviews focused on Digital Supply Chain or related topics have been conducted (e.g., Ben-Daya *et al.* 2019; Muñoz-Villamizar *et al.* 2019; Büyüközkan and Göçer 2018). However, none of them are focused exclusively on existing connectivity grids and how they improve the connectivity on Supply Chains. This article aims to fulfil this gap by showing and discussing how digitalization could be embedded in the supply chains.

The rest of this paper is divided as follows. Section 2 presents the related literature, while Sect. 3 describes the methodology used for gathering information. Section 4 presents the discussion of the reviewed papers and Sect. 5 concludes this paper.

2 Related Studies

Digital Supply Chain is a trending and important topic that has been implemented in different contexts because it allows companies to have a global reach and improve their logistic performance (Rozman *et al.* 2019). This section summarizes previous literature reviews on DSC.

Büyüközkan and Göçer (2018) analyse contributions in DSC from books, industrial reports, and academic literature. The authors identify advantages of DSC and key technologies such as Big Data, Cloud Computing and IoT. Finally, they propose a roadmap for implementing new technologies in the supply chain. The new framework integrates Supply Chain Management, Technology Implementation and Digitalization as main enablers to Digital Supply Chain.

Muñoz-Villamizar *et al.* (2019) focus the review on the relationship between sustainability and digital supply chains through a bibliometric analysis. The authors identify two main clusters: Supply Chain Management and Energy-related technologies. Moreover, the authors identify the timeline evolution of contributions and the most prolific year, i.e., 2017.

Makris *et al.* (2019) explore key trends technologies like Big Data, Cloud Computing, and 3D printing for Supply Chain 4.0. The authors consider the available literature, questionnaires, and interviews of different industries like pharmaceutical, retail, high tech, and food. This research concludes that Big Data is the most implemented tool while 3D printing is not commonly used by companies.

Considering Cloud Computing in DSC, Jede and Teuteberg (2015) develop a bibliometric analysis identifying keyword co-occurrence, disciplinary areas and regional importance. The analysis concludes that cost reduction and IT security are the main driven factors for implementing DSC.

In terms of Big Data Analytics (BDA) and DSC, Chehbi-Gamoura *et al.* (2019) conduct a comprehensive literature review. Authors recognize that future researches must focus on the procurement phase to explore full utilization of BDA and highlight the importance of cleaning data to avoid the extra cost. Finally, they conclude that the main reason for companies to implement BDA is the expected profits.

Finally, Ben-Daya *et al.* (2019) study the application of the Internet of Things in the supply chain and identify the importance of technologies like radio-frequency identification (RFID). Likewise, Sarac *et al.* (2010) explore the use of RFID in the supply chain evaluating the required costs and the return-on-investment.

3 Methodology

The literature review was gathered from Scopus Database and was done until December 2019. Although Industry 4.0 can be traced from 2011 (Kagermann *et al.* 2011), previous works show a high level of connectivity between logistic actors (Chow *et al.*, 2006; Ding *et al.* 2008). The search string structure was divided into two principal sections to get relevant information about connectivity in the supply chain (see Fig. 1).

The first part is focused on the title and abstract, looking for any combination between supply chain and digitalization, and using some synonymous. The second part considers the connection between DSC and connectivity, and the most related keywords, i.e., Cyber-Physical spaces, Smart Factories, 5G or RFID. Subject area limitations were also applied to refine results. As a result, 461 papers were obtained with a growing trend since 2012 as shown in Fig. 2. After a detailed analysis and filtering process, 40 papers were selected in this literature review. Finally, Fig. 3 classifies selected papers regarding the most representative categories in the literature.

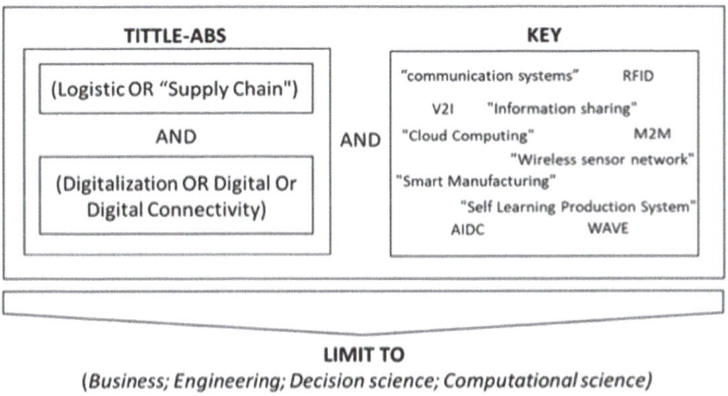

Fig. 1. Structured search string

Fig. 2. Number of publications per year

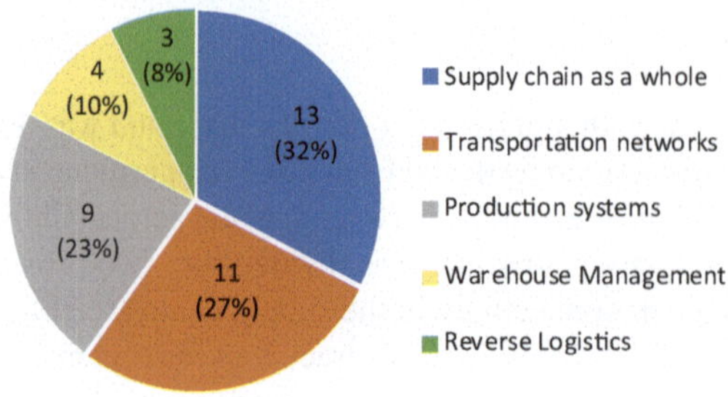

Fig. 3. Distribution of papers per category

4 Discussion

The keyword analysis depicted in Fig. 4 gives a preview of the results and shows the keywords clusters obtained. As we can see, there are two big groups of words clearly defined. This review is going to focus on the left group, where the red cluster seems to be the main cluster showing Industry 4.0 technologies like Cloud Computing, IoT, and Machine Learning. The full list of the reviewed articles and the corresponding IoT technologies considered to create connectivity networks is available at https://bit.ly/sohoma_LR_DSC. The following subsections will analyse the contributions considering the area of application in the supply chain, i.e., warehouse management, production systems, transportation networks, reverse logistics and supply chain management, and review the emerging trends in DSC.

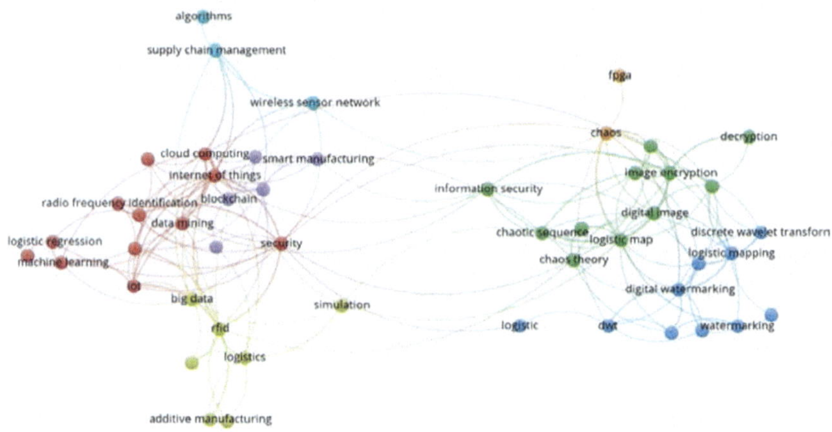

Fig. 4. Co-occurrence of author keywords

4.1 Warehousing Management

In warehousing management (WM), both storages of raw materials and final products are included. Considering WM, connectivity and spaces organization could improve the warehouse performance reducing picking times. For this reason, Ding *et al.* (2008) elaborated a complete structure of 4 layers, seeking a warehouse automation system that achieves connectivity using RFID reference tags, providing location information, and deploying strategic points to avoid data collision. The system is potentialized with the adoption of the Wi-Fi system, improving signal transmission and providing almost an exact position of required material. A similar study is made in a port side by Park *et al.* (2006), where container loading movements are analysed to deploy a Real-Time Location System complemented with RFID system to increase twin-lift operation efficiency. The proposed structure was modelled using the ARENA software where results show improvements in loading times.

In cases where warehouses have small areas and large number of goods, tag readers, data transfer and RFID became critical for warehouse management. For instance, Chow *et al.* (2006) implemented the RFID system based on a resource management system (RMS). The RMS is composed of two modules, i.e., the Front-end module responsible for data collection and Ultra-Wide Band for data transfer between readers and tags. The data processing is made by the Back-end module which receives all the information in the Resource Tracking Phase to transmit only useful information to Resource Management Engine and improve the decision-making process.

Considering large quantities of RFID tags, more than one reader antenna is necessary to capture all the information. Motivated by the reduction of infrastructure-related cost, Fuertes *et al.* (2018) focus on the efficiency of RFID, implementing an optimization model based on a Genetic Algorithm to identify the best position to place the antenna, considering warehouse dimension and layout.

4.2 Production Systems

Independent monitoring systems in production lines are constantly investigated. However, the connectivity between the production process and other logistics nodes is not very clear since most of the existing approaches consider them as two independent systems generating a lack of collaboration (Zhang *et al.* 2018). Therefore, Shrouf *et al.* (2014) study the impact of the Internet of Things in Smart factories, highlighting the Factory visibility as a key characteristic that must be achieved through a connection network, capable of identifying production problems and shutting down machines rapidly to avoid the production of defected goods.

The diverse benefits of sharing information are studied by Sahin and Robinson (2005) in a make-to-order network. Results show that when manufacturers share their planned replenishment schedules, the operational cost reaches a reduction of up to 2.33%. In the same line, a cost reduction of up to 30.69% is achieved when the manufacturer coordinates its scheduled orders with transportation decisions, achieving a 47.58% cost reduction when a complete information sharing system is implemented.

Considering decision making support, sensors are the most useful tool to gather information in different production environments; however, connectivity networks require data unification. Zhang *et al.* (2011) developed a data flow diagram in a steel-making factory, gathering real-time information from different technologies such as electronic tags and sensors. Information is conducted through wired and wireless networks, achieving a 99.8% of automated collecting rate. Thus, production managers can improve the operation level as well as product quality while reducing iron and steel consumption. Some authors deploy more sensors nodes seeking a continuous tracking system. Nevertheless, large data increases the complexity level of connectivity platforms that must deal with different information sources. For example, Wang *et al.* (2017) investigated the advantages and disadvantages of data-on-tag and data-on-network architectures to propose a hybrid-data-on-tag system, with a decentralized control node as part of a Smart machining shop floors (SMSF). The SMSF is composed of production resources, sensors, and machine information.

Wide connectivity grids are also studied by Zhong *et al.* (2015). The authors consider in a shop floor RFID tags on each batch of raw material, stationary readers on machines, and workers with handheld RFID readers. Thus, each actor generates and transmits information, becoming a Smart Manufacturing Object. Then, the resulted Big data is cleaned with an RFID-Cuboids for facilitating the data mining process and improving decision-making.

Cloud computing is also deployed among IoT technologies to improve the performance level and production management. Zhang *et al.* (2018) use a CPS to create a Smart Production Logistic System, where multi-source and real-time information of machine status and material handling can be perceived and stored in the cloud platform for future decision-making. Furthermore, the Analytical Target Cascading model is formulated to minimize the total weighted manufacturing cost, the total manufacturing time, and total energy consumption. Data is also analysed by Self-Learning Production Systems (SLPS). SLPS are huge connectivity platforms that usually employ a specific module to collect a vast number of data from different sources in a particular situation to identify the fittest solution for future similar contexts. Bittencourt *et al.* (2011) studied a detailed SLPS composed of many modules like the extractor, the adapter, and the evaluator. The authors conclude that continuous observation done by the extractor makes it possible to predict future machine utilization and reduce energy consumption. Based on the previous taxonomy, Candido *et al.* (2013) made a deeper research focusing on Adapter behaviour to improve predictive skills while improving idle-time performance and increasing energy savings.

4.3 Transportation Networks

Transport represents the most representative cost in logistics (Oliveira *et al.* 2015). Thus, many researchers try to integrate new technologies to get a higher degree of visibility and monitor goods inside and outside the shop floor. Yuvaraj and Sangeetha (2016) study a basic continuous monitoring system based on RFID for Indoor-tracking of goods and GPS for Outdoor-tracking of vehicles. An improved decision support system tracks real-time Cargo and vehicle location at the same time. Gnimpieba *et al.* (2015) present the idea of a vehicle with a General Packet Radio Service (GPRS) system embedded to monitor geographical location with RFID tags on each pallet to track and trace cargo. Shamsuzzoha and Helo (2011) explain some differences between track and trace and highlight the importance of real-time tracking systems. The authors suggested a friendly interface software to maximize delivery efficiency and improve customer service. Furthermore, and arguing that satellite-based positioning systems must fail for containers or freight vehicles, Hillbrand and Schoech (2007) propose a continuous track and trace network for packaged goods, based on a device-based positioning system implemented with cellular radio technology. Due to the lower frequency range of Global systems for mobile communication, the authors divide the covered geographic area into radio cells, seeking precision and coverage gains.

Among other works, Tjahjono *et al.* (2017) developed different platforms using existing technologies. Musa *et al.* (2014) proposed the integration of an intelligent RFID tag capable of measuring environmental factors, being monitored with GPS, communicated via Wi-Fi, and equipped with a high-speed microprocessor seeking to add valuable information about cargo status and position. Electronic Product Code (EPC) technology is also used by Zhengxia and Laisheng (2010) who developed an innovative EPC code into the RFID. The proposed tag can involve many static or single EPC into one dynamic EPC when goods are packaged together and returned to a single EPC when goods are unpacked.

Delivery efficiency could also be improved through IoT adoption. Papatheocharous and Gouvas (2011) proposed an e-tracer platform where transport vehicles are equipped with GPS technology and have a Mobile Logistics Station Network used to detect objects during loading. Moreover, warehouses, stations and hubs carry RFID readers and can detect all the RFID tags that enter and exit a facility. Oliveira *et al.* (2015) developed a similar network called SafeTrack capable of monitoring cargo movements and locating the truck position. In this platform, vehicles and stores are also equipped with RFID and a friendly interface allows managers to check wrong and correct deliveries, as well as detours of planned routes, optimizing deliveries, and reducing human mistakes.

Connectivity and visibility in transport logistics allow managers to gather valuable information that must be analysed in the decision-making process for economic benefits. Efforts to minimize distribution costs are focused on the development of routing algorithms, seeking near-optimal distribution plans (Zeimpekis *et al.* 2008). However, planned scenarios are affected by external variables like traffic jams or vehicle collisions, forcing managers to re-schedule fleet programming. Zeimpekis *et al.* (2008) studied three main types of accidents in transport logistics and proposed a structured system for dealing with vehicle delay, supported by a Decision Support Module that increased customer service by 25% . Similarly, Ngai *et al.* (2012), based on user requirements like context evaluation and real-time information, developed a Context-aware fleet management system that facilitates the managers' decision-making process. In the same way, Incident Response Logistics (IRL) aims to reduce the impact of unexpected situations on the supply chain performance. Thus, Zografos *et al.* (2002) present a real-time decision support system in charge of facing unplanned scenarios based on four mathematical models to minimize the incident response time, reduce the required response units and maximize covering location and service level.

4.4 Reverse Logistics

Reverse Logistics begins to receive deserved attention due to economic impact, mandatory legislation, and business outreach (Nunes *et al.* 2006). Thus, Xing *et al.* (2011) take advantage of IoT technologies to reduce the blind spots of the product recovery process, implementing an e-Reverse logistics network (e-RL) based on the connectivity of three main logistic partners: Collector, Remanufacture-to-Order companies, and the end-user. In the proposed approach, when products reach the end of its current life cycle, the end-user sends a disposal request to product collectors via his mobile device, using Near Field Communication (NFC) tags.

RFID is a constant tool used in the previous categories and Reverse logistics is not the exception. For example, Nunes *et al.* (2006) studied some RFID approaches to track and trace goods, and concluded that Product Embedded Information Devices associated with firmware and software components are very important for product lifecycle information in the decision-making, and for improving the recycling rate due to the higher quality of the input material and lower logistic costs. Besides, some others like Garrido-Hidalgo *et al.* (2019) consider product disposal after it was collected, using a RFID tag embedded to products in the production process in order to identify and classify potential re-usable parts or products to store them with a Smart Container connectivity system.

4.5 Supply Chain

Reviewed studies centralized their contributions in a specific task of the supply chain; however, this literature review would be incomplete if the connectivity networks for the entire supply chain are excluded. For example, Geerts and O'Leary (2014) states that supply chain management requires an in-depth understanding of its physical flow and implements the EAGLET ontology in order to provide a full information network inside the logistics process, as well as creating a Highly Visible Supply Chain (HVSC). The EAGLET method set ontological primitives (Event, Agent, Location, Equipment, and Thing) to study interactions between them and facilitate monitoring schemes. Relations between Supply Chain nodes are also studied by Kim *et al.* (2008) using a Fuzzy Cognitive map to build a complex supply chain network composed by the supplier, manufacturer, and retailer. This research uses the state vectors obtained from the RFID system, to make a bi-directional what-if and cause-effect analysis, seeking to identify the impact of nodes behaviour in the related actors.

Advanced technologies and their successful integration can accrue potential benefits concerning strategic SC partnership management (Kim and Shimn 2019). For that reason, Susilo and Triana (2018) explored the effect of Cloud Computing (CC) and Data encryption into Supply Chain management to state that introducing blockchain technology into the logistics process could improve tasks like data verification, artificial Intelligence or Machine Learning. Other blockchain benefits like Information Transparency and Information Immutability are deeply studied by Kim and Shimm (2019), regarding its relationship with partnership efficiency and partnership growth, and consequently their impact on financial and operational performance. In the same way, Rozman *et al.* (2019) proposed a Smart Contract platform based on Blockchain technologies with a friendly Application Programming Interface (API). Users access this platform in order to communicate or cooperate with other nodes, while providers publish their offers. All the information is saved in a genesis node and a rating node developed to exclude bad actors from the system.

4.6 Emerging Trends in DSC

The supply chain has experienced unprecedent vulnerabilities in lead times, order quantities, disruptions in network structures, and severe demand fluctuations due to the COVID-19 pandemic (Ivanov and Dolgui 2020). Considering these disruptions, connectivity became a critical service to clients. Thus, the implementation of technologies such as IoT, Cloud Computing, 5G, Artificial Intelligence, blockchain technologies, edge computing and analytics, cyber-security systems, 3D printing, and robotics need to be accelerated for improving visibility, agility, automation, synchronization, collaboration, and optimization in the digital supply chain management and the logistic processes.

5 Conclusions

This article presented a comprehensive literature review for better understanding the connectivity technologies throughout the principal Supply Chain operations. The literature was classified and analysed to provide a clear view of the connectivity benefits. After the analysis, we can conclude that there is a clear relation between connectivity and visibility in the Supply Chain. Data cleaning and information analysis are the most important phases to transform raw data into knowledge for improving the effectiveness and efficiency of the decision-making process. Also, results show that Radio Frequency Identification tags are the most used tool to gather information. However, this system is usually complemented with other technologies to construct a stronger monitoring system. Differently, Blockchains and Cloud Computing technologies show a huge potential for future works.

References

Ben-Daya, M., Hassini, E., Bahroun, Z.: Internet of Things and supply chain management: a literature review. Int. J. Prod. Res. **57**, 4719–4742 (2019)

Bittencourt, J.L., Bonefeld, R., Scholze, S., Stokic, D., Uddin, M.K., Lastra, J.L.M.: Energy efficiency improvement through context sensitive self-learning of machine availability. In: IEEE International Conference on Industrial Informatics (INDIN), pp. 93–98 (2011)

Büyüközkan, G., Göçer, F.: Digital supply chain: literature review and a proposed framework for future research. Comput. Ind. **97**, 157–177 (2018)

Candido, G., Di Orio, G., Barata, J., Bittencourt, J.L., Bonefeld, R.: Self-learning production systems (SLPS)-energy management application for machine tools. In: IEEE International Symposium on Industrial Electronics, pp. 1–8 (2013)

Chehbi-Gamoura, S., Derrouiche, R., Damand, D., Barth, M.: Insights from big data analytics in supply chain management: an all-inclusive literature review using the SCOR model. Prod. Plan. Control **31**, 355–382 (2019)

Chow, H.K.H., Choy, K.L., Lee, W.B., Lau, K.C.: Design of a RFID case-based resource management system for warehouse operations. Expert Syst. Appl. **30**(4), 561–576 (2006)

Ding, B., Yuan, H., Chen, D., Chen, L.: Application of RTLS in warehouse management based on RFID and Wi-Fi. In: 2008 International Conference on Wireless Communications, Networking and Mobile Computing, pp. 1–5 (2008)

Deloitte: COVID-19: managing supply chain risk and disruption (2020). https://www2.deloitte.com/content/dam/Deloitte/ca/Documents/finance/Supply-Chain_POV_EN_FINAL-AODA.pdf
Doyle, F., Duarte, M., Cosgrove, J.: Design of an embedded sensor network for application in energy monitoring of commercial and industrial facilities. Energy Procedia **83**, 504–514 (2015)
Fuertes, G., Alfaro, M., Soto, I., Carrasco, R., Iturralde, D., Lagos, C.: Optimization model for location of RFID antennas in a supply chain. In: 7th International Conference on Computers Communications and Control (ICCCC), pp. 203–209 (2018)
Garrido-Hidalgo, C., Olivares, T., Ramírez, F.J., Roda-Sánchez, L.: An end-to-end Internet of Things solution for reverse supply chain management in Industry 4.0. Comput. Ind. **112**, 103–127 (2019)
Geerts, G.L., O'Leary, D.E.: A supply chain of things: the EAGLET ontology for highly visible supply chains. Decis. Support Syst. **63**, 3–22 (2014)
Gnimpieba, D., Nait-Sidi-Moh, A., Durand, D., Fortin, J.: Using Internet of Things technologies for a collaborative supply chain: application to tracking of pallets and containers. Procedia Comput. Sci. **56**, 550–557 (2015)
Hillbrand, C., Schoech, R.: Shipment localization kit: an automated approach for tracking and tracing general cargo. In: International Conference on the Management of Mobile Business (ICMB 2007), p. 46 (2007)
Ivanov, D., Dolgui, A.: OR-methods for coping with the ripple effect in supply chains during COVID-19 pandemic: managerial insights and research implications. Int. J. Prod. Econ. **232**, 107921 (2020). https://doi.org/10.1016/j.ijpe.2020.107921
Kayikci, Y.: Sustainability impact of digitization in logistics. Procedia Manuf. **21**, 782–789 (2018)
Kagermann, H., Lukas, W., Wahlster, W.: Industrie 4.0: Mit dem Internet der Dinge auf dem Weg zur 4. industriellen Revolution, VDI Nachrichten (2011)
Kim, M.C., Kim, C.O., Hong, S.R., Kwon, I.H.: Forward-backward analysis of RFID-enabled supply chain using fuzzy cognitive map and genetic algorithm. Expert Syst. Appl. **35**(3), 1166–1176 (2008)
Kim, J.S., Shin, N.: The impact of blockchain technology application on supply chain partnership and performance. Sustainability **11**, 61–68 (2019)
Makris, D., Hansen, Z.N.L., Khan, O.: Adapting to supply chain 4.0: an explorative study of multinational companies. Supply Chain Forum **20**(2), 116–131 (2019)
Muñoz-Villamizar, A., Solano, E., Quintero-Araujo, C., Santos, J.: Sustainability and digitalization in supply chains: a bibliometric analysis. Uncertain Supply Chain Manag. **7**(4), 703–712 (2019)
Musa, A., Gunasekaran, A., Yusuf, Y., Abdelazim, A.: Embedded devices for supply chain applications: towards hardware integration of disparate technologies. Expert Syst. Appl. **41**(1), 137–155 (2014)
Ngai, E.W.T., Leung, T.K.P., Wong, Y.H., Lee, M.C.M., Chai, P.Y.F., Choi, Y.S.: Design and development of a context-aware decision support system for real-time accident handling in logistics. Decis. Support Syst. **52**(4), 816–827 (2012)
Nunes, K.R.A., Schnatmeyer, M., Thoben, K.D., Valle, R.A.B.: Using RFID for waste minimization in the automotive industry. IFAC Proc. Vol. **12**, 221–226 (2006)
Oliveira, R.R., Cardoso, I.M.G., Barbosa, J.L.V., Da Costa, C.A., Prado, M.P.: An intelligent model for logistics management based on geofencing algorithms and RFID technology. Expert Syst. Appl. **42**(15–16), 6082–6097 (2015)
Papatheocharous, E., Gouvas, P.: ETracer: an innovative near-real time track-and-trace platform. In: 2011 Panhellenic Conference on Informatics, pp. 282–286 (2011)
Park, D.-J., Choi, Y.-B., Nam, K.-C.: RFID-based RTLS for improvement of operation system in container terminals. In: 2006 Asia-Pacific Conference on Communications, pp. 1–5 (2006)
Rozman, N., Vrabič, R., Corn, M., Požrl, T., Diaci, J.: Distributed logistics platform based on blockchain and IoT. Procedia CIRP **81**, 826–831 (2019)

Sahin, F., Robinson, E.P.: Information sharing and coordination in make-to-order supply chains. J. Oper. Manag. **23**(6), 579–598 (2005)

Sarac, A., Absi, N., Dauzere-Pérès, S.: A literature review on the impact of RFID technologies on supply chain management. Int. J. Prod. Econ. **128**, 77–95 (2010)

Strategy&: Industry 4.0: how digitization makes the supply chain more efficient, agile, and customer-focused (2016). https://www.strategyand.pwc.com/gx/en/insights/2016/industry-4-digitization/industry40.pdf

Shamsuzzoha, A., Helo, P.T.: Real-time tracking and tracing system: potentials for the logistics network. In: 2011 International Conference on Industrial Engineering and Operations Management, pp. 22–24 (2011)

Shrouf, F., Ordieres, J., Miragliotta, G.: Smart factories in Industry 4.0: a review of the concept and of energy management approached in production based on the Internet of Things paradigm. In: IEEE International Conference on Industrial Engineering and Engineering Management, pp. 697–701 (2014)

Susilo, F.A., Triana, Y.S.: Digital supply chain development in blockchain technology using Rijndael algorithm 256. IOP Conf. Ser.: Mater. Sci. Eng. **453**, 1–11 (2018)

Jede, A., Teuteberg, F.: Integrating cloud computing in supply chain processes: a comprehensive literature review. J. Enterp. Inf. Manag. **28**(6), 872–904 (2015)

Tjahjono, B., Esplugues, C., Ares, E., Pelaez, G.: What does Industry 4.0 mean to supply chain? Procedia Manuf. **13**, 1175–1182 (2017)

Wang, C., Jiang, P., Ding, K.: A hybrid-data-on-tag-enabled decentralized control system for flexible smart workpiece manufacturing shop floors. Proc. Inst. Mech. Eng. Part C: J. Mech. Eng. Sci. **231**(4), 764–782 (2017)

World Economic Forum: Data-driven logistics businesses can optimize operations, reduce emissions and cut costs (2016). http://reports.weforum.org/digital-transformation/setting-data-at-the-heart-of-logistics/

Xing, B., Gao, W., Battle, K., Nelwamondo, F., Marwala, T.: e-RL: the Internet of Things supported reverse logistics for remanufacture-to-order. In: Fifth International Conference on Advanced Engineering Computing and Applications in Sciences, pp. 84–87 (2011)

Yuvaraj, S., Sangeetha, M.: EESCA: energy efficient structured clustering algorithm for wireless sensor networks. In: 2016 IEEE International Conference on Computing, Analytics and Security Trends, pp. 523–527 (2016)

Zeimpekis, V., Giaglis, G.M., Minis, I.: Development and evaluation of an intelligent fleet management system for city logistics. In: Proceedings of the 41st Annual Hawaii International Conference on System Sciences, p. 72 (2008)

Zhang, G., Zhao, G., Vi, Y., Wang, Z.: Research and application of steelmaking workplace crane logistics tracking system. In: IEEE Proceedings of International Conference on Electronics and Optoelectronics, pp. 29–31 (2011)

Zhang, Y., Guo, Z., Lv, J., Liu, Y.: A framework for smart production-logistics systems based on CPS and industrial IoT. IEEE Trans. Industr. Inf. **14**(9), 4019–4032 (2018)

Zhengxia, W., Laisheng, X.: Modern logistics monitoring platform based on the Internet of Things. In: 2010 International Conference on Intelligent Computation Technology and Automation, vol. 2, pp. 726–731 (2010)

Zhong, R.Y., Huang, G.Q., Lan, S., Dai, Q.Y., Chen, X., Zhang, T.: A big data approach for logistics trajectory discovery from RFID-enabled production data. Int. J. Prod. Econ. **165**, 260–272 (2015)

Zografos, K.G., Androutsopoulos, K.N., Vasilakis, G.M.: A real-time decision support system for roadway network incident response logistics. Transp. Res. Part C: Emerg. Technol. **10**(1), 1–18 (2002)

Cyber-Physical Systems in Logistics and Supply Chain

Erika Suárez-Riveros[1(✉)], Álvaro Mejia-Mantilla[1], Sonia Jaimes-Suárez[1], and Jose-Fernando Jimenez[2]

[1] Industrial Engineering Department, Escuela Colombiana de Ingenieria "Julio Garavito", Bogotá, Colombia
{erika.suarez-r,alvaro.mejia}@mail.escuelaing.edu.co, sonia.jaimes@escuelaing.edu.co

[2] Industrial Engineering Department, Pontificia Universidad Javeriana, Bogotá, Colombia
j-jimenez@javeriana.edu.co

Abstract. A Cyber-Physical System (CPS), as one of the most significant advances of information and communication technologies, is an interactive arrangement of digital and software elements engineered together for managing and controlling the functioning of physical elements. Even if this concept has been rapidly developed for manufacturing systems, there has been an increasing interest in recent years for developing and implementing it in a number of other domains: logistics, health, transport or education. Extensive research has shown that a straightforward application of CPS is in logistics or supply chain systems. However, research to date has not yet determined entirely the unambiguous characteristics of this technology within the logistic/supply chain domain. This paper provides an overview of cyber-physical systems applied to the logistics/supply domain. The approach of this paper is developed in three phases. First, it analyses and proposes a definition of cyber-physical logistics systems. Second, it explores the current literature approaches of cyber-physical logistics systems organized by the CPS maturity model. Third, it highlights the challenges and perspectives of cyber-physical logistics systems towards an industrial implementation.

Keywords: Cyber-Physical Systems · Logistic · Supply chain · Distributed systems

1 Introduction

A logistic system is set of products, raw materials, warehouses, organizations, people, activities, capital and information that are structurally and functionally organized to source, produce and deliver finished goods and/or services to customers within a global market. Nowadays, the increasing demands for customized products and services as well as frequent market fluctuations have posed challenges to the management and control of logistics systems (Y. Zhang *et al.* 2018). Traditional logistics managements systems lack features that respond efficiently to these challenges, causing operational inefficiency toward the logistics objectives. However, technological advances from the

D. Trentesaux et al. (Eds.): SOHOMA 2021, SCI 987, pp. 260–271, 2021.
https://doi.org/10.1007/978-3-030-80906-5_18

digital transformation age have changed the characteristics and dynamics of logistics systems (Cichosz *et al.* 2020). Cloud computing, Data-driven analytics, Autonomous Transportation Vehicles, Horizonal and Vertical integration and Internet of Things are some of the technological drivers that potentiate the efficiency and competitiveness of industries supply chain (Maqueira *et al.* 2019). These technologies have contributed to the improvement of the supply chain (SC) process achieving an adequate customer orientation process, reaching operational efficiencies and featuring adaptability characteristics in an increasing competitive global environment (Hübner *et al.* 2013). However, in addition to the implementation and orchestration of these drivers, the management and control of logistics systems are key aspects to achieve the expected objectives, especially for autonomous or semi-autonomous logistics systems. Therefore, considering the challenges of global competitive environment, it is critical to have a smart control system that monitors, manages and pilots cooperatively each of the components of a logistic system in order to obtain efficiency, adaptability, scalability, resiliency, safety and security needed for achieving the logistics operational objectives.

Due to the digital transformation revolution, the last two decades have seen a growing trend towards the implementation of technological advances in the manufacturing domain. The fourth industrial revolution, as the digital revolution is called in the manufacturing domain, has brought several technological advances towards the productivity and adaptability in production process such as collaborative robots, smart factories and 3D printing. Considering the stated challenges in logistics systems, one of the main technological drivers that orchestrates the technological drivers and serves as a control system for the manufacturing system may be extended to the management and control of supply chains. This driver, called cyber-physical system, is composed of collaborating computational entities which are in intensive connection with the surrounding physical world and the on-going processes, providing and using at the same time data-accessing and data-processing services available on the Internet (Monostori *et al.* 2016). In other terms, a CPS is a collection of communicating digital devices and algorithms that interact with the physical world via sensors and actuators in a feedback loop in order to a achieve certain objectives of the physical process. Certainly, the manufacturing domain had deployed vastly the use of CPS, known as Cyber-Physical Production Systems (CPPS), due to the need of orchestrating efficiently the products, resources, conveyors and operators towards the shop-floor productivity (Lee 2008). However, although some research has been carried out on CPS technology applied to logistics, there has been little development and discussion about the adoption of CPS in logistics and the potential benefits that this technological driver could bring. This indicates a need to understand the characteristics of the CPS in Logistics and the particularities that exist for the implementation of this technological driver in the Logistics field.

In this paper, an exploratory review of the literature concerning Cyber-Physical Systems for Logistics (CPLS) it is conducted. This paper is a milestone of a wider research aiming the construction of experimental platform of a CPS in logistics for assessing the cognitive characteristics of a human decision-making operator within a supply chain system. The objective of this research is the evaluation of a cognitive model for decision-making of a human supervisor enclosed in a human-CPS interaction framework for urban logistics. Therefore, this paper is organized as follows. Section 2 analyses a set of contributions that defines CPS in logistics and proposes a single definition to establish the scope of this term within the CPS research field. Then, Sect. 3 reviews a set of literature of CPS in logistics, organizing the contributions according to the CPS maturity model proposed by the RWTH Aachen University laboratory of Machine tools and production engineering, presented by Monostori *et al.* (2016). The challenges of the development and implementation and the research perspectives of CPS in logistics are presented in Sect. 4. Finally, Sect. 5 rounds up the paper with the conclusions and defines the next steps of this research.

2 Cyber-Physical Logistics Systems (CPLS)

In recent years, the logistics field has been undergoing great changes due to advances in technological innovation. Cyber Physical Systems, IoT, Big Data, Cloud Computing, Mining Data, Machine-to-Machine (M2M), among others, have led logistics to advance in big steps, modernizing its processes and making it increasingly flexible, autonomous and intelligent. The need to have information in real time and make timely decisions, make Cyber-Physical Systems a key element that adds value to logistics processes. These systems, based on a strong interconnection between objects (real world) and actors (virtual world) through ICT, aim to achieve stability, performance, reliability, and efficiency in managing physical systems for many application domains (Chen 2017).

Historically, manufacturing and industrial production in general have been the most investigated topics for Cyber-Physical Systems, their ease of relating to the environment and expanding the physical world through control technology and communication being important for different industries (Baheti and Gill 2011).

A CPLS is the union of human resources, equipment and elements establishing one or more interaction interfaces formulated cyber physically, where the physical and cybernetic worlds connect. The CPLS is used to monitor and control logistics operations, and to take advantage of the knowledge generated by human resources and equipment during the logistics process, together with the knowledge generated in different phases of the related supply chain. This internal knowledge is used at different time scales to continually improve operations depending on logistics activities in the chain, generating information on transport, storage and distribution. CPLS's results will also generate value for external systems that in turn provide feedback to CPLS and help improve it.

A case study that exemplifies the operation of a CPLS was proposed by Kong *et al.* for two SMEs in the city of Honk Kong in China. This paper proposes a cloud platform designed to solve problems in e-commerce logistics chains, capable of obtaining information from a physical world and generating useful information for decision-making in real time. As a result of these case studies, increases in productivity, greater efficiencies, greater speed, precision, and scalability are applied (2020).

3 CPS Maturity Model for Cyber-Physical Logistics Systems

A model of maturity is a valuable technique used to measure a process or an organization. This represents the path that must be followed towards a more organized and systematic way of carrying out an action or running a business. Normally, this consists of five levels, with level 1 being the most basic and level 5 the most advanced and can vary from one model to another depending on the domain and the concerns that motivate it (Proença and Borbinha 2016). According to Sen *et al.*, (2012) maturity is a concept that progresses from an initial state to a final state or a more advanced one.

The model of CPS depicted in Fig. 1 was proposed by Laboratory for Machine Tools and Production Engineering of University RWTH Aachen. This model was divided into five levels: 1) general conditions indicating setting basics, 2) information generation to create transparency, 3) information processing for increasing understanding, 4) information linking indicating improving decision making, and 5) interacting cyber-physical systems for self-optimizing (Yan *et al.* 2019).

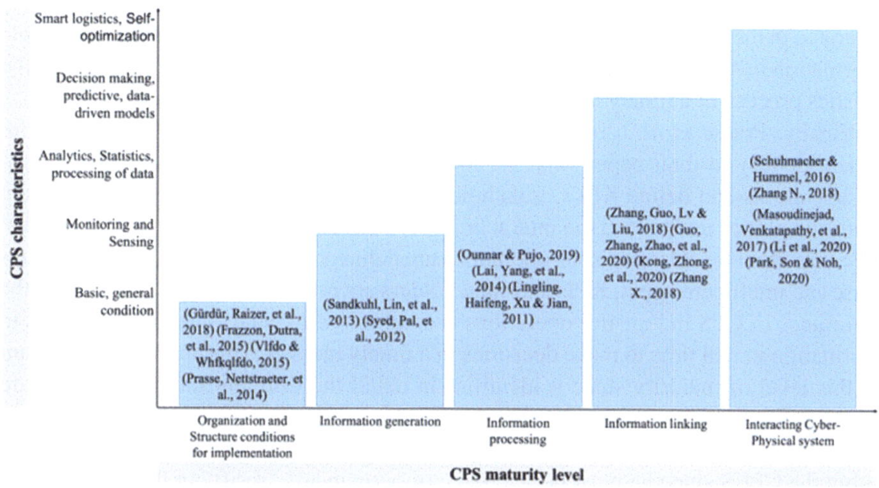

Fig. 1. CPS maturity model

"Setting basics" refers to the creation of organizational and structural conditions for the implementation of the CPLS. "Creating transparency" reflects the need for real-time data availability for all related CPLS activities. "Increasing understanding" indicates the existing aggregation instruments in order to deduce new knowledge. "Improving decision making" refers to the collaboration-based adaptation of CPLS process and "Self-optimizing or interacting Cyber Physical Logistic System" indicates the independent problem-solving capabilities of collaborative CPLS (Li *et al.* 2020).

3.1 Organizational and Structural Conditions for Implementation

The organizational and structural conditions for the implementation of the cyber-physical logistic system correspond to the first level of maturity according to the model created by the Machine Tools and Production Engineering Laboratory of University RWTH Aachen. This maturity level refers to the general conditions that are required to implement a cyber physical logistics system (CPLS).

An example of this level of maturity is set by Vlfdo *et al.* (2015) who identify in their paper which are the trends and emerging IT infrastructure within manufacturing, logistics and the supply chain. Among the mentioned technologies, there is Iot (Internet of Things), BD (Bid Data Analytics), IoS (Internet of Services), DSN (Digital Social Networks), CPS (Cyber Physical Systems) and their enabling technologies. Additionally, some fields of potential interest that may make use of these technologies in the future are mentioned. This information is relevant as it shows the use of these technologies and how they can contribute to logistics operations.

Another example is raised by Frazzon *et al.* (2015) who discuss the applicability of Big Data in CPLS and propose a conceptual model for the acquisition, application and analysis of Big Data as a contribution to decision-making. In his document, the authors offer an information analysis model that it called BogLogDatix Module. This model allows organizing large volumes of data (unstructured, useful, reliable, standardized) generated in the CPLS to later process them, generating structured and operationalizable information that allows decision makers to use much of the information produced in the logistics process in a timely manner.

Finally, Prasse *et al.* (2014) also carried out research at this level. Gürdür *et al.* (2018) identify in their paper what are the necessary requirements to carry out logistics operations and define KPI's, stakeholders and different data visualizations to help interested parties understand the interactions between entities easier and faster. For this, three important levels in logistics operations are defined: supply chain, automated warehouse and intelligent agent; KPIs and stakeholders are established for each level, and the importance of CPS in logistics operations is emphasised in what concerns the access to information in real time to make decisions in a timely manner. This approach is relevant for this level of maturity since it identifies in detail the different interactions that are generated in the logistics process and allows one to imagine how the physical Cyber System would be structured. Additionally, it considers the human as an important actor within the CPLS since he must be in charge of permanently verifying the KPIs in such a way that they support their decision-making.

3.2 Information Generation: Monitoring and Sensing

A system that generates information aims to understand and analyze, for example the impact of implementing new technologies in the processes of a company at administrative level. Its main component is information, since through monitoring and detection it guides decision-making. It also reflects in the performance indicators the operation of a system of any type. In a CPLS it is important to establish the means by which the information is obtained, whether this is in real time and the means of data output and visualization.

A CPLS is based on good communication, computing and control to obtain an efficient use of logistics infrastructure resources. Bearing in mind the above, Sandkuhl et al. (2013) presented a structure called logistics-as-a-Service (LaaS) that aims at representing elements of logistics networks as services. In this case, the system shares the information collected to facilitate decision-making about transportation, considering routes and user preferences. Although it focuses on representing the physical elements of the system, the work also considers the management of logistics competencies and helps in the search management of the available resources to make appropriate use according to the needs. This research proposes using ontology-based profiles to represent individual and organizational competencies that contribute to finding and managing resources in LaaS. It also include the human as an active resource of the LaaS network, involved in efficiently performing the established tasks.

Another example of this level of maturity is the work developed by Syed *et al.* (2012), where they used an accelerometer attached to a moving vehicle as a means of transferring information into the system. The data collected is in relation to the vertical vibration of the vehicle and by receiving a higher frequency already determined when driving on a bad road, a more detailed analysis of the received signal can be made. Models were developed to simulate different types of cars and road surfaces. To experiment with the model, the authors used a small car driven at a constant speed, and a smartphone that received the sensor signal; The results of the experiment were compared with those obtained in the simulation. Finally, they determined that it is feasible to use the functions created to detect irregularities on a road.

3.3 Information Processing: Analytics, Statistics and Data Processing

This level of maturity refers to the system's capacity to process data collected from the physical environment through sensors, with the purpose of influencing the state variables of the system in a desired way. This function allows a system to be more intelligent and flexible relative to the changing conditions of the environment in which it operates (Westermann *et al.* 2016).

An example related to this maturity level is presented in the paper by Huang *et al.* (2014) who propose an intelligent transport CPS based on mobile agents. In his document, the authors propose an architecture organized on three layers. In the first layer - called physical layer - it is possible to capture information from the physical environment where cars operate, not simply limiting to the vehicle but accessing more signals such as elements or traffic, e.g., sensor lights that ensure a more robust and intelligent transport system. In the second layer - called network layer - a means of transmitting the received information received is defined through a network architecture based on mobile agents, where each mobile object shares its information and can receive information from other objects instrumented with equipment of similar networks. In the third layer - application layer - an interface allows transmitting information and interacting with people in a comfortable and pleasant way. The work developed by Ounnar and Pujo (2019) is also classified at this level.

Another example is the vehicle monitoring system based on the IoT framework developed by Lingling *et al.* (2011) in which they developed an intelligent management of vehicles through control in a virtual information space, providing information feedback about the vehicle. driver and traffic in real time. The so-called Intelligent Vehicle Monitoring System (IVMS) is composed of five layers: the object layer, the detection layer, the network layer, the data layer and the application layer. The Vehicle Information System is designed to manage vehicle information and share it with other interested parties in the data. The authors implemented IVMS in Nanjing and observing the demand for network control and counting on an integration with public security work, they extended the work to smarter monitoring bases throughout the city.

3.4 Information Linking: Decision Making, Predictive, Reactive, Operational Research, and Data-Driven Models

CPS are more than complex and large-scale systems; they can also be decentralized and distributed, with components in physical world and in the network, heterogeneous and semi-autonomous (Villalonga Jaen *et al.* 2020); in ranking by maturity level, these systems are referred to as information links. They can process information and run models to predict patterns, be reactive, and make decisions. Although they can do all these functions on their own, they require some sort of initial indication as a specific task that they are assigned to perform. A simple example refers to cars that do not require a human driver, but only the initial indication of the place to be reached and this makes autonomous decisions about the route, the navigation objectives and the space to park (Thekkilakattil Dodig-Crnkovic 2015).

A data-driven model based on analytical goals has been developed to implement the self-organizing setup with case study at a Chinese engine manufacturer. The model poses challenges in excessively long lead time and energy waste in the integration in the Industrial Internet of Things (IIoT) of production and logistics. The results show that with a reasonable calculation time on the part of the model, manufacturing time and energy consumption are reduced; as a conclusion, the use of this type of applications could be viable for intelligent modelling of manufacturing resources, and intelligent production logistics systems with self-organized configuration. This system is composed of an intelligent modelling layer where IIoT and sensors are integrated with key resources they monitor in real time. In the smart production-logistics systems layer, a knowledge base is built with the data collected to build a smart decision-making system for machines and the material handling. Finally, in the self-organizing configuration layer, when production orders or exceptions are entered, self-organization is activated: production and logistics tasks are decomposed to be processed by the model; logistics is put into operation when a machine places a task in the cloud; tasks are assigned in the chain between logistics and production in a bidirectional way exploiting the cooperation between machines, materials and humans (Zhang 2018).

Also focusing on logistics in production, the work developed in 2020 by two of the authors of the afore mentioned document brings new contributions. The CPS has sensor-based layers for the reception of information as the previous work. The difference consists in a self-adaptive collaborative control (SCC) mode that improves the flexibility and resilience of the system, leading to collaborative optimization problems that are solved using the analytical objective cascade (ATC) method. The nodal SCC, local SCC and global SCC levels of collaborative control are introduced. The results show that the proposed SCC mode outperforms the non-SCC mode in reducing standby time, duration and power consumption (Guo *et al.* 2020).

In implementation cases, the logistics CPS for e-commerce studied in the city of Hong Kong in China is of importance. This study proposes a cloud platform enabled for a multilayer CPS to achieve virtualization of assets and their control, process execution, reconfiguration and synchronization of simultaneous processes in small and medium-sized companies. At the most global layer, the means are sought to improve the utilization rate of space and resources, while reducing waiting and waste. This platform and the different scenarios where it was tested, showed that it can carry out the modularization of the application to obtain an improvement in productivity, which is important in the logistics of electronic commerce. In this research the authors show that human capacities cannot be easily substituted; they involve the human in the system and try to potentiate its reach through technological tools such as drones; the iCoordinator software for intelligent operations is developed to handle service requests, to collect and distribute notifications in real time on top of the cloud layer (Kong *et al.* 2020).

3.5 Interacting Cyber-Physical System (Smart Logistics)

A top-level CPS is considered to perform a self-optimization process, where the autonomous and continuous system configuration adapts to the profile of the situation it faces in the environment. This type of intelligence eliminates the need for human intervention during the process. CPS interacting are smart solutions that are considered important to meet the needs of the industry of the future (Masoudinejad *et al.* 2017).

Schuhmacher and Hummel (2016) demonstrated that decentralized control systems are able to respond to the need for logistics management in production in an increasingly fast-paced and changing world, in addition to the high demand for individualized products and services, as they allow autonomous and local decision-making. Using an ESB Logistics Learning factory (LLF) platform, Reutlingen University educates and trains students and professionals in the specific competencies of the industry, making this platform an important research tool since it includes a production system, a framework for decentralized control systems, a decentralized control of intralogistics processes, as well as a collaborative and decentralized trailer train transportation system.

Continuing with the development of smart platforms, Zhang's proposal for an intelligent route decision solution is based on the logistics route decision model of CPS, Internet of Things and data storage technology of the cloud platform, which are interconnected with a data processing layer that uses several algorithms: ant colony, simulated annealing and the genetic algorithm in optimizing the logistics route (X. Zhang, 2018).

Because a supply chain that works with manufacturing to order is characterized by redundant inventory and operational capacity, a smart system is needed for the strength of the operations. To solve this problem, the CPS must be engaged; therefore, Park *et al.* (2020), proposed a CPLS that cooperates with the CPPS in a multilevel CPS structure. This multi-level design was intended to provide the necessary control functions for a supply chain with the above mentioned characteristics. Finally, the authors of this research deduced that the application of the proposed CPLS and the multilevel CPS allows the control of the supply chain with better resilience, compared to operating only the CPS of independent applications, alleviating the whip effects.

As well as in the research of Li *et al.* (2020), none of the works reviewed at this level of maturity suppress the human from interacting with the CPLS. The information collected, processed and displayed is organised to better serve people.

4 CPLS: Challenges and Research Perspectives

With the investigations of the CPS in different areas challenges have arisen. Concerning those that include logistics we can mention the difficulty to integrate physical resources with computerized ones and achieve improvement: in the capacity of the intelligence of the system, in the dynamic representation of the physical elements of the system at the control point, and in the different degrees of exceptions that must be dealt with in order to strengthen the resilience of the system (Guo, Zhang *et al.* 2020). Other important issues are the security, privacy, regulations and data exchange handled by the system, considering the dissemination and exchange of data through the network (Sowe *et al.* 2019).

Another important challenge that arises from the implementation and commissioning of a CPLS is related to the variety and management of large volumes of data generated by the system. The processing and analysis of the information generated by the components of the system is of vital importance in decision-making. Researchers such as Frazzon *et al.* (2015) have stated that the challenge consists not only in having high-tech elements that allow storing a large amounts of information, but also in the adaptation of processes to the change of management and data analysis.

From a CPS perspective, the challenge of integrating humans and intelligent robots is very important to enable all CPLS actors to achieve better cooperation, collaboration and organization to overcome complex tasks (Lai *et al.* 2014). The success of the interaction between the CPLS and the Human will consist in allowing to define the means, requirements, methods and tools to be carried out and guide the intervention of humans in the system in such a way that the potential of connecting the physical world, the sensory perceptions and dexterity of the operator, the detection and monitoring capabilities of the devices with digital technology is exploited (Fantini *et al.* 2016).

For companies, one of the biggest challenges is related to determining if the cost-benefit of implementing a cyber-physical logistics system is favourable and if it will bring important revenues for the business; making logistics operations more efficient and automated through CPLS needs large capital investments.

5 Conclusions

From the literature review on CPS looking for research works that focus on logistics and were classified in five levels of maturity: organizational and structural conditions for implementation, information generation, information processing information, linking and interacting Cyber-Physical System it can be concluded first that most of the research is focused on production and manufacturing, and fewer papers address other areas of CPS application. Because of this open space for research, there are few approaches to a definition of CPLS; in addition several of the documents included in this analysis are related to logistics within production. Secondly, there is research at different levels of maturity for CPLS, and it is important to note that in recent years research has been targeting maturity levels 4 and 5. At this last maturity level called Interacting Cyber-Physical Systems, it can be seen that the systems that belong to it are developed for teaching and research.

Finally, a review was made on the challenges that CPLS bring due to recent advances in technology and how they can be potentiated if the enabling technologies of the fourth industrial revolution are used appropriately. The implementation of technologies that store information in the cloud using the internet deserves special attention concerning the integrity, privacy and security of the information given the risk of information leakage when being exposed to cyber-attacks or information theft. This study ends by showing the great challenge of achieving a harmonious connection between human and CPLS to fulfil the organization's objectives.

References

Baheti, R., Gill, H.: Cyber-physical systems. Impact Control Technol. **12**(1), 161–166 (2011)

Chen, H.: Applications of cyber-physical system: a literature review. J. Ind. Integr. Manage. **2**(03), 1750012 (2017)

Cichosz, M., Wallenburg, C.M., Michael Knemeyer, A.: Digital transformation at logistics service providers: barriers, success factors and leading practices. Int.J. Logistics Manage. **31**(2), 209–238 (2020)

Fantini, P., et al.: Exploring the integration of the human as a flexibility factor in CPS enabled manufacturing environments: Methodology and results. In: IECON 2016 - 42nd IEEE Conference on Industrial Electronics Society, pp. 5711–5716 (2016), https://doi.org/10.1109/IECON.2016.7793579

Frazzon, E.M., Dutra, M.L., Vianna, W.B.: Big data applied to cyber-physical logistic: systems conceptual model and perspectives. Braz. J. Oper. Prod. Manag. **12**(2), 330 (2015)

Guo, Z., Zhang, Y., Zhao, X., Song, X.: CPS-based self-adaptive collaborative control for smart production-logistics systems. IEEE Trans. Cybern. **51**(1), 188–198 (2021). https://doi.org/10.1109/TCYB.2020.2964301

Gürdür, D., Raizer, K., El-Khoury, J.: Data visualization support for complex logistics operations and cyber-physical systems. In: Proceedings of 13th International Joint Conference on Computer Vision, Imaging and Computer Graphics Theory and Applications, 200–211(2018)

Huang, W., Shang, J., Chen, J., Huang, Y., Li, G.: Mobile Agents for CPS in Intelligent Transportation Systems. Advanced Technologies, Embedded and Multimedia for Human-Centric Computing. Lect. Notes Electr. Eng. **260**, 721–729 (2014). https://doi.org/10.1007/978-94-007-7262-5_82

Hübner, A.H., Kuhn, H., Sternbeck, M.G.: Demand and supply chain planning in grocery retail: an operations planning framework. Int. J. Retail Distrib. Manage. **41**(7), 512–530 (2013)

Kong, X.T.R., et al.: Cyber physical ecommerce logistics system: an implementation case in Hong Kong. Comput. Ind. Eng. **139**, 106170 (2019). https://doi.org/10.1016/j.cie.2019.106170

Lai, M., Yang, H., Yang, S., Zhao, J., Xu, Y.: Cyber-physical logistics system-based vehicle routing optimization. J. Ind. Manage. Optim. **10**(3), 701–715 (2014). https://doi.org/10.3934/jimo.2014.10.701

Lee, E.A.: Cyber physical systems: design challenges. In: Proceedings 11th IEEE Symposium on Object Component Service-Oriented Real-Time Distributed Computing, ISORC 2008, (August), pp. 363–369 (2008). https://doi.org/10.1109/ISORC.2008.25

Li, Q., Li, X., Li, J.: Research on logistics vehicle scheduling optimization based on cyber-physical system. Proce. Int. Conf. Comput. Eng. Appl., ICCEA **2020**, 537–542 (2020). https://doi.org/10.1109/ICCEA50009.2020.00119

Lingling, H., Haifeng, L., Xu, X., Jian, L.: An intelligent vehicle monitoring system based on internet of things. In: 7th International Conference on Computational Intelligence and Security, pp. 231–233 (2011). https://doi.org/10.1109/CIS.2011.59

Manuel Maqueira, J., Moyano-Fuentes, J., Bruque, S.: Drivers and consequences of an innovative technology assimilation in the supply chain: cloud computing and supply chain integration. Int. J. Prod. Res. **57**(7), 2083–2103 (2019)

Masoudinejad, M., Venkatapathy, A.K.R., Emmerich, J., Riesner, A.: Smart sensing devices for logistics application. In: Magno, M., Ferrero, F., Bilas, V. (eds.) S-CUBE 2016. LNICSSITE, vol. 205, pp. 41–52. Springer, Cham (2017). https://doi.org/10.1007/978-3-319-61563-9_4

Monostori, L., et al.: Cyber-physical systems in manufacturing. CIRP Ann. **65**(2), 621–641 (2016)

Ounnar, F., Pujo, P.: Pilotage auto-organisé de l'Internet physique via des systèmes logistiques cyber-physiques: cas du transport Chine-France. Logistique. Manage. **27**(4), 255–263 (2019). https://doi.org/10.1080/12507970.2019.1635049

Park, K.T., Son, Y.H., Noh, S.D.: The architectural framework of a cyber physical logistics system for digital-twin-based supply chain control. Int. J. Prod. Res., **0**, 1–22 (2020). https://doi.org/10.1080/00207543.2020.1788738

Prasse, C., Nettstraeter, A., Hompel, M.: How IoT will change the design and operation of logistics systems. In: International Conference on the Internet of Things (IOT), pp. 55–60 (2014). https://doi.org/10.1109/IOT.2014.7030115

Proença, D., Borbinha, J.: Maturity models for information systems - a state of the art. Procedia Comput. Sci. **100**, 1042–1049 (2016). https://doi.org/10.1016/j.procs.2016.09.279

Sandkuhl, K., Lin, F., Shilov, N., Smirnov, A., Tarasov, V., Krizhanovsky, A.: Logistics-as-a-service: Ontology-based architecture and approach. Inv. Oper. **34**(3), 188–194 (2013)

Schuhmacher, J., Hummel, V.: Decentralized control of logistic processes in cyber-physical production systems at the example of ESB logistics learning factory. Procedia CIRP **54**, 19–24 (2016). https://doi.org/10.1016/j.procir.2016.04.095

Sen, A., Ramamurthy, K., Sinha, A.P.: A model of data warehousing process maturity. IEEE Trans. Softw. Eng. **38**(2), 336–353 (2012). https://doi.org/10.1109/TSE.2011.2

Sowe, S.K., Fränzle, M., Osterloh, J.-P., Trende, A., Weber, L., Lüdtke, A.: Challenges for Integrating Humans into Vehicular Cyber-Physical Systems (2019) https://doi.org/10.1007/978-3-030-57506-9_2

Syed, B., Pal, A., Srinivasarengan, K., Balamuralidhar, P.: A smart transport application of cyber-physical systems: Road surface monitoring with mobile devices. In: 6th Int. Conf. on Sensing Technology, pp. 8–12 (2012). https://doi.org/10.1109/ICSensT.2012.6461796

Thekkilakattil, A., Dodig-Crnkovic, G.: Ethics aspects of embedded and cyber-physical systems. In: Proceedings - International Computer Software and Applications Conference, vol. 2, pp. 39–44 (2015) https://doi.org/10.1109/COMPSAC.2015.41

Villalonga Jaen, A., Castaño Romero, F., Haber, R., Beruvides, G., Arenas, J. (2020). El control de sistemas ciberfísicos industriales: revisión y primera aproximación. pp. 916–923 (2020). https://doi.org/10.17979/spudc.9788497497565.0916

Vlfdo, E.H.U., Dv, V., Whfkqlfdo, W.K.H.: Cyber-physical systems as the technical foundation for problem solutions in manufacturing, logistics and supply chain management.In: 5th International Conference on the Internet of Things (IOT), pp. 12–19(2015). https://doi.org/10.1109/IOT.2015.7356543

Westermann, T., Anacker, H., Dumitrescu, R., Czaja, A.: Reference architecture and maturity levels for cyber-physical systems in the mechanical engineering industry. In: IEEE International Symposium on Systems Engineering (ISSE), pp. 1–6 (2016). https://doi.org/10.1109/SysEng.2016.7753153

Yan, J., Zhang, M., Zimin, F.: An intralogistics-oriented Cyber-Physical System for workshop in the context of Industry 4.0. Procedia Manuf. **35**, 1178–1183 (2019). https://doi.org/10.1016/j.promfg.2019.06.074

Zhang, N.: Smart logistics path for cyber-physical systems with internet of things. IEEE Access **6**, 70808–70819 (2018). https://doi.org/10.1109/ACCESS.2018.2879966

Zhang, X.: Automatic task matching and negotiated vehicle scheduling for intelligent logistics system. Int. J. Eng. Model. **31**(4), 43–55 (2018). https://doi.org/10.31534/engmod.2018.4.si.04g

Zhang, Y., Guo, Z., Lv, J., Liu, Y.: A framework for smart production-logistics systems based on CPS and industrial IoT. IEEE Trans. Industr. Inf. **14**(9), 4019–4032 (2018). https://doi.org/10.1109/TII.2018.2845683

Managing Disruptions in Supply Chains

Jairo R. Montoya-Torres(✉)

School of Engineering, Universidad de La Sabana,
Km 7 autopista norte de Bogota D.C., Chia, Colombia
jairo.montoya@unisabana.edu.co

Abstract. Globalized supply chain networks are prone to high levels of turbulence and uncertainty. This globalization induces new challenges that go far beyond the typical concern on demand and supply uncertainties. The management of disruptions in globalized supply chains requires a fully interdisciplinary approach. The COVID-19 outbreak of 2020 resulted in international border closures and shutdowns of many critical facilities and, thus, has exposed the fragility of the global supply network. The COVID-19 outbreak has made many organizations realize the importance of supply chain resilience, and many firms across industries will consider reconfiguring their supply chains to be ready for disruptions. The aim of this paper is to review relevant academic literature published between 2004 and 2020 examining issues related with disruptions in supply chains. Papers were selected following criteria established by the systematic approach for literature reviews, ensuring it is auditable and repeatable. This paper contributes to the knowledge on business continuity and in particular to the management of disruptions as it relates to supply chain management. The analysis of the literature allows to suggest future research opportunities in this area.

Keywords: Supply chain management · Disruption · Vulnerably · Resilience

1 Introduction

The environment of supply chain operations has changed drastically in recent years. Due to globalization, short product life-cycles and pressure from lean production, supply chain has become more vulnerable [46]. These issues have put almost all firms in a very complex commercial network in which supply chain managers have to proactively deal with disruptions that might affect the complicated supply networks characterizing modern enterprises. Globalized supply chain networks are prone to high levels of turbulence and uncertainty since the customer order fulfilment process is no longer controlled by a single, integrated organization, but by a number of decentralized and independent firms collaborating together. Theorists have focused on the transaction and production costs for explaining the shifting modes of production. But global sourcing has contributed to enhanced supply chain risks and vulnerabilities in the supply

D. Trentesaux et al. (Eds.): SOHOMA 2021, SCI 987, pp. 272–284, 2021.
https://doi.org/10.1007/978-3-030-80906-5_19

chain [40]. Indeed, the impact of this globalization is at least twofold [22]. Firstly, there appear huge opportunities for cost reduction and access to labour/talent pools, capital and markets. Secondly, this increase in supply chain scope induces new challenges that go far beyond the typical concern on demand and supply uncertainties.

In the best possible and idealized world, disruptions never occur. However, in real world, disruptions do and will always occur. Hence, business plans should anticipate and prepare organizations for these events. This is especially critical when supply chains are globalized and firms are cooperating with international suppliers and impacting international markets. According to the literature, a disruption in a supply chain is a situation where there is a physical problem with product delivery to final market. In order to understand the concept of a supply chain disruption, it is first necessary to understand the concepts of supply risk, as well as other two interconnected concepts: supply chain vulnerability and supply chain resilience.

Supply chain risk is defined as the "variation in the distribution of possible supply chain outcomes, their likelihoods, and their subjective values" [20,35]. This definition highlights the two dimensions characterizing risk: impact and likelihood of occurrence [11]. Supply chain disruption is one of the major categories of supply chain risk [28]. The study of supply risk dates back to the early 1990's with the study of inventory models and multiple sourcing policies [1,24]. More recently, the study of supply chain risk include the study of assessment tools, risk perceptions, risk management, financial effects, etc. For more detailed analysis of current literature on supply chain risk, we refer the reader to [37,45,49,50] or the book of Zsidisin and Ritchie [51].

Closely related with risk is the concept of supply chain vulnerability. It is defined as the "existence of random disturbances that lead to deviations in the supply chain from normal, expected or planned activities, all of which cause negative effects or consequences" [43]. The academic literature has confirmed the strong relationship between these two concepts [7,19,36]. Finally, supply chain resilience is defined as the "ability of a system to return to its original (or desired) state or move to a new, more desirable state after being disturbed" [7], while resilience is defined as "the adaptive capability of the supply chain to prepare for unexpected events, respond to disruptions, and recover from them by maintaining continuity of operations to the desired level of connectedness and control over structure and function" [36]. On the day-to-day management of a supply chain, resilience is viewed as an inherent strength of an operational configuration [31].

The academic literature has witnessed an increased level of publications covering a wide range of issues in supply chain risk management, employing different research methodologies, such as modelling and optimization, simulation, questionnaire surveys, event studies or case studies [46]. The majority of these researches focus on the economic impact of disruptive events and analyze the continuity of businesses and the economic survivability of firms. Even the recent works about the high impacts of COVID-19 pandemic on globalized supply chains focus on the economic impacts, as pointed out in the recent review [29]. The management of disruptions in globalized supply chains requires a fully interdisciplinary approach, embracing recent developments from Decision Theory, Operational Research and the Management Sciences, finance, organisational theory, information systems, among others disciplines. The aim of this paper is to review relevant academic literature that links the management of disruptions in supply chains with the evaluation of performance metrics. According to Fink [12] and Badger *et al.* [2], from a methodological point of view, a literature review is a systematic, explicit, and reproducible approach for identifying, evaluating, and interpreting the existing body of documents. This paper follows the principles of systematic literature reviews (SLR), in contrast to narrative reviews, by being more explicit in the selection of the studies and employing rigorous and reproducible evaluation methods [9].

Many scholars have proposed classifications in the form of typologies and/or taxonomies that are often labelled as supply chain risk sources [7,13–15,18, 21,28,35,41–43]. Throughout this paper, we adopt the classification proposed in [28]. The current paper builds upon those previous reviews by assessing the extend at which quantitative and qualitative research approaches have employed to evaluate the impact of major disruptions on the performance of supply chains.

The rest of this paper is organized as follows. Section 2 presents the methodology employed to carry out the review. Section 3 analyzes the findings. The paper ends in Sect. 4 by presenting the conclusions and drawing some opportunities for further research.

2 Review Methodology

As stated earlier, this paper reviews the relevant literature available in order to study the impact of major disruptions on the performance of supply chains. As pointed out in [2], a key requirement in a literature review is that each stage of the process has to be defined in a protocol "intended to guide the whole, thereby reducing the possible sources of bias which arise from authors making idiosyncratic decisions at different stages of the review process". Under these guidelines, as with any research, a primary decision to make is to formulate a clear and answerable question. For the purpose of this paper, the research question is formulated as follows: *What approaches have been employed in the academic literature to assess the impact of major disruptions on the performance of supply chains?*

The previous question will be addressed by identifying:

1. The types of disruptions in supply chains that have been studied in the academic literature, and
2. The performance metrics that have been considered in such academic works.

In order to answer the previous research questions, a determination was made to collect and identify the research papers from the Scopus database. We used the web-based interface. The rationale for this choice is that Scopus "is the largest abstract and citation database of peer-reviewed literature: scientific journals, books and conference proceedings. Delivering a comprehensive overview of the world's research output in the fields of science, technology, medicine, social sciences, and arts and humanities, Scopus features smart tools to track, analyze and visualize research" [10].

Various sets of keywords and phrases were employed to search in titles, abstracts, and keywords using logical connectors: disruption* AND supply chain*, disruption* AND logistics, resilience AND supply chain*, resilience AND logistics, disaster management AND supply chain*, disaster management AND logistics, ecological discontinuities AND supply chain*, ecological discontinuities AND logistics, disruption* AND (optimization OR optimisation), resilience AND (optimization OR optimisation), business continuity AND supply chain*, business continuity AND logistics.

We recognize that this was less efficient than just focusing on the keywords but ensured that we captured as many relevant papers as possible. The initial collection of references counted publications appeared between 2004 and 2020. The rationale for considering articles published during this timeframe is that the link between supply chain management issues, in a large extent, and disruption/risk management as a field of study has only relatively recently been addressed. Thus, a 17-year literature review allows for a sufficiently exhaustive analysis of the scientific research on this area.

Each short-listed paper was reviewed in its entirety. In order to maintain the necessary degree of consistency in this review, a database was completed for each paper. This database included information about: study (authors and year), method (conceptual, case-based, empirical, analytical), definition of disruption (if present in the paper), supply chain structure (serial, dyadic, convergent, divergent, network), location and source of disruption (natural, human-made, demand, supply, transport, etc.), performance metrics (economic, environmental, social), and focus/findings (the main findings of the study). Finally, we critically analyzed the classified articles so as to examine how the research has evolved over the timeframe. The analysis also provided an ability to identify research opportunities in the existing literature. This analysis is given in following sections.

3 Findings

This section summarizes the findings of the literature review. The methodology employed to assess the impact of disruptions in supply chain performance metrics is one of the main issues of interest. Inspired from previous works in the literature (e.g., [23]), two main research methodologies were identified: qualitative and quantitative approaches. While the use of multiple methodologies within an article was possible, each article was classified as to the primary methodology employed. Papers using qualitative methods were categorised as theoretical/conceptual papers, case studies, or surveys, while quantitative or modelling papers were categorised as analytical/mathematical programming, simulation-based, or algorithmic papers. From our review, we observe that majority of works used quantitative approaches (78% of papers), and mainly analytical or mathematical modelling techniques (54%); 12% are theoretical or conceptual papers, 15% are papers based on survey-based research techniques, 14% of papers considers a case-study, 15% of papers employs computer simulation approaches, and 9% of papers develops an algorithmic approach.

When focusing on the structure of the supply chain, we observed that most of papers developing a mathematical analysis mainly considered simple supply chains structures (dyadic, convergent or divergent), due to their simplicity to deliver analytical insights. When the structure is much more complex (e.g., network configuration), either computer simulation or a qualitative approach is preferred. As a matter of fact, an important issue of interest in our analysis of literature was the structure of supply chains considered in the papers. The serial structure is the typical structure studied in the literature in which supplier, manufacturer, distributor and retailer are considered. This structure is in fact obtained by cascading several dyadic structures. The dyadic structure consists of two business entities. In the study, we observe that 11% of the papers considered in this review focused on a serial structure, while dyadic was studied in 18% of reviewed papers. A divergent structure is used to represent a more realistic supply chain in which one entity (e.g. supplier) distributes stock to several downstream entities. In a convergent structure, several entities (e.g. several suppliers) deliver components to a single manufacturer or to a distribution centre. Finally, the network structure is a complex supply chain with a combination of divergent and convergent structures. In the study, we observed that divergent and convergent structures were analysed each in 14% of reviewed papers, while a network configurations is considered in 38% of the references.

In order to answer the research questions presented previously in this paper, an important output of this review is to identify where the disruption occurred in the supply chain. As explained before, broadly speaking, disruption can be categorised as supply-side or demand-side. However, we noticed during our review process that it was important to specify if disruptions concern the demand, the supply or the transport between members of the supply chain. Figure 1 presents statistics about the number of times that demand, supply and/or transport disruptions were considered in the analysed literature. We observe that disruptions in demand and supply have been the most studied issues (85% of the total number of paper analysed in this review), and transportation issues being of interest only in a smaller proportion during the last years.

In addition to the type of disruption, an important issue is to identify the source of such disruption(s) in the supply chain. We observed however that 46% of reviewed works do specify the source of such disruption, and simply assume that something is happening that disrupts the flow of material between the members of the supply chain (see Fig. 2); only few papers state the source of disruption. Natural disasters, human / organisational and production issues represent 43% of causes in reviewed papers.

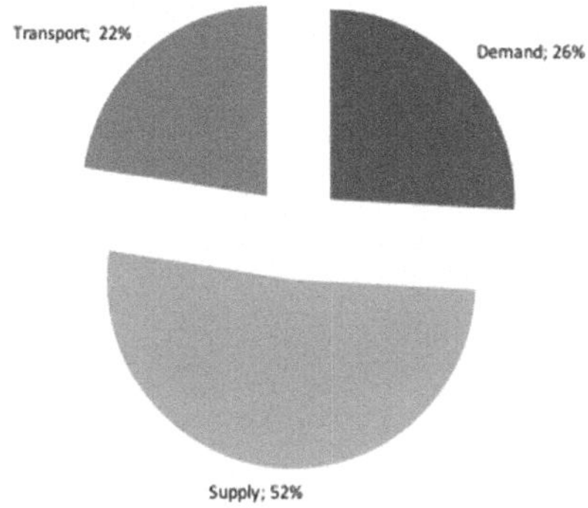

Fig. 1. Frequency of demand, supply and/or transport disruptions

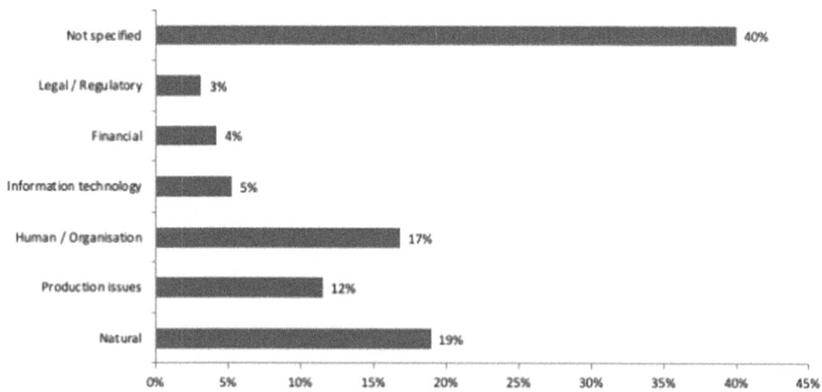

Fig. 2. Source of disruptions

Research papers included in this review were also classified according to the level of impact (severity) and frequency of disruptions. When these two elements were not presented in the reviewed paper, the framework proposed in [32] and adapted in [33] was employed. This framework allows the decision-maker to assess and prioritize disruptions in order to choose management actions appropriate to the situation. The method consists in comparing events by assessing their probabilities and consequences and put them in a risk map/matrix. Table 1 presents this matrix in terms of the number of published papers that were evaluated in the current review.

Finally, an important outcome of this analysis is the performance metrics measured by the existent literature when evaluating the impact of disruptions in supply chains. Our review revealed that various indicators at strategic, tactical and operational decision-making levels have been employed to measure economic-, financial- and productivity-oriented metrics, as shown in Fig. 3. This is a very interesting outcome in the current context of the growth of sustainability issues in supply chain management and design. Defining sustainability in supply

Table 1. Severity (impact) versus frequency of disruptions in reviewed papers

		Frequency		
		Low	Medium	High
Impact	Low	12%	12%	23%
	Medium	11%	51%	8%
	High	37%	18%	10%

chains as the simultaneous optimisation of economic, social and environmental metrics during the decision-making process, our review revealed, however, that the economic dimension has been the only dimension studied when dealing with disruptions in supply chains.

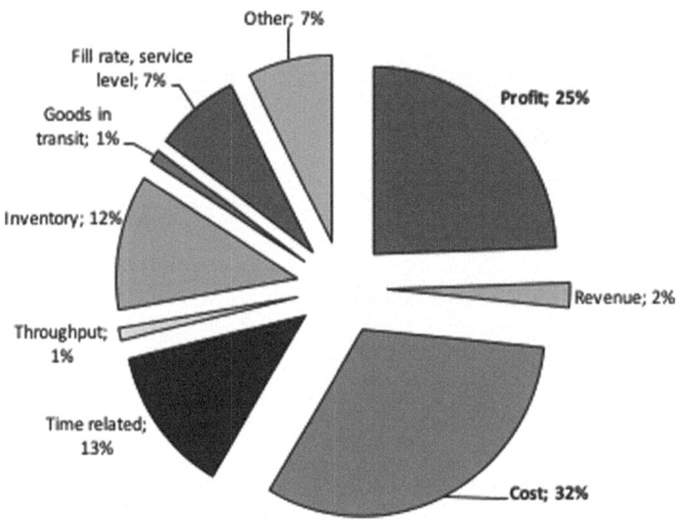

Fig. 3. Performance criteria

4 Discussion, Conclusions and Opportunities for Future Work

As pointed out in [39], there is abundant anecdotal and empirical evidence that supply chain disruptions negatively affect firm performance [16,39,40]. Given these costs, it is understandable that disruptions may also negatively affect the value of the firm. Developing an accurate assessment of the impact of operational disruptions on firm value is of practical importance because it produces managerial decisions on supply chain structure [25], mitigation investments, and supply chain insurance [38]. From the findings of our systematic review, we conclude that the implementation of supply chain wide risk management is a very complex task. A "real" extensive and effective supply chain wide with all supply chain partners risk management is ideal. An entire implementation of the concept in supply chain practice seems to be impossible due to intersections and the length of today's supply chains. However, an implementation based, for example, on collaboration between the focal company and its tier-1 and tier-2 suppliers and customers is possible and worthwhile. For this purpose, companies have to fulfil the requirement of having precise knowledge of their supply chain and the nature of relation to their partners.

As explained earlier in this paper, the notion of supply chain performance is nowadays a more complex concept that needs to be concerned with environmental as well as social and economic issues. An implicit assumption, in current state-of-the art models of supply chain management is that the supply chain is never affected by any kind of major disruption (but is rather driven by efficiency). The responsiveness of a supply chain to unexpected major disruptions must be modelled in an integral manner such that applying the model in practice will allow the comparison of the pre- and post-disruption effects on economic, social and environmental supply chain dimensions.

Consider a supply chain where a focal organization (node) has an important part to play in channelling goods through the network, e.g. it may be a manufacturing plant. Such a node is subject to major disruptions in its immediate vicinity and the degree of impact on the node depends on such as proximity and other characteristics of the disruption (type, frequency and severity). However, the impact also depends on the resilience of the individual organization arising from its characteristics such as size. For example, a flood of the plant could cause a major disruption to the supply chain. The impact on the supply chain would also depend upon whether an alternative path could be switched to, thus avoiding the impacted node. However, a wider spread disruption (e.g. a continental-wide pandemic such as the COVID-19) could impact on a variety of nodes and therefore in that case the vulnerability of the supply chain depends more on the network topology rather than the robustness of the individual node. Figure 4 proposes a conceptual framework for disruption management in supply chains. Its elements are explained next:

- Characteristics of the supply chain. These determine the way major external disruptions affect performance within the supply chain, such as: transport mode, structure or typology, dynamics and complexity of supply chains. Indeed, as pointed out in extant literature on supply chain complexity (e.g. [5,6,26]), horizontal, vertical and spatial complexity of the supply chain all three increase the frequency of disruptions and intensifies the effects of the other two in synergistically [4].
- Characteristics of the organizations within a supply chain. These influence the way major external disruptions affect organization performance within the supply chain, such as: size, age, location, industry sector and position in supply chain (e.g. warehouse, distributor, 3PL).
- Major external disruptions. These are infrequent disruptions, typically external to the organization, which seriously affect the distribution channels of

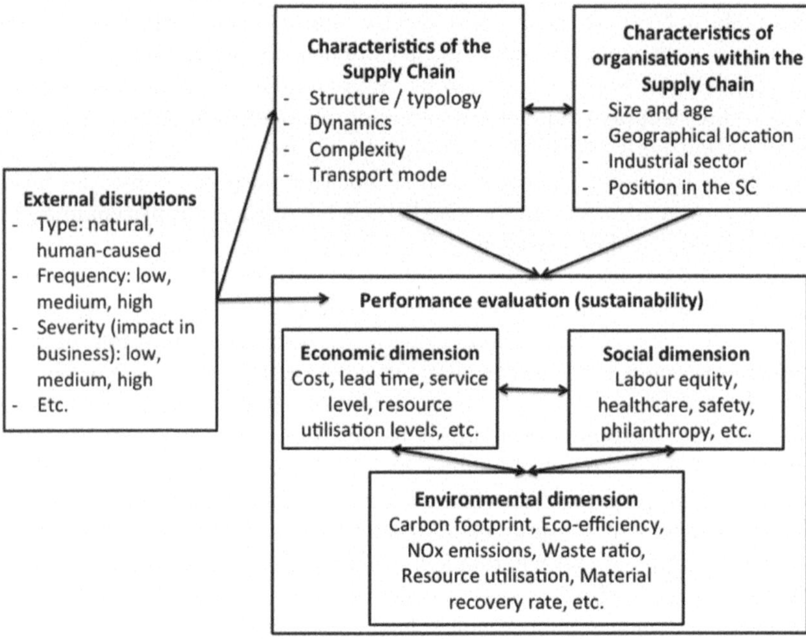

Fig. 4. A framework for studying the impact of major disruptions on sustainability in supply chains

the supply chain. These could be human-caused (terrorist attacks, nuclear power station failures, and air-traffic control accidents), or natural (floods, hurricanes, earthquakes, etc.).

- Performance evaluation. This refers to the supply chain performance impacted by the major disruptions. The performance is measured across the three sustainability dimensions (economic, environmental, and social), as a multi-dimensional performance measure taking into account the following:
  - Economic dimension: Taking into account the typical five operations performance objectives: cost, speed, dependability, quality and flexibility. Affected factors would be cost, lead time (related to dependability and speed) and service level (related to quality and flexibility). Several authors have conducted critical reviews and proposed frameworks on how to measure supply chain performance, while others have focused their effort on approaches improving supply chain design (e.g. [3,8]).
  - Environmental dimension: There is a variety of environmental sustainability metrics that organizations report, mainly related to resource use and waste generation [44], while other relate to the wide supply chain [30]. Those considered in the framework could be: carbon footprint, eco-efficiency, NOx emissions, waste ratio, and material recovery rate.
  - Social dimension: Analyzing social issues in any decision-making problem containing multiple stakeholders requires a multi-disciplinary approach.

In practice it is difficult to measure all social aspects in a single decision-making problem. Given the importance and growth of social issues in business, a number of standards have been developed to support the planning and implementing of corporate social responsibility [34]. To achieve a standard framework, the International Standard Organization (ISO) developed the "International Guidance Standard on Social Responsibility - ISO 26000". This standard classifies the social issues of firms in to various groups. As reported in [17] the measures of social sustainability as they refer to supply chain management can be classified into: human health, equity, quality of life and philanthropy.

Based on the analysis of reviewed papers, several opportunities for future research appear. We observ in that majority of works studies simple structures of supply chains (dyadic, convergent, divergent or serial with two echelons). This is explained by the fact that these structures are mathematically simpler and tractable. Agent-based and discrete-event simulation approaches are preferred for more complex structures [47]. However, much more than modelling or simulating the structure of a supply chain and evaluating the impact on disruption on a given set of performance metrics, the knowledge and understanding of supply chain structures are important elements of supply chain resilience. An appropriate structure of the supply chain can facilitate resilience and a faster or even to a certain extent a proactive response to disruptions [48]. This is actually a very important issue since most supply chain design models reported in the academic literature do not usually address the practical global supply chain design problem in its entirety [27].

Regarding supply chain performance metrics, an interesting outcome of this review is that all reviewed papers evaluate economic- and financial-related metrics. This includes both costs, profit and revenue-based metrics and service level as well as productivity related metrics. Despite the international trend to also consider environmental and social impacts of management decisions, the academic literature on disruption management in supply chains has not yet included social and environmental metrics. Facing disruptions in logistics and supply chain management plays a key role in sustaining dynamic capabilities of all and each one of the supply chain actors. Sustainability is a key enabler for resilience of supply chains. The understanding of sustainability issues of each supply chain to what the focal enterprise belongs to does help managers make better decisions. However, there is no evidence of this despite the fact that there is a clear relation between the disruptions as causes of effects on sustainability metrics.

Finally, an important topic concerns internal risks and disruptive events. As observed in the literature review, all reviewed papers consider that an event outside the company and outside the supply chain occurred and hence impacts the flows of products along the supply chain. However, internal risks must not be unobserved as they may also have an important impact on the different actors of the supply chain. This is another interesting line for further research.

References

1. Anupindi, R., Akella, R.: Diversification under supply under uncertainty. Manag. Sci. **39**(8), 944–963 (1993)
2. Badger, D., Nursten, J., Williams, P., Woodward, M.: Should all literature reviews be systematic? Eval. Res. Educ. **14**(3–4), 220–230 (2000)
3. Beamon, B.M.: Measuring supply chain performance. Int. J. Oper. Prod. Manag. **19**(3), 275–292 (1999)
4. Bode, C., Wagner, S.M.: Structural drivers of upstream supply chain complexity and the frequency of supply chain disruptions. J. Oper. Manag. **36**, 215–228 (2015)
5. Bozarth, C.C., Warsing, D.P., Flynn, B.B., Flynn, E.J.: The impact of supply chain complexity on manufacturing plant performance. J. Oper. Manag. **27**(1), 78–93 (2009)
6. Choi, T.Y., Krause, D.R.: The supply base and its complexity: implications for transaction costs, risks, responsiveness, and innovation. J. Oper. Manag. **24**(5), 637–652 (2006)
7. Christopher, M., Peck, H.: Building the resilient supply chain. Int. J. Log. Manag. **15**(2), 1–13 (2004)
8. Cuthbertson, R., Piotrowicz, W.: Performance measurement systems in supply chains - a framework for contextual analysis. Int. J. Prod. Perform. Manag. **60**(6), 583–602 (2011)
9. Delbufalo, E.: Outcomes of inter-organizational trust in supply chain relationships: a systematic literature review and a meta-analysis of the empirical evidence. Supply Chain Manag: An Int. J. **17**(4), 377–402 (2012)
10. Elsevier: About Scopus (2020). https://www.elsevier.com/en-in/solutions/scopus. Accessed 10 Nov 2020
11. Faisal, M.N., Banwet, D.K., Shankar, R.: Supply chain risk mitigation: modelling the enablers. Bus. Process Manag. J. **12**(4), 535–552 (2006)
12. Fink, A.: Conducting Research Literature Reviews: From Paper to the Internet. Sage, Thousand Oaks (1998)
13. Fahimnia, B., Tang, C.S., Davarzani, H., Sarkis, J.: Quantitative models for managing supply chain risks: a review. Eur. J. Oper. Res. **247**, 1–15 (2015)
14. Giannakis, M., Papadopoulos, T.: Supply chain sustainability: a risk management approach. Int. J. Prod. Econ. **171**, 455–470 (2016)
15. Hallikas, J., Karvonen, I., Pulkkinen, U., Virolainen, V.-M., Tuominen, M.: Risk management processes in supplier networks. Int. J. Prod. Econ. **90**(1), 47–58 (2004)
16. Hendricks, K.B., Singhal, V.R.: An empirical analysis of the effect of supply chain disruptions on long-run stock price performance and equity risk of the firm. Prod. Oper. Manag. **14**(1), 35–52 (2005)
17. Hutchins, M.J., Sutherland, J.W.: An exploration of measures of social sustainability and their application to supply chain decisions. J. Clean. Prod. **16**(15), 1688–1698 (2000)
18. Jüttner, U.: Supply chain risk management - understanding the business requirements from a practitioner perspective. Int. J. Logist. Manag. **16**(1), 120–141 (2005)
19. Jüttner, U., Maklan, S.: Supply chain resilience in the global financial crisis: an empirical study. Supply Chain Manag.: An Int. J. **16**(4), 246–259 (2011)
20. Jüttner, U., Peck, H., Christopher, M.: Supply chain risk management: outlining an agenda for future research. Int. J. Logist.: Res. Appl. **6**(4), 197–210 (2003)
21. Kamalahmadi, M., Parast, M.M.: A review of the literature on the principles of enterprise and supply chain resilience: Major findings and directions for future research. Int. J. Prod. Econ. **171**, 116–133 (2016)

22. Kouvelis, P., Dong, L., Boyabatli, O., Li, R.: Handbook of Integrated Risk Management in Global Supply Chains. Wiley-Blackwell, Hoboken (2011)
23. Kotzab, H., Seuring, S., Muller, M., Reiner, G.: Research Methodologies in Supply Chain Management. Physica-Verlag, Heidelberg (2007)
24. Lee, H.L., Padmanabhan, V., Whang, S.: Information distortion in a supply chain: the bullwhip effect. Manag. Sci. **43**(4), 546–558 (1997)
25. Leone, M.: A case for fatter supply chains. Technical report, CFO.com (2006). http://www.cfo.com/article.cfm/5491636. Accessed 19 May 2015
26. Manuj, I., Sahin, F.: A model of supply chain and supply chain decision-making complexity. Int. J. Phys. Distrib. Logist. Manag. **41**(5), 511–549 (2011)
27. Meixell, M.J., Gargeya, V.B.: Global supply chain design: a literature review and critique. Transp. Res. Part E **41**, 531–550 (2005)
28. Melnyk, S.A., Rodrigues, A., Ragatz, G.L.: Using simulation to investigate supply chain disruptions. In: Zsidisin, G.A., Ritchie, B. (eds.) Supply Chain Risk – A Handbook of Assessment, Management, and Performance. International Series of Operations Research & Management Sciences, vol. 124, pp. 103–122. Springer (2009)
29. Montoya-Torres, J.R., Muñoz-Villamizar, A., Mejia-Argueta, C.: Mapping research in logistics and supply chain management during COVID-19 pandemic. Submitted manuscript (2020)
30. Moreno-Camacho, C.A., Montoya-Torres, J.R., Jaegler, A., Gondran, N.: Sustainability metrics for real case applications of the supply chain network design problem: a systematic literature review. J. Clean. Prod. **231**, 600–618 (2019)
31. Munoz, A., Dunbar, M.: On the quantification of operational supply chain resilience. Int. J. Prod. Res. **53**(22), 6736–6751 (2015)
32. Norrman, A., Jansson, U.: Ericsson's proactive supply chain risk management approach after a serious sub-supplier accident. Int. J. Phys. Distrib. Logist. Manag. **34**(5), 434–456 (2004)
33. Oke, A., Gopalakrishnan, M.: Managing disruption risks in supply chains: a case study of a retail supply chain. Int. J. Prod. Econ. **118**(1), 168–174 (2009)
34. Pishvaee, M.S., Razmi, J., Torabi, S.A.: Robust possibilistic programming for socially responsible supply chain network design: a new approach. Fuzzy Sets Syst. **206**, 1–20 (2012)
35. Pfohl, H.-C., Köhler, H., Thomas, D.: State of the art in supply chain risk management research: empirical and conceptual findings and a roadmap for the implementation in practice. Logist. Res. **2**, 33–44 (2010)
36. Ponomarov, S.Y., Holcomb, M.C.: Understanding the concept of supply chain resilience. Int. J. Logist. Manag. **20**(1), 124–143 (2009)
37. Rao, S., Goldsby, T.J.: Supply chain risk: a review and typology. Int. J. Logist. Manag. **20**(1), 97–123 (2009)
38. Reynolds, D.: The Dawn of Supply Chain Insurance. Risk & Insurance, Brand Studio, Horsham (2011)
39. Schmidt, W.: Supply chain disruptions and the role of information asymmetry. Decis. Sci. **46**(2), 465–475 (2015)
40. Sheffi, Y.: The Resilient Enterprise: Overcoming Vulnerability for Competitive Advantage. MIT Press, Cambridge (2007)
41. Snyder, L.V., Atan, Z., Peng, P., Rong, Y., Schmitt, A.J., Sinsoysal, B.: OR/MS models for supply chain disruptions: a review. IIE Trans. **48**(2), 89–109 (2016)
42. Spekman, R.E., Davis, E.W.: Risky business: expanding the discussion on risk and the extended enterprise. Int. J. Phys. Distrib. Logist. Manag. **34**(5), 414–433 (2004)

43. Svensson, G.: A conceptual framework for the analysis of vulnerability in supply chains. Int. J. Phys. Distrib. Logist. Manag. **30**(9), 731–749 (2000)
44. Székely, F., Knirsch, M.: Responsible leadership and corporate social responsibility: metrics for sustainable performance. Eur. Manag. J. **23**(6), 628–647 (2005)
45. Tang, C.: Perspectives in supply chain risk management. Int. J. Prod. Econ. **103**(2), 451–488 (2006)
46. Tang, O., Matsukawa, H., Nakashima, K.: Supply chain risk management. Int. J. Prod. Econ. **139**(1), 1–2 (2012)
47. Tordecilla, R.D., Juan, A.A., Montoya-Torres, J.R., Quintero-Araujo, C.L., Panadero, J.: Simulation-optimization methods for designing and assessing resilient supply chain networks under uncertainty scenarios: a review. Simul. Model. Pract. Theory **106**, 102166 (2020)
48. Trkman, P., McCormack, K.: Supply chain risk in turbulent environments - a conceptual model for managing supply chain network risk. Int. J. Prod. Econ. **119**(2), 247–258 (2009)
49. Wieland, A.: Selecting the right supply chain based on risks. J. Manuf. Tech. Manag. **24**(5), 652–668 (2013)
50. Williams, Z., Lueg, J.E., LeMay, S.A.: Supply chain security: an overview and research agenda. Int. J. Logist. Manag. **19**(2), 254–281 (2008)
51. Zsidisin, G.A., Ritchie, B.: Supply Chain Risk – A Handbook of Assessment, Management, and Performance. International Series of Operations Research & Management Sciences, vol. 124. Springer (2009)

Weighted Sum Method for Multi-objective Optimization LP Model for Supply Chain Management of Perishable Products in a Diary Company

Sara Manuela Alayón Suárez(✉)

Pontificia Universidad Javeriana, Bogota, Colombia
salayon@javeriana.edu.co

Abstract. The Inventory Routing Problem (IRP) allows companies and systems to take decisions in two important areas: Inventory Management and Routing. Nevertheless, the IRP models developed for perishable products are limited due to the strong restrictions that this type of product presents. In this paper, a lineal mathematical model is proposed in the context of a deductive decision making problem to tackle this atypical kind of problem. This article proposes a weighted sum method for multi-objective optimization linear model that takes into account logistics costs, customer service level and CO_2 emissions. The mathematical model aims to be a contribution relative to the studies that currently report problems related to supply chain management of perishable products and analyses the sensitivity of the problem of having a restricted horizontal transshipment versus an open one. The results showed that the approach with continuous variables to make deductive decisions is quite close to the integer problem and that the problem is quite sensitive to the condition of lateral transshipment presenting gaps up to -25.76%, in the total objective function. These results invite to develop in future research non-exact models such as metaheuristics and heuristics to find solutions to larger instances and allow to conclude that a condition such as transshipment is a sensitive variable in this kind of problem.

Keywords: Inventory Routing Problem · Distribution Requirement Planning · Perishable products · Supply Chain Management · CO_2 emissions · Transport · Inventory

1 Introduction

In recent decades, the problems that analyse inventory and routing decisions together (Inventory Routing Problem - IRP) have become very important in the management of the supply chain [3]. This type of problem has several classifications in which lateral-transshipment is considered. In this type of problem, it is possible to send products between agents at the same level of the supply chain. This approach has achieved excellent results in inventory systems policies [11]. These types of problems have been considered of great interest, especially when they are associated with perishable products

D. Trentesaux et al. (Eds.): SOHOMA 2021, SCI 987, pp. 285–298, 2021.
https://doi.org/10.1007/978-3-030-80906-5_20

[10] since they are an important part of the supply of the world population [7]. The management of inventories had to face the analysis of methods for their planning and control, the optimal balance of inventory levels and the minimization of costs. However, the consideration of additional restrictions such as the management of perishable products [7], has led to numerous attempts to adapt the classical models that have been of great interest in the literature.

Inventory models for this type of products have been studied by multiple authors where problems with different characteristics and objectives are considered. Although, this investigations did not consider the multi-product approach, lead-time management and multi-links, only one article was found associated with transshipment of these types of products [5]. In 2017, a first study of a real problem associated with inventory management and supply chain management of a dairy company was carried out [2]. The study addressed a problem of perishable products taking into account a multi-product approach, with lead-time variation and multi-period with three links in the supply chain.

Based on this first model, the research reported in this article tries to develop a model with a multi-target function that will not only consider logistics costs but also the reduction of CO2 emissions throughout the supply chain and the evaluation of customer satisfaction through demand satisfaction. Likewise, restrictions that were not taken into account in the first model will be added. These restrictions consider the variety of trucks and their transport capacities and the handling of products for storage and transport.

The model was tested in 20 different instances, based on some of the previously developed one. In these instance the number of agents in each link and the number of raw ingredients and products will be varied by analysing the behaviour for two different planning horizons. The other restrictions and parameters associated to the problem will be kept constant.

This paper proposes an exact method solution in a deductive decision making problem, to solve the described problem and as an input to literature for the case of perishable products as a problem insufficiently analysed. The paper describes a first part of a project that aims to have also a predictive model in the future.

Two hypotheses are considered in relation to these approaches. First, it is believed that the model will only be able to solve the instances by having all the variables continuous. Likewise, it is considered that having an open transshipment the model will not present major changes taking into account that it is a complex problem affected by several capacity restrictions.

2 State of the Art

Inventory models for perishable products have been studied by multiple authors where problems with different characteristics and objectives are considered, such as [6] in which a hybrid tabu search based heuristic for the periodic distribution inventory problem with perishable goods is proposed. This article has as objective to minimize the transport cost between a depot and the customers. Furthermore, a multi-period IRP mode for a fresh tomato distribution centre is presented in [14] taking into account constraints such as truck capacity. The objective was to minimize the sum of transportation, waste and inventory costs. In [9] a column generation-based solution approach was developed taking into account an IRP for one product and considering as nodes or links only the customers.

However, the solution above does not consider the multi-product approach, lead-time management and multi-links and the conditions of lateral transshipment presented in the case study of this article. In previous work [2], this condition was taking into account leading to interesting results in terms of logistics cost; some restrictions such as the number of trucks available and the procurement were not taken into account.

In recent approaches of this type of problem for perishable products other types of restrictions have been taken into account. These approaches include the carbon dioxide emissions associated with transportation [1], the multi-period IRP accounting for CO_2 emissions [13], the multi-period IRP with products having fixed shelf life [12], the IRP model with single product transshipment [4], and the multi-stage production systems with shipment to distributors [15]. The above-mentioned works represent a progress with respect to previous works, and provide opportunities to develop models that include multiple constraints. These restrictions are not only logistical, but also associated with the reduction of pollution and other aspects related to green logistics [8].

To conclude, this study adds to the literature on IRP for perishable products by: 1) considering a multi-product approach with lead-time management and multi-links to which the possibility of having lateral transshipment between some links at the same level of the supply chain is added; (2) presenting a weighted sum method for multi-objective optimization mathematical model that takes into account not only logistics costs associated with warehouse, transport, picking, and backorders, but also CO_2 emissions and customer satisfaction.

3 Methods

3.1 Problem Description

The problem in this study is defined by a supply chain of a dairy product company where F are the suppliers, P correspond to the production plants and Ito the national (CDN) and regional (CDR) distribution centres. The company produces K products that have a specific shelf life (perishable products), which can be of cold or dry type. These products must be delivered to customers through CDRs taking into account their demand $D_{i,k,t}$ for each period, T within the planning horizon H. This demand may have backorders across the planning horizon, but has a maximum number of backorders at the end of the planning horizon.

The company currently seeks not only to reduce logistics costs but also CO_2 emissions and increase customer satisfaction. The company has multiple suppliers for the multiple raw material ingredients M needed, and stores them according to their type (cold or dry) in the limited warehouses of each production plant. Not all suppliers ship all raw ingredients to all plants and have associated restricted shipping capacity and lead times for product delivery. Each plant transforms the raw ingredients into products according to the company's recipes. Each product in each plant has a maximum production capacity, production time associated and the amount of emissions of CO_2 in ppm generated for each ton produced.

Each plant has its own CDN relative to which it can make lateral transshipment of all products. These CDNs take the products to the CDRs according to the established lead-times. All distribution centres have limited storage capacities for each type of product. The company has a limited fleet of diesel and electric vehicles V, with specific transport capacities between the distribution centres. Each type of truck emits a quantity of ppm of CO_2 for each kilometer travelled.

The problem consists in determining the quantities of product and material (in tons) to be sent between the distribution centres, plants and suppliers respectively and the necessary freight to distribute the finished product. It also defines inventory levels of raw materials and products, as well as production levels in each node of the supply chain. These variables are associated with the amount of CO_2 emissions generated throughout the chain and with the satisfaction of customer demand.

3.2 Mathematical Model

This section presents the mathematical formulation for the studied problem. Table 1 presents the notation for the model.

We now present the model formulation, starting with the objective function.

$$\begin{aligned}
\text{Minimize } Z = 2 + \sum_{t\in T} co_2T_t + 6 * \frac{\left(\sum_{i\in CDR}\sum_{k\in K}\sum_{t\in T} BB_{ikt}/H\right)}{\sum_{i\in CDR}\sum_{k\in K}\sum_{t\in T} D_{ikt}} + 6 \\
* \Big(\sum_{i\in I}\sum_{j\in I}\sum_{k\in K}\sum_{t\in T} X_{ijkt} * CP_{ik} \\
+ \sum_{p\in P}\sum_{m\in M}\sum_{t\in T} INVM_{pmt} * CIMP_{mp} + \sum_{p\in P}\sum_{k\in K}\sum_{t\in T} S_{pkt} * CPR_{pk} \\
+ \sum_{i\in I}\sum_{j\in I}\sum_{v\in V}\sum_{t\in T} Y_{ijvt} * CF_{ijv} * DT_{ij}\Big) \\
+ \sum_{f\in F}\sum_{f\in P}\sum_{m\in M}\sum_{t\in T} W_{fpmt} * CMP_{ifmp}\Big) \\
+ \sum_{i\in I}\sum_{k\in K}\sum_{t\in T} BB_{ikt} * CB_{ik} \qquad (1)
\end{aligned}$$

The objective function comprises three parts: i) expected CO_2 emissions from the production plants and the vehicles (note that CO_2T_t is derived from constraints (7)) ii) expected customer satisfaction based on average backorders and demand on the planning horizon, and iii) expected logistics costs. Each part was assigned a corresponding weight. Part i) and ii) have a weight of 2 respectively, and the sum of logistics a weight of 6.

$$\sum_{m\in M} INVM_{pmt} * TMP_m \le CapFB_p, \forall p \in P, \forall t \in \{T|t > 0\} \qquad (2)$$

$$\sum_{m\in M} INVM_{pmt} * (1 - TMP_m) \le CapSB_p, \forall p \in P, \forall t \in \{T|t > 0\} \qquad (3)$$

$$\sum_{p\in P} W_{fpmt} \le CapProv_{fm}, \forall f \in F, \forall m \in M, \forall t \in \{T|t > 0\} \qquad (4)$$

$$\sum_{m \in M} INV_{ikt} * TP_k \leq CapFCD_i, \qquad \forall i \in I, \forall t \in \{T|t > 0\} \tag{5}$$

$$\sum_{m \in M} INV_{ikt} * (1 - TP_k) \leq CapSCD_i, \forall i \in I, \forall t \in \{T|t > 0\} \tag{6}$$

$$CO_2T_t = \sum_{i \in I} \sum_{j \in I} \sum_{v \in V} Y_{ijvt} * CO_{2_v} * DT_{ij} + \sum_{p \in P} \sum_{k \in K} S_{pkt} * CO_{2_k}, \forall t \in \{T|t > 0\} \tag{7}$$

$$S_{pkt} \leq CapProd_{pk} * BI_{pk}, \qquad \forall p \in P, \forall k \in K, \forall t \in \{T|t > 0\} \tag{8}$$

Constraints (2, 3) relate the inventory levels of raw materials with the warehouse capacities of each type of ingredients in each plant. Also, constraint (4) relates the inventory levels of products to warehouse capacities in each distribution centre. Constraints (5, 6) limit the amount of raw ingredients that each supplier sends according to their capacity, and constraints (7–8) relate the production level with the capacity of each plant if it is allowed that the plant produces all related product.

$$\sum_{i \in I} \sum_{j \in I} Y_{ijvt} \leq CaV_v, \qquad \forall v \in V, \forall t \in \{T|t > 0\} \tag{9}$$

$$\sum_{t \in T} X_{ijkt} * TP_k \leq CapV_1 * Y_{ij1t}, \forall i, \forall j \in I, \forall t \in \{T|t > 0\} \tag{10}$$

$$\sum\nolimits_{t \in T} X_{ijkt} * (1 - TP_k) \leq CapV_2 * Y_{ij2t}, \forall i, \forall j \in I, \forall t \in \{T|t > 0\} \tag{11}$$

Constraints (9–11) relate the capacity and amount of freights that the company can make daily; (12, 13) ensure that raw ingredients and products are sent in valid areas.

$$W_{fpmt} \leq M * B_{fpm}, \forall f \in F, \forall p \in P, \forall m \in M, \forall t \in \{T|t > 0\} \tag{12}$$

$$X_{ijkt} \leq M * A_{ijk}, \forall i, \forall j \in I, \forall k \in K, \forall t \in \{T|t > 0\} \tag{13}$$

Constraints (14–17) define inventory variables of raw ingredients and products respectively in plants and centres of distribution and limit the amount sent by the distribution centre according to the inventory level. The shelf-life (18) compares in a given period all the system inventory for each product with what was produced in that period minus the shelf-life of the product, considering the lead-times and ensuring that there are no expired products [2].

$$\sum_{j \in I} X_{ijkt} \leq INV_{ikt-1} + \sum_{j \in I} X_{ijkt-Lji} + S_{pkt}, \quad \forall i \in I, \forall k \in K, \forall p \in \{P|p = i\}, \forall t \in \{T|t > 0\} \tag{14}$$

$$INVM_{pmt} = INVM_{pmt-1} + \sum_{f \in F} W_{fpmt-LFP_{fpm}} - \sum_{k \in K} S_{pkt} * R_{mk}, \quad \forall m \in M, \forall p \in P, \forall t \in \{T|t > 0\} \tag{15}$$

$$INV_{ikt} - BB_{ikt} = INV_{ikt-1}BB_{ikt-1} + \sum_{q \in I} X_{qikt-L_{qi}} + S_{pkt-TPK_k} - \sum_{q \in I} X_{iqkt},$$
$$\forall i \in CDN, \forall k \in K, \forall p \in \{P|p = i\}, \forall t \in \{T|t > 0\} \quad (16)$$

$$INV_{ikt} - BB_{ikt} = INV_{ikt-1} - BB_{ikt-1} + \sum_{q \in I} X_{qikt-L_{qi}} - D_{ikt},$$
$$\forall i \in CDR, \forall k \in K, \forall t \in \{T|t > 0\} \quad (17)$$

$$\sum_{i \in CDN} \sum_{j \in I} \sum_{d=t-V_k}^{d=t-L_{ij}} X_{ijkd} + \sum_{p \in P} S_{pkt} \geq \sum_{i \in I} INV_{ikt}, \forall k \in K, \forall d \in \{T|t \geq 1\} \quad (18)$$

Finally, constraints initialising the model by not allowing production or distribution at negative or zero periods, establishing initial inventories of raw ingredients and products and establishing the backorders levels in plants, CDN and CDR were considered in the model.

Table 1. Parameters and decision variables.

Meaning	Meaning	Meaning
I: Set of nodes that represents de CDR and CDR	*T:* Set of time periods	*V:* Set of types of vehícles
K: Set of products	*M:* Set of raw ingredients	*P:* Set of pruduction plants
F: Set of suppliers	*CDN*: Set of CDN	X_{ijkt} : Amount of product $k \in K$ sent between distribution centres $i, j \in I$ in each period $t \in T$
Y_{ijvt}: Amount of freights between the pair of distribution centres $i, j \in I$ in vehicle $v \in V$ in each period $t \in T$	W_{fpmt} : Amount of raw ingredient $m \in M$ sent from supplier $f \in F$ to plant $p \in P$ in each period $t \in T$	INV_{ikt} : Amount of product $k \in K$ inventory in the distribution centre $i \in I$ in each period $t \in T$
R_{mk} : Amount of raw ingredient $m \in M$ needed to produce $k \in K$	$INVM_{pmt}$: Amount of raw ingredient $m \in M$ inventory in plant $p \in P$ in each period $t \in T$	$CapSCD_i$: Capacity of distribution centre $i \in I$ warehouse for "dry" raw ingredients
MinBB: Min. level of backorders	$CapV_v$: Capacity of vehicle $v \in V$, ton	CaV_v: Number of vehicles $v \in V$

(*continued*)

Table 1. (*continued*)

Meaning	Meaning	Meaning
CMP_{fmp} : Raw ingredient $m \in M$ cost from supplier $f \in F$ to plant $p \in P$	S_{pkt} : Amount of product $k \in K$ produced in plant $p \in P$ in each period $t \in T$	LFP_{fpm} : Leadtime from each supplier $f \in F$ to each plant $p \in P$ for a raw ingredient $m \in M$
BB_{ikt} : Amount of back-orders of product $k \in K$ in the distribution centre $i \in I$ in each period $t \in T$	CM_{ik} : Product $k \in K$ warehouse cost in distribution center $i \in I$	$IT0F_{pm}$: Initial inventory of raw ingredient $m \in M$ in plant $p \in P$
CDR: Set of CDR	CF_{ijv} : Freight cost between distribution centres $i, j \in I$ in vehicle $v \in V$	$IT0I_{ik}$: Initial inventory of product $k \in K$ in distribution centre $i \in I$
$CapProv_{fm}$: Capacity of supplier $f \in F$ of each raw ingredient $m \in M$	CP_{ik} : Picking cost for product $k \in K$ in distribution centre $i \in I$	CO_{2t}: Amount of CO_2 produced in period $t \in T$
$CapFCD_i$: Capacity of distribution centre $i \in I$ warehouse for "cold" raw ingredients	CB_{ik} : Backorders cost for product $k \in K$ in distribution centre $i \in I$	$CIMP_{mp}$: Raw ingredient $m \in M$ warehouse cost in plant $p \in P$
$CapSF_p$: Capacity of plant $p \in P$ warehouse for "dry" raw ingredients	TP_k : Binary parameter that assumes value 1 if the product $k \in K$ is "cold" type	CO_2 : Emissions associated to the execution of product $k \in K$
B_{fpm}: Binary parameter that determines if the supplier $f \in F$ can send the raw ingredient $m \in M$ to the plant $p \in P$	TMP_m : Binary parameter that assumes value 1 when the raw ingredient type is "cold" $m \in M$	$CapBF_p$: Capacity of plant $p \in P$ warehouse for "cold" raw ingredients
DT_{ij} : Distance between each pair of distribution centres $i, j \in I$	CO_{2V} : Emissions associated to the vehicle $v \in V$	D_{ikt} : Demand in period of CDR for product
H: Planning horizon	*M*: Big number	VUP_k : Shelf life of product $k \in K$
$CapProd_{pk}$: Capacity of plant production $p \in P$ for product $k \in K$	L_{ij} : Leadtime between each distribution centre $i \in I$	CPR_{pk} : Production cost for product $k \in K$ in plant $p \in P$
A_{ijk} : Binary parameter that indicates if each pair of distribution centres $i, j \in I$ are connected for product $k \in K$	TPK_k : Production time of product $k \in K$, days	BI_{pk} : Binary parameter that determines if the plant $p \in P$ produces product $k \in K$

4 Experimental Protocol

4.1 Purpose

This research has two fundamental objectives with their corresponding hypotheses:

1. To validate the proposed mathematical model and to observe if it is possible, under exact methods, to obtain a solution in the large instances with limited run time. For this question, it is assumed that, with at least one variable set as integer, the exact method will not be able, within the 1 h run time limit, to show a solution to the problem for the largest instances.
2. Determine how sensitive is the proposed configuration regarding a restricted lateral transshipment between CDN and an open lateral transshipment. In the first case, for each product not all the CDN's are allowed to send products between them. In the second case there are not restrictions for the CDNs. It is considered that having an open transshipment, the model will not present major changes taking into account that it is a complex problem affected by several capacity restrictions.

4.2 Materials

The proposed experiment has three main components that are presented below: experimental instances, implementation software and the stop criteria of the runs.

Experimental Instances

In order to validate the model, 20 instances were generated, based on those elaborated in the previous work and under the conditions defined in the Fig. 1. The first 10 instances were generated with a planning horizon of 7 days. The instances from 11 to 20 were generated with a planning horizon of 14 days. The instances allow to measure the behaviour of the model over a deductive context.

ID	Suppliers	Ingredients	Production Plants	CDN	CDR	Products
1 -11	2	6	1	1	2	3
2 - 12	3	9	2	2	3	4
3 -13	3	12	3	3	4	6
4 - 14	5	13	4	4	5	7
5 -15	5	15	5	5	6	9
6 -16	7	20	6	6	7	15
7 - 17	10	30	6	6	7	25
8 - 18	12	35	6	6	7	40
9 - 19	15	40	6	6	7	50
10 -20	15	40	7	7	7	50

Fig. 1. Instances parameters

The parameters related with warehouse and production capacities, demand, shelf life, type of products, amount of trucks and costs were generated taking into account the reference values and information obtained in the first article that tracks parameters of a real case. Parameters such as connections between links and initial inventories for the products follow the guidelines below:

- *Generation of connections between links*: Connections were generated where plants only send products to their own CDN, and CDR do not send products to any other centre. Additionally, one CDN does not send to all CDN a product. The assignments were made taking into account that the plants that make the products are the ones that send to the other plants.
- *Generation of connections between suppliers and plants*: These connections were generated according to the products to be made and their raw material needs, ensuring that at least half of the suppliers could provide raw materials to the plant.
- *Initial inventories*: The initial inventories were randomly, following the parameters given by the real bounds, but so that there are no backorders at initial times.

LP Software Implementation
The instances were tested on the GUSEK software with open source LP/MILP IDE for Win32; it limits running time and generates optimal Lineal Programming Models.

Stop Criteria for Integer Solutions
The integer model was run with a stop criteria of one hour.

4.3 Methods

To create each instance according to the limits and guidelines set defined in the previous section, a master file was generated for the largest instance - no. 20. It was confirmed that each scenario is feasible and there are no cases such as a CDN whose plant does not manufacture a product distributed to other centres, adhering to the policy that the company applies in real cases.

A first test was carried out with this instance to validate the feasibility of the parameters generated; based on this feasible solution the other smaller instances were adjusted. Likewise, in all instances, the parameters associated with the trucks (quantity, capacity and CO_2 emissions produced per km driven) remained constant.

Once the first run time test was carried out and defined with a maximum duration of 1 h, the whole experimentation was carried out in all instances. Taking into account the first hypothesis, the same instances were run in parallel with the variable quantity of freight. Likewise, in order to fulfil the second objective, the instances were run with all the continuous variables (relaxed), changing exclusively the control variable.

For each one of the instances there were obtained the values of total run time, total objective function, total amount of CO_2 emissions generated, level of customer satisfaction, total logistics costs and the value of each logistics cost in order to analyze its incidence in the supply chain.

4.4 Controls

By establishing as second objective of the research the measurement of model sensitivity for the restriction of lateral transshipment between CDNs, the variation of the parameter $A_{i,j,k}$ was considered as a control variable in the experiment - specifically in the configuration of the parameter for the connections between CDNs for each product. This parameter was altered allowing the existence of lateral transshipment for all the products between all the centres. The behaviour of the model was observed in contrast with the results of the relaxed instances.

4.5 Data Interpretation

The collected data of each executed instance, including those associated with the control variable, were analysed in terms of running time and behaviour of each variable associated with the objective function. The percentage gap between the reference model and the model established for comparison was established as gap measure, where R_e are the results of the model establish for comparison and R_{ref} are the results of the reference model:

$$gap = \frac{R_e - R_{ref}}{R_{ref}} \tag{19}$$

Two main comparisons were made: one between the whole model and the relaxed model in order to check the first hypothesis, and the another between the relaxed model and the model with the control variable for the analysis and review of the second hypothesis.

5 Results

The results of the experimentation protocol are shown in the following section. Based on the results, we will proceed to present as results the comparison between the integer a relaxed model in both open and restricted transshipment variables. Each comparison presents the results in terms of running time, behaviour of the objective function and its components (CO_2 emissions, customer satisfaction and total logistics cost), and behaviour of each of the logistic cost that the model considers.

5.1 Comparison Between Integer and Relaxed Models - Open Transshipment

The results of the comparison between integer model and relaxed model allow solving the problem about the capacity of the exact integer model in large instances. In this part of the experiment it was not possible to obtain results for four integers during the established run time.

In terms of the behaviour of the objective function, the average gap for the weighted function considering the three components was of $-16{,}03\%$, However, in most of the instances the gap ranged from $-0{,}64\%$ to $-0{,}07\%$, having atypical behaviour in instances 13 and 14 as a consequence of a substantial difference in the amount of emissions produced and logistical costs having lower values in the relaxed model. Respecting customer satisfaction was in average of 69,2% and 70,8%.

It is important to mention that the customer satisfaction indicator was calculated using the formula:

$$1 - \frac{(\sum_{i \in CDR} \sum_{k \in K} \sum_{t \in T} BB_{i,k,t}/H)}{\sum_{i \in CDR} \sum_{k \in K} \sum_{t \in T} D_{i,k,t}} \tag{20}$$

These results lead to the inference that the relaxed model is an excellent approximation to the optimal model and can be taken as a baseline for future studies with the instances that did not obtain results in their entire version.

Performing the analysis of logistics costs, it can be seen that the average gap for each cost was between −5.18% and 4.60. The costs with more variation between models are those associated with inventories, followed by production cost ones. The cost with the greatest weight is the one associated with backorders, representing an average of 76,5% for the relaxed model and 72,8% for the entire model. This difference can be explained with the higher production and lower inventory cost.

This last result of backorders cost offers an interesting perspective for the next phase of the experiment where special attention will be given to what happens with this variable when the parameter associated to the transshipment changes.

Results Analysis

With the results obtained in this part of the experiment, the first hypothesis could be verified, seeing that for large instances the exact method could not lead to a solution in the established running time. Likewise, it was possible to observe that, given the generated parameters, the most significant logistic cost in the proposed supply chain is the cost associated to backorders. Although the service indicator associated with average backorders offers in both cases (the integer model and the relaxed model) a service level close to 70%, it generates at the cost level a great impact on the supply chain. This invites to think that the system offers improvement possibilities to reach better levels for the indicator as well as to reduce costs. Likewise, this point of the research imposes taking into account this indicator and the costs as a priority point for the second part of the experimentation that considers the open transshipment.

At the level of CO_2 emissions, we can observe that its association with the productions costs can have great implications when making decisions in the model.

5.2 Comparison Between Integer and Relaxed Models - Restricted Transshipment

The results of the comparison between the integer model and the relaxed model with an open transshipment allow solving the sensitivity problem of the model for the consideration of this variable. Based on the results of the first comparison, the analysis will take into account how the cost of backorders and the level of service evolves.

With the change applied to an open transshipment model, it is evident that the instances take three or four times more time compared to the relaxed model with open transshipment. A different behaviour is observed in terms of gaps between the planning horizons, in which for the planning horizon of 7 days the gap oscillates between 0% and −0.26%, and for the 14-day planning horizon the gap ranges from 0% to 25.76%. This results implies a greater sensitivity of the model to the change of open transshipment.

A large variability in the behaviour of each component of the objective function was obtained for all 20 instances. Regarding CO2 emissions, an average variation of 105% was obtained, explained by the greater number of trips made and tons produced. In terms of customer satisfaction, for the instances with a planning horizon of 7 days, the values remained between 66.2% and 73.4%. For the instances with planning horizons of 14 days, this indicator was improved with respect to the results of the relaxed solution, with values ranging from 82.9% to 90.6%. This presents a considerable enhancement and shows the sensitivity of the model in instances with long time horizons.

Finally, the total logistical costs showed a reduction in all the instances, ranging from 0.07% to 25.76%. Taking into account the previous results of the comparison between the integer model and the relaxed model, it can be observed that for 14 days planning horizon, having an open transshipment causes a significant incidence in the model results and changes the participation of the cost in the model. This reduces the backorders cost and increases the freight and production cost that support the positive growth of the customer satisfaction indicator.

Results Analysis
From the results presented, it can be seen that the second hypothesis is partially fulfilled. For the instances with a planning horizon of 7 days, having an open transshipment versus a restricted one, even though it reduces some costs, the difference is not so significant compared to the results of the instances with a planning horizon of 14 days.

The CO_2 emissions, the level of customer satisfaction, and the total logistics costs had a considerable impact with the changes made. For logistics costs and customer satisfaction the impacts were positive, greatly reducing the costs associated with backorders and obtaining levels of up to 90%. However, given the increase in production and the amount of freight, the average CO_2 was tripled compared to the instances of the relaxed model.

6 Conclusions

In this paper an IRP problem with lateral transshipment was modelled and analysed taking into account a supply chain including production plants and distribution centres for perishable products. The model evaluated a multi-objective function with a weighted sum method that aimed to assess not only total logistics costs, but also the amount of CO_2 emissions produced and the level of customer satisfaction from the point of average backorders. This model is a contribution to the current research literature taking into account not only the particular case of perishable products but also the multi-period, multi-agent and variable lead-times within an IRP that allow transshipment between its nodes. The proposed model can be used to make logistics decisions in companies like the one presented in the case study and integrate green logistics aspects.

The results obtained showed that through accurate models solutions can be generated for small instances; however, for large instances unless a relaxation of the variables is presented, it is not possible to obtain results in running times of one hour. Likewise, it was proved that the relaxed model presents a very good solution against the integer one because it has an average gap of -1% and can be taken into account for comparisons of future developments for the same case study. This relaxed model presents its major difference in inventory costs (5.2%) and production costs (4.6%).

Evaluating the second hypothesis, it was possible to refute partially the initially presented hypothesis and to observe that the model is very sensitive to the change from a restricted transshipment to an open one, especially in instances with 14-day planning horizon, obtaining an average gap of (-16.03%). In the scenarios with open transshipment it was observed in general that logistics costs were reduced and the level of customer satisfaction increased; however, the costs of CO_2 emissions increased based on an increase in production and internal distribution of the products.

6.1 Research Perspectives

Based on the results presented for future research, the sensitivity to other variables such as different storage and production capacities can be reviewed for both the restricted and open transshipment scenario. Likewise, the objective function can be planned, not under a proposal of goal weighting, but under a proposal of programming by goals with minimum values of fulfilment of service level to the client and maximum values for CO_2 emissions and logistic costs.

In addition, the model presents opportunities to develop non-solution methods, such as heuristics and metaheuristics, considering the same scenarios and control variables. Additionally, the model presented in the article can extend the nodes to retailers and approaches with stochastic demand. This model can be also used for future studies about predictive decision making and not only as the deductive perspective presented.

References

1. Al Shamsi, A., Al Raise, A., Muhammad, A.: Pollution-inventory routing problem with perishable goods. In: Proceedings of the Conference on Logistics Operations, Supply Chain Management and Sustainability, pp. 585–596 (2014)
2. Alayón, S., Castro, A., Restrepo, N., Montoya, C.: Diseño de un aplicativo que determine las políticas de inventarios en una empresa de productos lácteos. Degree Thesis, Institutional Repository - Pontificia Universidad Javeriana (2017)
3. Amorim, P., Almada-Lobo, B.: The impact of food perishability issues in the vehicle routing problem. Comput. Ind. Eng. **67**(1), 223–233 (2014)
4. Azadeh, A., Elahi, S., Farahani, M.H., Nasirian, B.: A genetic algorithm-Taguchi based approach to inventory routing problem of a single perishable product with transshipment. Comput. Ind. Eng. **104**, 124–133 (2017)
5. Cheong, T.: Joint inventory and transshipment control for perishable products of a two-period lifetime. Int. J. Adv. Manuf. Technol. **66**(9–12), 1327–1341 (2013)
6. Diabat, A., Abdallah, T., Le, T.: A hybrid tabu search based heuristic for the periodic distribution inventory problem with perishable goods. Ann. Oper. Res. **242**(2), 373–398 (2014). https://doi.org/10.1007/s10479-014-1640-4
7. Filina-Dawidowicz, L., Postan, M.: Optimal inventory control for perishable items under additional cost for deterioration reduction. Logforum **12**(2), 147–156 (2015)

8. Jedliński, M.: The position of green logistics in sustainable development of a smart green city. Procedia. Soc. Behav. Sci. **151**, 102–111 (2014)
9. Le, T., Diabat, A., Richard, J.-P., Yih, Y.: A column generation-based heuristic algorithm for an inventory routing problem with perishable goods. Optim. Lett. **7**(7), 1481–1502 (2012). https://doi.org/10.1007/s11590-012-0540-2
10. Muriana, C.: An EOQ model for perishable products with fixed shelf life under stochastic demand conditions. Eur. J. Oper. Res. **255**(2), 388–396 (2016)
11. Paterson, C., Kiesmüller, G., Teunter, R., Glazebrook, K.: Inventory models with lateral transshipments: a review. Eur. J. Oper. Res. **210**(2), 125–136 (2011)
12. Shaabani, H., Kamalabadi, I.N.: An efficient population-based simulated annealing algorithm for the multi-product multi-retailer perishable inventory routing problem. Comput. Ind. Eng. **99**, 189–201 (2016)
13. Soysal, M., Bloemhof-Ruwaard, J.M., Haijema, R., Van Der Vorst, J.: Modeling an Inventory Routing Problem for perishable products with environmental considerations and demand uncertainty. Int. J. Prod. Econ. **164**, 118–133 (2015)
14. Vahdani, B., Niaki, S., Aslanzade, S.: Production-inventory-routing coordination with capacity and time window constraints for perishable products: heuristic and meta-heuristic algorithms. J. Clean. Prod. **161**, 598–618 (2017)

Optimization in Manufacturing and Services

A Mixed-Integer Linear Model for Solving the Open Shop Scheduling Problem

Daniel Morillo-Torres[1(✉)] and Gustavo Gatica[2]

[1] Pontificia Universidad Javeriana - Cali, Cali, Colombia
daniel.morillo@javerianacali.edu.co
[2] Universidad Andres Bello, Santiago, Chile
ggatica@unab.cl

Abstract. This paper addresses the Open Shop Scheduling Problem with non-identical parallel machines. In this context, a finite set of jobs must be processed by a finite set of machines in any order. However, the machines can only process a single job at a time. The objective is to minimize the maximum completion time of the jobs, known as $C_{\max}$ or *makespan*. In this paper, a mixed-integer linear programming model is presented for this problem; it uses time-based decision variables and disjunctive constraints. The model allows each job to have a different number of operations. Computational results are tested with the Gurobi solver and the three best-known benchmark libraries from the literature. These results show that the mathematical model proposed efficiently solves open shop scheduling problems with 6 machines and 6 jobs and its optimal value has a maximal deviation of 6.18%.

Keywords: Open Shop Scheduling · MILP · Job scheduling

1 Introduction

Scheduling problems can be defined as the allocation of scarce resources over time for processing a finite set of activities that make up a project [1]. In a decision-making process, the objective is to optimize a measure of performance. Generally, it is to maximize the profit or minimize project execution time. At an operational level, this allocation can be viewed as a sequencing problem.

Job scheduling became more important at the end of the last century [14]. Companies realized that tactical decisions related to scheduling had an impact on processing times, delayed deliveries and customer service. This problem is further highlighted by production philosophies such as just-in-time or lean-manufacturing [12].

Job scheduling problems are a classic in the Operations Research community given their complexity and relevance in industrial environments [3]. In addition, new metaheuristics based on artificial intelligence contribute to its resolution in smart manufacturing environments as required by Industry 4.0 [24].

D. Trentesaux et al. (Eds.): SOHOMA 2021, SCI 987, pp. 301–310, 2021.
https://doi.org/10.1007/978-3-030-80906-5_21

In the industry, most scheduling problems can be defined by the processing routes and special constraints of the jobs. The three main production systems are: Flow Shop, Job Shop and Open Shop [17]. The first two that have been given higher priority in the literature [16]. In a Flow Shop, all jobs have the same processing route through the machines. In a Job Shop, each job can have a specific processing route through the machines. In an Open Shop, there is no defined processing route through the machines. That means each job can be scheduled in any processing route (any sequence).

In these three systems, the main constraint is that each machine can process only one job at a time. Nevertheless, the Flow Shop and Job Shop models have only one decision level: finding the optimal job sequence. On the other hand, the Open Shop has two decision levels: finding an optimal processing route and an optimal job sequence [17].

Most scheduling problems are highly complex; even for small instances, the problems are classified as NP-Hard [7]. For this reason, researchers around the world have studied new intelligent solutions in recent decades [12].

In practice, several real-world applications can be modelled as an Open Shop Scheduling Problem (OSSP). Furthermore, formulating mathematical models to highly complex problems as OSSP is a contribution of classical scheduling problems. For these reasons, in this paper, a Mixed-Integer Linear Programming (MILP) is proposed for solving the OSSP, where every jobs can have a different number of operations. Computational experimentation was carried out on well -known state-of-the-art benchmarks: the first set, with 60 instances [21], the second set, with 52 more difficult instances [4] and the third set, with 80 instances [10].

The remainder of this paper is organized as follows: Sect. 2 formally introduces the OSSP, Sect. 3 describes the main methodologies in the existing literature for solving the OSSP, Sect. 4 presents the MILP proposed for this problem, then Sect. 5 presents computational experiments and their results. Finally, in Sect. 6 the concluding remarks and areas open to future work are presented.

2 Problem Description

Formally, in the OSSP a set J of n jobs must be processed on a set I of m machines (or stages). Each job consists of N activities (operations or tasks), each of which is executed on a machine [17]. If a job needs to be processed by all machines, $N = m$. Additionally, the problem includes a processing time (P_{ji} for all $j \in J$ and $i \in I$). In deterministic scheduling, all parameters in the problem are known [3].

The activities of each job can be done in any order (processing route), but only one at a time. Similarly, each machine only can process one task at a time. Finally, only non-preemptive schedules are allowed. The objective is to find both an optimal processing route for each job and an optimal job sequence in order to minimize the total processing time of the whole project (known as $C_{\max}$ or *makespan*).

Following the standard classification of scheduling problems introduced by [8] ($\alpha \mid \beta \mid \gamma$), the OSSP can be defined as $O_m \mid\mid C_{\max}$ where m is the number of machines. The $O_2 \mid\mid C_{\max}$ can be solved by the Longest Alternate Processing Time First rule (LAPT) in a polynomial time. For $m \geq 3$ ($O_3 \mid\mid C_{\max}$), the problems are NP-Hard [6].

As an example of this problem, the instance *tai04_01* proposed by [21] is depicted in Table 1. Additionally, a feasible solution and an optimal solution for this instance are presented in Fig. 1.

Table 1. Processing times of the instance *tai04_01*.

	M1	M2	M3	M4
Job 1	34	15	38	95
Job 2	2	89	19	7
Job 3	54	70	28	34
Job 4	61	9	87	29

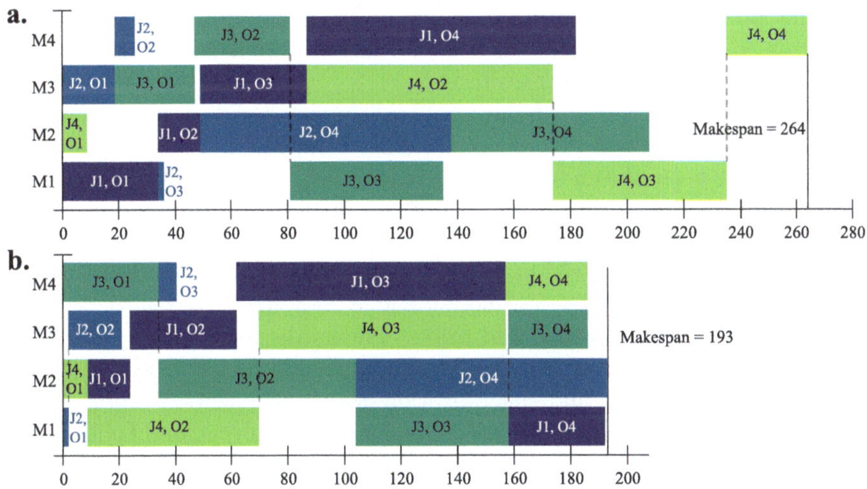

Fig. 1. (a) A feasible solution and (b) an optimal solution of the instance *tai04_01*.

3 Related Works

Despite the large number of publications in the scheduling field, most of them are focused on Job Shop and Flow Shop Problems. The development of efficient methodologies for solving Open Shop Problems, on the other hand, has received less attention [16].

The main solving methods for this problem can be classified into two categories: exact and approximate approaches. The former achieve an optimal solution but with a high computational cost. The latter are very time efficient but sacrifice accuracy to find an optimum solution [12].

One of the first exact methods was proposed by [4]. It consists of a branch and bound algorithm based on a disjunctive graph model. Its main contribution is the branching scheme and the heuristics for calculating the upper bounds. The results show that some of the problems with seven machines cannot be solved despite its efficiency. Later, [9] improved on the [4] branch and bound approach through a technique based on intelligent backtracking. It reduces the number of backtracks in the algorithm, but it doubles the processing time of each node.

[5] improved the branch and bound approach by focusing on constraint propagation-based methods that reduce the search space. Their algorithm was the first to solve many benchmark-problem instances to optimality in a short amount of time. However, the computation time was unfeasible for real-size problems. In [22] proposed a new method to encode Constraint Optimization Problems (COP) with integer linear constraints into Boolean Satisfiability Testing Problems (SAT). This approach proved to be highly efficient, optimally solving all of the benchmark instances. The Constraint Programming (CP) approach has also been developed for the OSSP. [15] proposed a CP model that integrates constraint propagation and the best branching techniques enhanced with randomization and restart. The results show that their model is a very efficient approach for most of the instances tested.

Due to the NP-hardness of the OSSP, the exact methods are not able to achieve optimal solutions in polynomial time. Considering that approximate algorithms have been proven to efficiently reach a near-optimal solution, their use is justified [3]. However, the formulation of new MIP models is still of great relevance in science, mainly due to three reasons: first, multiple small and moderate size NP problems can be solved by MIP efficiently; second, MIP models represent the base for exact decomposition methods able to address large scale problem namely branch and cut, branch and price, Benders decomposition, among others. Finally, a novel MIP can be solved by commercial optimization solvers, and facilitate the modification and generalization of the optimization problem.

Finally, a novel MIP can be solved by optimization solvers, and facilitate the modification and generalization of the optimization problem.

Approximate algorithms can be classified as heuristic or metaheuristic. The former are mostly used to build a feasible initial solution while the latter improve the first solution through intelligent search strategies. [23] were among the first to propose a Genetic Algorithm (GA) for solving the OSSP. Later, [18] improved the performance of the GA through two key features: a population in which each individual has a different makespan, and a reordering procedure that increases the efficiency of the crossing operator. [2] proposed a hybrid Ant Colony Optimization, which improves the building solutions with a Beam Search (a tree search method). The results showed that this approach is competitive. A Particle Swarm Optimization was proposed by [20]; they modified the particle position representation and the particle movement. [16] analyzed the unwanted phenomenon in building solutions using metaheuristic. It is called redundancy on the encoding scheme and it causes different encodings to produce identical real solutions.

More recently, [11] proposed an adapted Simulated Annealing, which uses two neighborhood transformations to efficiently explore the solution space. [19] analyzed of the effects of the crossover and mutation operation on GAs and proposed an improved GA with hybrid selection. This proposal achieves one of the best results in the literature.

Despite significant improvements in metaheuristic development to solve the OSSP, research on exact methods in the scheduling field also continues to progress steadily. This paper contributes to the field by proposing a new linear mixed-integer formulation.

4 Proposed Model

In this section, the proposed MILP model is presented; it uses time-based decision variables and disjunctive constraints. The model is based on the model proposed by [13] for the Job Shop Problem. The proposed model permits jobs to have a different number of operations (less than or equal to) the number of machines. This is useful in practical scenarios where jobs may not require processing on all available machines.

Below is the definition of the sets, followed by the parameters and decision variables used. Finally, a complete mathematical model is given.

Sets

- $J = \{1, ..., j, ..., n\}$ set of jobs to be scheduled.
- $I = \{1, ..., i, ..., m\}$ set of machines.

Parameters

- P_{ji} : execution time of job $j \in J$ processed on machine $i \in I$.
- $r_{ji} = \begin{cases} 1, & \text{if job } j \in J \text{ must be processed on machine } i \in I. \\ 0, & \text{otherwise.} \end{cases}$
- N : maximum number of job operations.
- M : a big value regarding the other parameters.

Decision variables

- s_{jki} : start time of operation $k \in I$ of job $j \in J$ processed on machine $i \in I$.
- $x_{jki} = \begin{cases} 1, & \text{if operation } k \in I \text{ of job } j \in J \text{ is processed on machine } i \in I. \\ 0, & \text{otherwise.} \end{cases}$
- $Z_{j_1k_1j_2k_2i} = \begin{cases} 1, & \text{if operation } k_1 \in I \text{ of job } j_1 \in J \text{ precedes operation } \\ & k_2 \in I \text{ of job } j_2 \in J \text{ on machine } i \in I. \\ 0, & \text{otherwise.} \end{cases}$
- $C_{\max}$: the makespan of the project.

The proposed Mixed-Integer Linear Program

$$\text{minimize: } C_{\max} \tag{1}$$

Subject to:

$$x_{jki} \leq r_{ji}, \quad \forall j \in J, \forall k, i \in I \tag{2}$$

$$\sum_{k \in I} x_{jki} = r_{ji}, \quad \forall j \in J, \forall i \in I \tag{3}$$

$$\sum_{i \in I} x_{jki} \leq 1, \quad \forall j \in J, \forall k \in I \tag{4}$$

$$s_{jki} \leq M \cdot x_{jki}, \quad \forall j \in J, \forall k, i \in I \tag{5}$$

$$s_{jki_1} + P_{ji_1} \leq s_{j,k+1,i_2} + M \cdot (2 - x_{jki_1} - x_{j,k+1,i_2}), \quad \forall j \in J, \forall k \in \{1, ..., N-1\}, \forall i_1, i_2 \in I; i_1 \neq i_2 \tag{6}$$

$$s_{j_1k_1i} \geq s_{j_2k_2i} + P_{j_2i} - M \cdot (2 - x_{j_1k_1i} - x_{j_2k_2i}) - M \cdot Z_{j_1k_1j_2k_2i}, \quad \forall j_1, j_2 \in J; j_1 < j_2, \forall k_1, k_2, i \in I \tag{7}$$

$$s_{j_2k_2i} \geq s_{j_1k_1i} + P_{j_1i} - M \cdot (2 - x_{j_1k_1i} - x_{j_2k_2i}) - M \cdot (1 - Z_{j_1k_1j_2k_2i}), \quad \forall j_1, j_2 \in J; j_1 < j_2, \forall k_1, k_2, i \in I \tag{8}$$

$$C_{\max} \geq s_{jNi} + P_{ji}, \quad \forall j \in J, \forall i \in I \tag{9}$$

$$s_{jki} \geq 0, \quad \forall j \in J, \forall i, k \in I \tag{10}$$

$$x_{jki} \in \{0, 1\}, \quad \forall j \in J, \forall i, k \in I \tag{11}$$

$$Z_{jkj_2k_2i} \in \{0, 1\}, \quad \forall j, j_2 \in J, \forall i, k, k_2 \in I \tag{12}$$

Expression (1) represents the objective function that seeks to minimize the variable $C_{\max}$ (described by Constraint (9)). Constraint (2) states that operation allocation x_{jki} only can exist if the job j requires processing on machine i (represented by r_{ji}). Constraint (3) ensures that each job j be processed on the machines it requires. Constraint (4) guarantees that no more than one operation can be processed on each machine at any given time. Constraint (5) imposes the dynamics of the continuous variable s_{jki} and binary variable x_{jki}. Constraint (6) prevents operations k and $k+1$ (in any order) from running simultaneously. Expressions (7) and (8) are disjunctive constraints. They ensure that if operation k_1 of job j_1 precedes operation k_2 of job j_2 on the same machine i (i.e. $z_{j_1k_1j_2k_2i} = 1$), the start time of operation k_2 of j_2 must be greater than or equal to the start time of operation k_1 of job j_1 plus its processing time. In this case, the expression (8) is the active constraint. Similarly, in the other case when $z_{j_1k_1j_2k_2i} = 0$, the start time of operation k_1 of j_1 must be greater than or equal to the start time of operation k_2 of job j_2 plus its processing time. In this case, the expression (7) is the active constraint. Constraint (9) states that the *makespan* must be greater than or equal to the greatest finish time of the last operation of each job. Finally, constraints (10) to (12) specify the nature of the variables.

The proposed model has $n \cdot m^2(1+n \cdot m)$ binary variables, $n \cdot m^2+1$ continuous variables and $n \cdot m(n \cdot m^2 + 4)$ constrains. The polynomial behaviour close to an $\mathcal{O}(n^3)$, implies the need to explore new formulations and efficient algorithms in computational time. Without loss of generality, in Fig. 2 the behaviour of the number of binary variables is depicted.

5 Computational Experiments

The three classical benchmarks of OSSP instances are used in this section. The first, [21] set, consists of 60 symmetric instances ranging from 4 jobs and 4 machines (16 operations) to 20 jobs and 20 machines (400 operations). In this paper, this set is denoted by *tai**, where * is *n_m*, the number of jobs and machines, respectively. The second, [4] set, consists of 52 more difficult instances ranging from 3 jobs and 3 machines (9 operations) to 8 jobs and 8 machines (64 operations). This set is denoted by *j*-per*, where * is the number of jobs and machines. Finally, the third is [10] set, with 80 instances ranging from 3 jobs and 3 machines (9 operations) to 10 jobs and 10 machines (100 operations). This set is denoted by *GP**, where * again is the number of jobs and machines. The benchmarks and the optimal solutions reported are available at the following Google drive link: https://shorturl.at/tDKT6.

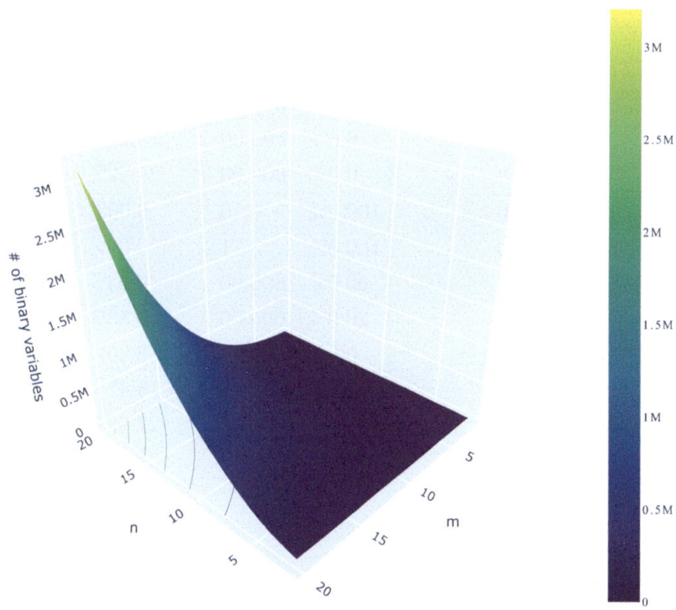

Fig. 2. The number of binary variables according to the number of jobs n and machines m.

The experiments were carried out on a Windows PC with an Inter(R) Core(TM) i7 - 3.40 GHz processor and 16 GB of RAM. The instances are solved using CPLEX 12.8.0.0 and the AMPL language. Table 2 presents a summary of the computational experimentation. It shows the three sets of instances, the number of instances of each subset, the proportion of optimal solutions, the execution time (limited to 600 s) and the gap value. This last value is calculated according to Eq. (13) where O^*_y is the optimal solution of the instance y, BF_y is the best solution found and Y_T is the total number of instances in a given subset.

$$gap_y = \frac{|O^*_y - BF_y|}{O^*_y}, \qquad \overline{gap} = \sum_{y=1}^{Y_T} \frac{1}{Y_T} \cdot gap_y * 100 \tag{13}$$

Table 2. Summary of the computational experimentation of the proposed model on the benchmarks of OSSP instances.

Subset	# instances	% opt	Time (s)	$\overline{gap}$
tai04	10	100.0%	2.33	0.00%
tai05	10	50.0%	558.33	0.53%
tai07	10	0.0%	600.00	8.41%
tai10	10	0.0%	600.00	35.26%
tai15	10	0.0%	600.00	120.71%
tai20	10	0.0%	600.00	124.07%
j3-per	8	100.0%	0.07	0.00%
j4-per	9	100.0%	2.08	0.00%
j5-per	9	55.6%	568.05	1.03%
j6-per	9	0.0%	600.00	6.18%
j7-per	9	0.0%	600.00	15.25%
j8-per	8	0.0%	600.00	25.22%
GP3	10	100.0%	0.10	0.00%
GP4	10	100.0%	3.21	0.00%
GP5	10	90.0%	600.00	0.03%
GP6	10	20.0%	600.00	0.53%
GP7	10	0.0%	600.00	2.96%
GP8	10	0.0%	600.00	13.47%
GP9	10	0.0%	600.00	27.62%
GP10	10	0.0%	600.00	51.41%

6 Conclusions and Future Work

In this paper, the Open Shop Scheduling Problem has been addressed, and a new mixed-integer linear programming model has been proposed for solving it. Decision variables are time-based, and they are complemented by binary variables. Disjunctive constraints define the dynamics of jobs, operations and the allocation of machines. One of the main contributions is that the proposed model allows jobs that require a number of operations less than or equal to the number of machines. The objective is to minimize the total execution time of the whole project (*makespan*). In particular, the disjunctive constraints can be modeled as indicator constraints to skip the trickle flow effect, in new releases of commercial solvers.

The computational experiments were carried out in three well-known benchmark sets proposed by [4,10,21], with a total of 204 instances ranging from 3 jobs and 3 machines to 20 jobs and 20 machines. The computational results show that the proposed model is efficient for solving problems with 6 jobs and 6 machines. Its maximal deviation from the optimal value is 6.18% and it has an average execution time of 10 min.

In further research, an algorithm based on a Logic-Based Bender Decomposition (LBBD) of the problem will be proposed to improve the computational time. Furthermore, the authors plan to design a metaheuristic algorithm to evaluate the performance of the LBBD.

References

1. Błażewicz, J., Ecker, K.H., Pesch, E., Schmidt, G., Weglarz, J.: Handbook on Scheduling. International Handbook on Information Systems. Springer, Heidelberg (2007). https://doi.org/10.1007/978-3-540-32220-7
2. Blum, C.: Beam-ACO - Hybridizing ant colony optimization with beam search: an application to open shop scheduling. Comput. Oper. Res. **32**(6), 1565–1591 (2005)
3. Bouzidi, A., Riffi, M.E., Barkatou, M.: Cat swarm optimization for solving the open shop scheduling problem. J. Ind. Eng. Int. **15**(2), 367–378 (2019)
4. Brucker, P., Hurink, J., Jurisch, B., Wöstmann, B.: A branch & bound algorithm for the open-shop problem. Discr. Appl. Math. **76**(1–3), 43–59 (1997)
5. Dorndorf, U., Pesch, E., Phan-Huy, T.: Solving the open shop scheduling problem. J. Sched. **4**(3), 157–174 (2001)
6. Gonzalez, T., Sahni, S.: Open shop scheduling to minimize finish time. J. ACM **23**(4), 665–679 (1976)
7. Gonzalez, T., Sahni, S.: Flowshop and jobshop schedules: complexity and approximation. Oper. Res. **26**(1), 36–52 (1978)
8. Graham, R.L., Lawler, E.L., Lenstra, J.K., Rinnooy-Kan, A.H.G.: Optimization and approximation in deterministic sequencing and scheduling: a survey. Ann. Discret. Math. **5**(1), 287–326 (1979)
9. Guéret, C., Jussien, N., Prins, C.: Using intelligent backtracking to improve branch-and-bound methods: an application to open-shop problems. Eur. J. Oper. Res. **127**(2), 344–354 (2000)
10. Guéret, C., Prins, C.: A new lower bound for the open-shop problem. Ann. Oper. Res. **92**, 165–183 (1999)

11. Harmanani, H.M., Ghosn, S.B.: An efficient method for the open-shop scheduling problem using simulated annealing. Presented at the (2016). https://doi.org/10.1007/978-3-319-32467-8_102
12. Huang, Z., Zhuang, Z., Cao, Q., Lu, Z., Guo, L., Qin, W.: A survey of intelligent algorithms for open shop scheduling problem. In: Procedia CIRP, vol. 83, pp. 569–574. Elsevier B.V., January 2019
13. Ku, W.-Y., Christopher Beck, J.: Mixed Integer Programming models for job shop scheduling: a computational analysis. Comput. Oper. Res. **73**, 165–173 (2016)
14. Lei, D., Cai, J.: Multi-population meta-heuristics for production scheduling: a survey. Swarm Evol. Comput. **58**, 100739 (2020)
15. Malapert, A., Cambazard, H., Guéret, C., Jussien, N., Langevin, A., Rousseau, L.-M.: An optimal constraint programming approach to the open-shop problem. INFORMS J. Comput. **24**(2), 228–244 (2012)
16. Naderi, B., Fatemi Ghomi, S.M.T., Aminnayeri, M., Zandieh, M.: A contribution and new heuristics for open shop scheduling. Comput. Oper. Res. **37**(1), 213–221 (2010)
17. Pinedo, M.L.: Scheduling - Theory, Algorithms, and Systems, 5th edn. Springer, Cham (2016). https://doi.org/10.1007/978-3-319-26580-3
18. Prins, C.: Competitive genetic algorithms for the open-shop scheduling problem. Math. Meth. Oper. Res. **52**(3), 389–411 (2000)
19. Rahmani-Hosseinabadi, A.A., Vahidi, J., Saemi, B., Sangaiah, A.K., Elhoseny, M.: Extended Genetic Algorithm for solving open-shop scheduling problem. Soft Comput. **23**(13), 5099–5116 (2019)
20. Sha, D.Y., Hsu, C.Y.: A new particle swarm optimization for the open shop scheduling problem. Comput. Oper. Res. **35**(10), 3243–3261 (2008)
21. Taillard, E.: Benchmarks for basic scheduling problems. Eur. J. Oper. Res. **64**(2), 278–285 (1993)
22. Tamura, N., Taga, A., Kitagawa, S., Banbara, M.: Compiling finite linear CSP into SAT. Constraints **14**(2), 254–272 (2009)
23. Yamada, T., Nakano, R.: A genetic algorithm applicable to large-scale job-shop problems, vol. 2, pp. 283–292 (1992)
24. Yin, Y., Stecke, K.E., Li, D.: The evolution of production systems from industry 2.0 through industry 4.0. Int. J. Prod. Res. **56**(1–2), 848–861 (2018)

On the Complexity of the Collaborative Joint Replenishment Problem

Carlos Otero-Palencia[1], Jairo R. Montoya-Torres[2(✉)], and René Amaya-Mier[3]

[1] Department of Civil Engineering and Environmental Engineering, Institute of Transportation Studies, University of California at Davis, One Shields Avenue, Ghausi Hall 3143, Davis, CA 95616, USA
cotero@ucdavis.edu

[2] School of Engineering, Universidad de La Sabana, km 7 Autopista Norte de Bogota D.C., Chia, Colombia
jairo.montoya@unisabana.edu.co

[3] Department of Industrial Engineering, Universidad del Norte, km 5 via Puerto Colombia, Barranquilla, Colombia
ramaya@uninorte.edu.co

Abstract. In this paper, we study the open question regarding the computational complexity of one newest version of the Joint Replenishment Problem, the Collaborative Joint Replenishment Problem. This problem has received attention due to his potential in practical settings. However, the complexity of the problem remains unresolved. In this paper, we provide insights to proof that the problem is indeed NP-complete. The complexity is also analyzed using computational experiments.

Keywords: Collaborative Joint Replenishment · Complex analysis

1 Introduction

The market intense dynamics and financial pressures imperatively demand the satisfaction of customer demand at a lessening cost. Under this scenario, companies have implemented different strategies to improve supply chain resilience, among which collaboration between companies appears as a viable alternative. Collaboration allows the generation of win-win situations that could become competitive advantages and which individually could not be achieved [21]. Collaborative and Cooperative practices date back to the 1990's with the implementation of strategies that sought synchronization between echelons in the same supply chain to reduce the bullwhip effect, a known over-cost generator due to the lack of coordination [22]. Later, strategies appeared that sought integration between echelons in companies belonging to different supply chains that might be or not antagonists [4, 17]. In general these strategies demonstrate benefits such as: cost reduction, better service levels, better inventory control, etc. [9].

In the inventory management context, Otero-Palencia *et al.* [19] introduced a model called the Stochastic Collaborative Joint Replenishment Problem (S-CJRP). The S-CJRP explores collaboration between companies in inventory replenishment as an alternative to reduce logistics costs through the joint exploitation of economies of scale.

D. Trentesaux et al. (Eds.): SOHOMA 2021, SCI 987, pp. 311–318, 2021.
https://doi.org/10.1007/978-3-030-80906-5_22

The simplest version of this coordination problem is known in the literature as the Joint Replacement Problem (JRP) introduced early in [16]. Since its appearance, the JRP has been recognized for its potential for application in real settings; however some of the assumptions of the model are controversial such as considering unlimited capacities, deterministic demand, and to neglecting lead times. The S-CJRP challenges those assumptions considering limited warehouse and transport capacities, stochastic demand, and non-zero lead times. Furthermore, it extends the model to a new context where multiple companies jointly replenish their inventory.

The S-CJRP's strategy consists of coordinating the inventory replenishment of multiple non-competing companies, essentially importers that face high fixed costs inherent in the international replenishment process. In particular, such companies may face capacity limitations. For instance, limited capacity of warehouses, transport units, and even budget [2,5]. Under these circumstances, the following question arises: What is the frequency and/or cost-efficient quantity in which each item must be replenished?

In simulated scenarios where the S-CJRP has been tested, it has demonstrated potential to reduce logistics cost; 32.3%. 28.2% and 32.7% for 3, 4 and 5 companies in collaborative agreement, respectively [19,20]. In addition, the model has shown to be a more convenient alternative for expanding warehouse capacity than individual investment; since the model favours a more efficient replenishment operation the storage requirements are reduced. Furthermore, required investments and implicit risks are shared between the cooperating companies [18]. Beyond, alternative extensions have been considered, for instance, to consider sustainable policies implying additional restrictions such as limited budget emissions or required green-friendly fleet composition (the one required to transport the cargo during replenishment) [11]. In this context, collaboration could be a good alternative to leverage the purchase of electric vehicles or automated vehicles in the imminent fourth revolution in transportation [10,12].

Given the benefits of S-CJRP and other JRP's extensions, it is expected that they will eventually migrate from theory to practice. Therefore, computational complexity is an indispensable requirement for three main reasons. Firstly, it allows a glimpse of which solution could be more suitable. Secondly, it allows estimating the required computational time to find an optimal value. Lastly, it contributes by providing insights and approaches to researchers working on the complexity of similar extensions.

The classic JRP is known to be NP-hard [1,15], reason why most authors have proposed heuristic procedures as a solution method [14,15]. Extensions could add or remove elements from the base model eventually reducing or increasing the complexity of the problem. In particular, the S-CJRP is composed of two parts. The first deal with an inventory replenishment model. i.e., an extension of the JRP model with reconsidered assumptions. The second part deals with the allocation among companies of benefits obtained from the exploitation of economies of scale thanks to the model from the first part. Such allocation is performed by means of the Shapley Value function. Which has the virtue of allocating benefits fairly. Fairness in this sense refers to the fact that companies would receive benefits proportional to their contributions to reduce the costs of all companies in the agreement. This paper focuses on the complexity of the first section since the Shapley Value function has a proven NP-complete complexity [7].

The JRP is a special case of the one-warehouse N-retailers problem with very high holding costs (i.e., a single warehouse receives goods from an external supplier and distributes to multiple retailers). In addition to the NP-hardness of the JRP with non-stationary demands [1, 15], other variants of the JRP have been proven to be NP-hard, such as the periodic JRP [6]. Nevertheless, the complexity of the S-CJRP is still an open question. This paper fits such gap.

Section 2 presents the framework to analyze the computational complexity, while a numerical analysis of such complexity is presented in Sect. 3 through a set of computational experiments. The paper concludes in Sect. 4 by presenting the conclusions and drawing some opportunities for further research.

2 Complexity Analysis

This section introduces the S-CJRP mathematical formulation and the proposed complexity analysis approach.

2.1 The JRP Extension in the S-CJRP

Next, the model notation is defined: TC is the total annual cost (ordering, holding and transportation cost), M_s is the set of family items with $M_s = 1,2,3,...,M$, N_s is the set of players with $N_s = 1,2,3,...,N$, n is the number of item families, being $N \times M$; I is the set of family items/players pair, $I = U_{y \in M_s, z \in N_s}(y,z) = \{(1,1),(1,2),...,(1,N),...,(2,1),(2,2)...(2,N),...,(M,N)\}$; D_i is the annual demand rate for the item $i, i \in I$; L is the vendor lead time; σ_i is the standard deviation for the item $i, i \in I$; Z_α is the security level; S is the major ordering cost; s_i is the minor ordering cost for the item $i, i \in I$; T is the time between two consecutive replenishments; h_i is the holding cost for the item family $i, i \in I$; k_i is a positive integer multiple of T for the item $i, i \in I$; A is the cost of a full transport/container unit; W is the maximum capacity of a transport unit; w_i is the weight/volume per unit of item $i, i \in I$; H is the maximum storage capacity available.

The JRP extension in the S-CJRP below has of three components. The first (from left to right) refers to the annual ordering cost, where for each period (T) a major ordering cost (S) is incurred plus an aggregated minor ordering cost s_i from the i-th item each k_i per T periods. The second represents the holding cost, defined as the cost of keeping the average of units stocked over the time horizon plus the annual holding cost including both cycle and security stock. The last refers to the total transportation cost, which is calculated as the sum of the cost to transport the total volume $w_i D_i k_i TA$ of item i, in transport units with capacity W a number of times $1/(k_i T)$ over the time horizon. The whole formulation is as follows:

$$\text{Minimize}: TC(T, k_1, k_2, ..., k_n)$$

$$TC = \frac{S + \sum_{i \in I} \frac{s_i}{k_i}}{T} + \frac{T}{2} \sum_{i \in I} D_i k_i h_i + \sum_{i \in I} Z_\infty \sigma_i h_i \sqrt{L + Tk_i} + \sum_{i \in I} \frac{w_i D_i A}{W}$$

subject to:

$$\sum_{i\in I} D_i T k_i w_i + \sum_{i\in I} Z_{\infty}\sigma_i w_i \sqrt{L+Tk_i} \leq H : T > 0; k_i \in N, \forall i \in I$$

The constraints above are concerned with the warehouse limited capacity. It should be noted that the objective function depends on the variables T and k_i.

2.2 Complexity Proof

Considering the formulation from the previous section, the complexity analysis approach is introduced now. For this purpose, additional notation and remarks are needed, however some parameters remain the same as specified in the previous section: n, i are indices for nodes or items, r, t are indices for the time period, N is the number of items, $T-1$ is the number of time periods in the time horizon, d_{nt} is the demand for item n in period t, h_n is the holding cost of item n per unit per time period, s_n is the minor cost of item n, and S is the major cost. A is the cost of a full cargo load, W is the maximum capacity of a full cargo load, and w_n is the unit weight of item n.

The period T is a dummy variable, meaning the last period. An order can be given in any time period between 1 and T. For the interval $1 < r < t < T$, an order placed in period s covers the demand up to $t-1$. This assumes that there is no lead time, since an order arrives immediately as required. However, if we want to consider constant lead time, as in our case, it would only be necessary to convert each period i, into the following: $i' = i + \text{lead time}$, for all i in the interval 1 to T. Therefore, assuming lead time or not is irrelevant. Then, the cost per period would be given by the following expression:

$$C_{nst} = S_n + h_n \sum_{i=r+1}^{t-1} (i-r)d_{ni} + \sum_{i=s+1}^{t-1} \frac{w_n d_{ni} A}{W} + \sum_{i=r+1}^{t-1} Z_{\alpha}\sigma_n h_{ni}\sqrt{i-r+L_n}$$

For each item n, consider a network such as the one shown in Fig. 1. From each node r, there is an arc to node $t > r$, with an associated distance defined as the cost C_{nst}.

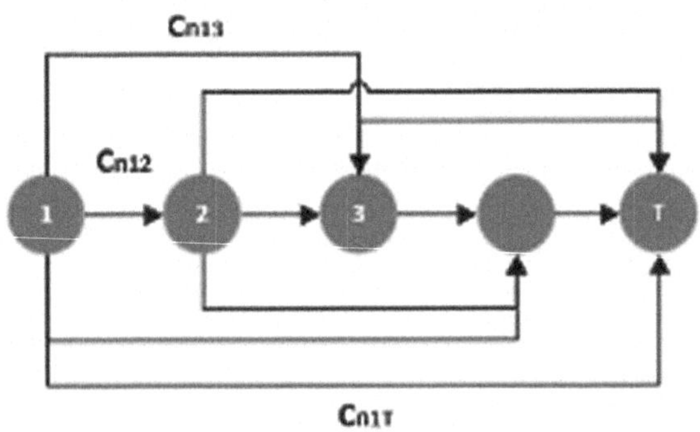

Fig. 1. Network for item n, each item is represented by one similar network.

Each path represents a feasible solution that covers the demand of all intermediate nodes. For example, if the C_{n1T} path is sealed, it is assumed that a replenishment is made in period 1 to cover all the periods up to $(T-1)$. Such a network exists for all items. The major cost S is incurred each time a node is selected. So, if there were two products and the solutions of each one were identical (for example, make a replenishment in S and then another in t) i.e. if there exists an arc from a 1 to r followed by another from r to t and then another from t to T (recall that the latter is identical, that is, it follows the same policy), then they share the major cost and the total cost will be the cost of the arc chosen by each one plus $3S$, which is $C_{11r} + C_{1rt} + C_{1tT} + C_{21r} + C_{2rt} + C_{2tT} + 2S$.

In order to prove that the S-CJRP is an NP-complete problem two conditions must be met: The problem must be NP and complete.

To prove the first condition recall that it can be considered that the replenishment of each product can be represented by the network displayed in Fig. 1. Then, the following question arises: Does there exist a set of paths from nodes $(n,1)$ to (n,T) for all n for which the sum of the setup costs incurred plus the path costs is not greater than B? The answer is yes; in fact there are algorithms that run in polynomial time for fixed T [23] or for fixed N [13] that solve such network problem.

Next, to prove completeness, the network problem in Fig. 1 can be reduced from the 3-SAT problem, which is a known NP-complete problem [8]. Such proof was already presented in [3]. Finally, it is proven that the S-CJRP is NP-Complete, a problem at least as complex as the JRP.

3 Computational Experiments

In order to have an overview of the computational complexity of the problem, a set of numerical experiments were run. Random data were generated for the parameters of each product, as follows. Demand was generated using a uniform distribution with parameters [1000, 100000]; major costs followed a discrete distribution S (100, 200, 300, 400); minor costs were generated s_i ($0.05 \times S$, $0.1 \times S$, $0.3 \times S$, ..., S), the standard deviation was determined considering changes in the coefficient of variation in the range [0.05, 0.15] uniformly distributed to avoid too large deviations. The other parameters were determined using uniform distributions as in the case of the aforementioned demand.

These problems were run with sizes from $n = 5$ to $n = 35$ using the BARON non-linear global optimization solver. Two different computers were used. Core i7/8 generation PC, 8 GB of RAM, and Neos Server (https://neos-server.org/neos/). Neos server does not deliver more results after $n = 30$, since the maximum computational time allowed per task is 8 h. In the case of the PC problems, $n = 35$ takes 15 h; problems with $n = 40$ were not possible to be solved because between the 4th and 5th day the PC stops working.

Figure 2 shows the evolution of this experimental computational time. Table 1 presents key indicators of the computational time, including minimum, maximum and average values.

Table 1. Summary of computational time

Size of N	5	10	15	20	25	30
Min	0.36	3.58	6.18	61.42	114.46	350.32
Avg	0.55	3.89	6.77	64.79	120.22	356.43
Max	0.71	4.23	7.2	70.15	126.67	359.35
Range	0.35	0.65	1.02	8.73	12.21	9.03

According to the calculations of the function that estimates the computational time, shown in Fig. 2, the estimated time required to obtain an optimal solution for an instance with $N = 5$ will be 4.6 days, while for $N = 50$, this time will be about 242 days. Note that the problem is restricted. However, the representation in Fig. 2 not consider such a restriction. Restricted problems were generated to verify how much the restriction affects computational time; and for moderate restrictions there was no apparent change. In both cases, it is impractical to try sizes with $n \geq 35$.

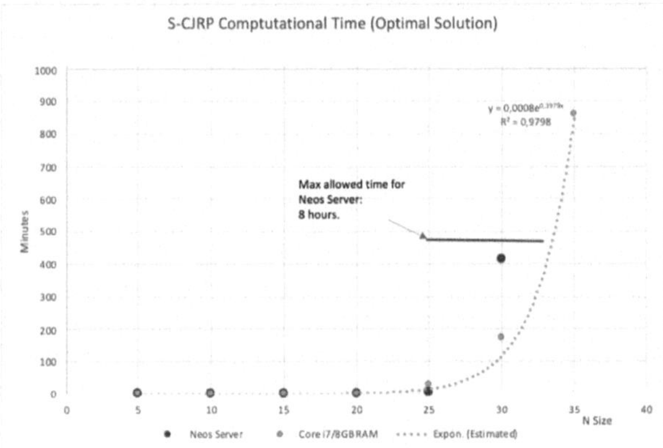

Fig. 2. Experimental CPU time

4 Conclusions and Further Work

This paper provided insights to demonstrate that the Stochastic Collaborative Joint Replenishment Problem (S-CJRP) is a problem at least as complex as the JRP; in fact it was proven that the S-CJRP is an NP-Complete problem.

The key to verify the S-CJRP's complexity lies in devising a network of paths as the one shown in Fig. 1. Later, it is easy to verify that such network problem can be solved in a polynomial time by known algorithms, so the S-CJRP is NP. Next, in order to prove completeness, the problem is reduced from the 3-SAT problem, a well-known

NP-Complete problem. Such reduction is proposed in [3] for the JRP; it can be extended to the S-CJRP considering the similarities of our approach to the aforementioned network problem.

Computational times for optimal solution analysis support the NP-completeness endowed. Through the computational experiments, it was possible to obtain a close fit ($r_2 = 0.9798$) for an exponential curve of the computational time versus the number of items to be shipped, as follows $CPU\,time = 0.0008exp^{(0,3379N)}$. It is always possible to deal with restricted versions of CJRP, yet it is impractical to try sizes of more than 35 items.

Acknowledgements. The work of the second author was supported under grant INGPHD-10-2019.

References

1. Aksoy, Y., Erenguc, S.S.: Multi-item inventory models with co-ordinated replenishments: a survey. Int. J. Oper. Prod. Man. **8**(1), 63–73 (1988)
2. Amaya Mier, R.A.: Intervención sobre Prácticas Integrativas en el Clúster de Logística del Atlántico: Cadenas Logísticas De Comercio Exterior. Ediciones Uninorte, Barranquilla, Colombia (2018)
3. Arkin, E., Joneja, D., Roundy, R.: Computational complexity of uncapacitated multi-echelon production planning problems. Oper. Res. Lett. **8**(2), 61–66 (1989)
4. Barratt, M.: Understanding the meaning of collaboration in the supply chain. Supply Chain Manag. **9**(1), 30–42 (2007)
5. Chopra, S., Meindl, P.: Supply Chain Measurement: Strategy, Planning and Operation. Prentice-Hall, Hoboken (2007)
6. Cohen-Hillel, T., Yedidsion, L.: The periodic joint replenishment problem is strongly NP-hard. Math. Oper. Res. **43**(4), 1269–1289 (2018)
7. Deng, X., Papadimitriou, C.H.: On the complexity of cooperative solution concepts. Math. Oper. Res. **19**(2), 257–266 (1994)
8. Garey, M.R., Johnson, D.S.: Computers and Intractability: A Guide to the Theory of NP-Completeness. Freeman, W.H., New York (1979)
9. Holweg, M., Disney, S., Holmström, J., Småros, J.: Supply chain collaboration: making sense of the strategy continuum. Eur. Manag. J. **23**(2), 170–181 (2005)
10. Jaller, M., Otero-Palencia, C., Pahwa, A.: Automation, electrification, and shared mobility in urban freight: opportunities and challenges. Transp. Res. Procedia **46**, 13–20 (2020)
11. Jaller, M., Otero-Palencia, C., Yie-Pinedo, R.: Inventory and fleet purchase decisions under a sustainable regulatory environment. Supply Chain Forum Int. J. **21**(1), 1–13 (2019)
12. Jaller, M., Otero, C., Pourrahmani, E., Fulton, L.: Automation, electrification, and shared mobility in freight, Sacramento, California (2020)
13. Kao, E.P.: A multi-product dynamic lot-size model with individual and joint set-up costs. Oper. Res. **27**(2), 279–289 (1979)
14. Kaspi, M., Rosenblatt, M.J.: The effectiveness of heuristic algorithms for multi-item inventory systems with joint replenishment costs. Int. J. Prod. Res. **23**(1), 109–116 (1985)
15. Khouja, M., Goyal, S.: A review of the joint replenishment problem literature: 1989–2005. Eur. J. Oper. Res. **186**(1), 1–16 (2008)
16. Starr, M.K., Miller, D.W.: Inventory Control: Theory and Practice. Prentice-Hall, Englewwod Cliffs (1962)

17. Naesens, K., Gelders, L., Pintelon, L.: A swift response tool for measuring the strategic fit for resource pooling: a case study. Manage. Decis. **45**(3), 434–449 (2007)
18. Otero-Palencia, C., Amaya-Mier, R., Montoya-Torres, J., Jaller, M.: Collaborative inventory replenishment: discussions and insights of a cost-effective alternative for noncompetitive small- and medium-sized enterprises. In: Yoshizaki, H.T.Y., Mejia Argueta, C., Mattos, M.G. (eds.) Supply Chain Management and Logistics in Emerging Markets, pp. 215–234. Emerald Publishing Limited, Bingley (2020)
19. Otero-Palencia, C., Amaya-Mier, R., Yie-Pinedo, R.: A stochastic joint replenishment problem considering transportation and warehouse constraints with gainsharing by Shapley value allocation. Int. J. Prod. Res. **57**(10), 3036–3059 (2018)
20. Otero-Palencia, C., Amaya, R.: A cost-effective collaborative inventory management strategy between non-competitor companies - a case study. In: Proceedings of the International Conference on Industrial Engineering and Operations Management, pp. 948–960 (2017). http://ieomsociety.org/bogota2017/papers/235.pdf
21. Simatupang, T.M., Sridharan, R.: The collaborative supply chain. Int. J. Logist. Manag. **13**(1), 15–30 (2002)
22. Småros, J., Lehtonen, J., Appelqvist, P., Holmström, J.: The impact of increasing demand visibility on production and inventory control efficiency. Int. J. Phys. Distr. Log. **33**(4), 336–354 (2003)
23. Veinott Jr., A.F.: Minimum concave-cost solution of Leontief substitution models of multi-facility inventory systems. Oper. Res. **17**(2), 262–291 (1969)

Hybrid Model for Decision-Making Methods in Wireless Sensor Networks

Martha Torres-Lozano(✉) and Virgilio González

Department of Electrical and Computer Engineering,
University of Texas at El Paso, El Paso, TX 79968, USA
mltorresloz@miners.utep.edu, vgonzalez3@utep.edu

Abstract. Technology selection for wireless sensor networks is becoming more difficult because of the number of technologies (alternatives are increasing) and their complexity. Multicriteria selection method describes the process of getting a choice from a group of alternatives. This process involves the requirements and inputs provided by the customer and the opinions of experts becoming the method for pure quantitative decision. The paper proposes a new methodology based on quantitative and qualitative parameters that makes more accurate and precise the technology selection. The paper describes the advantages and disadvantage of the proposed methodology compared with selection models currently used.

Keywords: WSN · Decision model · Hybrid model · AHP/ANP

1 Introduction

A fundamental factor for the social and economic development is the Telecommunication infrastructure, which allows transactions and communication to take place in either small or wide areas. One of the most important technologies used currently is represented by Wireless Sensor Networks (WSN). Academia and companies have been making efforts to achieve the best option for selecting the appropriate technology according to customer requirements.

Some factors for evaluating technology are technical, sociological, environmental, regulatory, infrastructural, and economic ones that make the selection process complicated [1]. Infrastructure factors are related to the presence of services such as Wi-Fi, cellular reception, microwave and satellite communications, and electrical power; also, building, offices, warehouses are included. Social factors are referred to as size of population, social structure communities, industrial and public installations [2, 3]. Geographical parameters, flora, fauna, weather, type of location (plain, mountainous) are examples of environmental aspects [1, 3]. Technical factors include type of data, reliability, flexibility, scalability, bandwidth velocity rate, and channel capacity. On the other hand, economic indicators are associated with operation cost, investment, and payback period. Regulatory parameters are related to licensing, rights of way, and spectrum availability [1, 4, 5]. Figure 1 presents different criteria for technology evaluation.

D. Trentesaux et al. (Eds.): SOHOMA 2021, SCI 987, pp. 319–330, 2021.
https://doi.org/10.1007/978-3-030-80906-5_23

In the selection process, one of the common mistakes made by companies is related to the decision process, when superficial characteristics are compared without any in-depth analysis and evaluation of risks that increase operation and/or implementation costs because of the issues encountered in real scenarios. The present research is focused on generating a quantitative and qualitative analysis of the techniques involved in the selection process. A basic methodology is also presented in this paper.

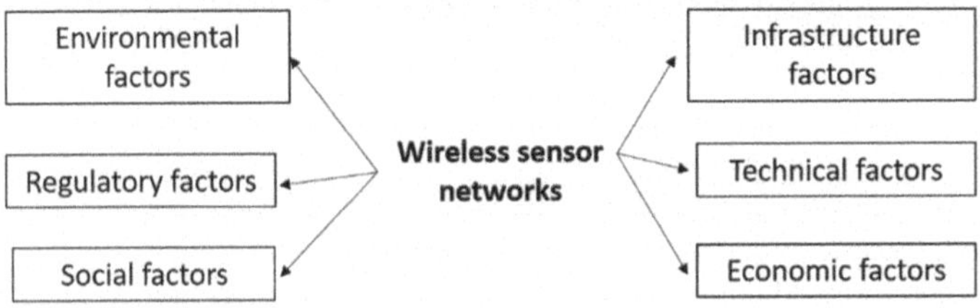

Fig. 1. Evaluating factors involved in the technology selection process

2 Descript Ion of Decision-Making Methods

A brief description of Multicriteria Decision Methods (MCDM) is provided in order to identify the gap that the current paper intends to close.

According to [6], MCDM methods can be divided in two different groups, weighted models and comparison-based methods. These groups include the most important decision models used for WSN for selecting the more appropriate technology application in a specific scenario. For the weighted method, a numerical weight is assigned to each evaluation parameter and a mathematical model is applied to get the final decision. In contrast, the methods by comparison is evaluate the parameters needed with respect to the infrastructure already installed (to assure cost reduction), to big cities close to the areas of interest, to the availability of carriers and others. Notice that weight is not assigned for the analysis.

2.1 Weighted Models

The strategy of the weighted models consists in identifying and clarifying the problem and goal, as well as defining the evaluation parameters based on expert opinions. Alternatives are analysed and weights are assigned to the parameters based on beneficial or non-beneficial criteria; a mathematical model is applied to make the final decision. These steps are described in Fig. 2; see also [7–10].

The most important models belonging to the weighted models are: Analytic Hierarchy Process, Analytic Network Process, and Vikor [11–15]. All these models are focused on theoretical fundaments, standards, literature review and information provided by vendors. However, none of the models offers a detailed risk analysis or a previous simulation of the proposed technology.

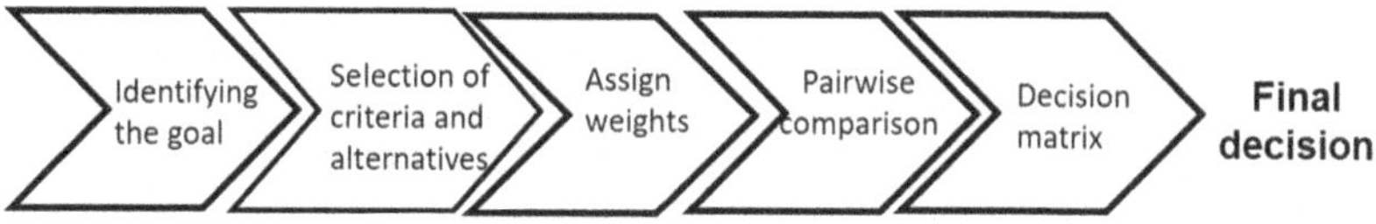

Fig. 2. Weighted model levels

2.2 Models Based on Comparison

In comparison-based models, the goal is defined according to the infrastructure already built, and applies nearby large cities where the infrastructure can be extended to reduce the investment and costs. Figure 3 illustrates the steps of the comparison-based models [6].

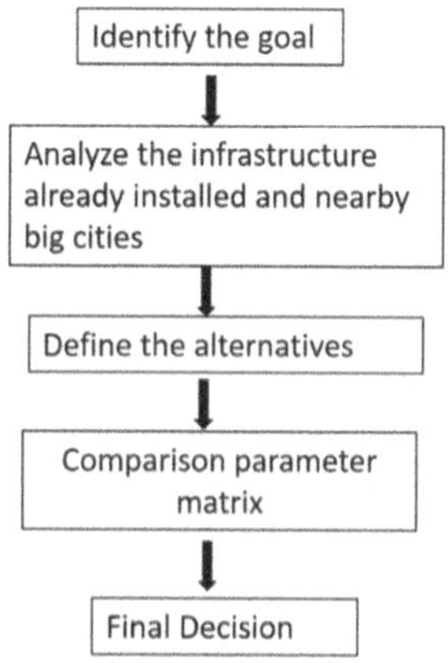

Fig. 3. The strategy of comparison-based models

In the first step, the problem and goal are clarified and identified according to the community's needs and/or customer requirements. The next step is to analyze the infrastructure installed in the area, or the existence of big cities close to the development project area in order to define possible alternatives that meet the client's expectations. Then, a comparison matrix is realized to compare the alternative attributes and take the final decision.

Business canvas model applied on technologies selection and Hub structure models are examples of comparison-based models. As for the weighted models, these models do not implement a simulation part where technologies proposed are evaluated and risks are analysed [6].

3 Hybrid Model for Multicriteria Decisions. The Methodology

One of the most important factors in communications, more specifically in wireless sensor networks, is the decision-making on the technologies that will be used in the implementation of projects in rural, urban, or remote areas. Currently, there are several decision-making methods such as AHP, ANP, Vikor. However, these methods only analyze the quantitative part of the alternatives, leaving the qualitative part outside the study. This problem can produce an cost increase in the project implementation, caused by environmental factors such as electrical storms, noise induction in the communication signal, quality factors, reachability, network congestion, coverage area, and others. Based on these factors, a methodology is proposed to reduce risks, avoid investment increase, and find the most viable technology for projects.

3.1 Hybrid Model - A New Methodology

The hybrid model is organized on three levels, as shown in Fig. 4. At the first level, the problem and the goal are defined and placed in a hierarchical structure related to the elements and decision factors. The next step of the first level is the evaluation of environmental and technological risks that may be present once the project is implemented. This initial evaluation is superficial and based on the theoretical information obtained from the client and literature review. Finally, the last step of level 1 is the comparison between the project requirements and risks in order to ensure that the technology meets the project specifications.

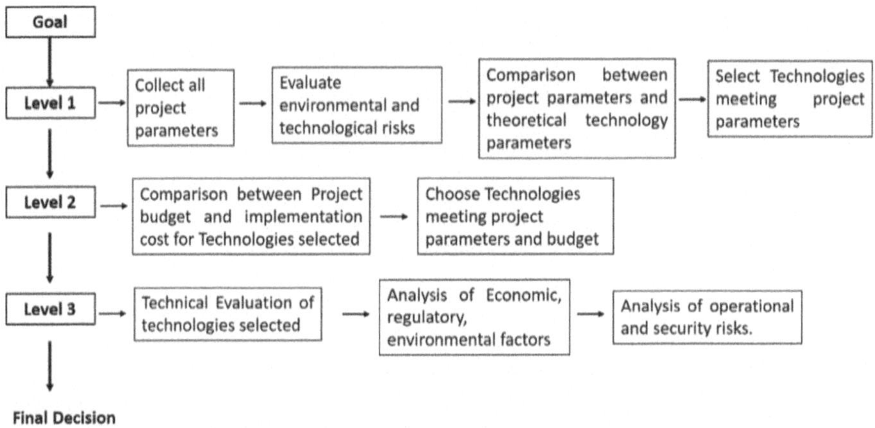

Fig. 4. The hybrid model proposed

If x_i is the customer requirement and x_j is the technology specification, then the result should be 1 if the specification covers the customer requirement, otherwise the result is 0.

$$x_{i,j} = \begin{cases} 1 \; if \;\; x_i \;\; meets \;\; x_j \\ 0 \; if \;\; x_i \;\; not \;\; meets \;\; x_j \end{cases} \tag{1}$$

After this evaluation the weight factor W_1 is calculated for all technologies according to Eq. 2.

$$W_1 = \sum x_{i,j} \tag{2}$$

A threshold α is used in order to discard the technologies that are not meeting the initial customer requirements. Then, if W1 < α, the technology is rejected from the project study.

On the second level, the budget plays the most important role in the decision method. A comparison between the technology chosen and the budget available is made in order to decide upon including or not the technologies in the list of the selected technologies for level 2. Equation 1 is rewritten as specified in Eq. 3.

$$y_{i,j} = \begin{cases} 1 \; if \; y_i \; meets \; y_j \\ 0 \; if \; y_i \; not \; meets \; y_j \end{cases} \tag{3}$$

where y_i is the customer requirement and y_j is the budget calculated for the technology. After this evaluation, the weight factor W_2 for level 2 is calculated for all technologies, according to Eq. 3.

$$W_2 = \sum y_{i,j} \tag{4}$$

If threshold value β for level 2 is accomplished, then the technology continues to be accepted in the selection process; therefore, $W_2 \geq \beta$.

For the third level, a technical evaluation is performed in order to identify critical factors such as poor coverage, network congestion, and wrong sensor topology. This evaluation is performed through network and traffic simulators such as SIMIO®, NS3®, and Opnet®. Then, an exhaustive economic, regulatory, and environmental analysis must be carried out to avoid increases in the necessary budget. Finally, an analysis of the operational and security risk is made. This risk study serves to validate all the data obtained at this level and, if necessary, to implement some strategies to improve the communication system and avoid failures, e.g., equipment redundancy and encryption.

The Architecture Trade-off Analysis Method (ATAM) is used for the risk analysis. The function of the ATAM model is to assess the consequences of an architectural design decision and its influence on the quality parameters [17, 18]. Analyzing different scenarios in terms of inputs and outputs, quality parameter could be clarified according to their responses. Figure 5 presents the ATAM method phases. To identify the sensitivity points, risk and trade-off points, the evaluator traces scenarios throughout the architecture, identifying the relevant components and connectors involved in the current scenario. The evaluators then ask specific questions about the system response.

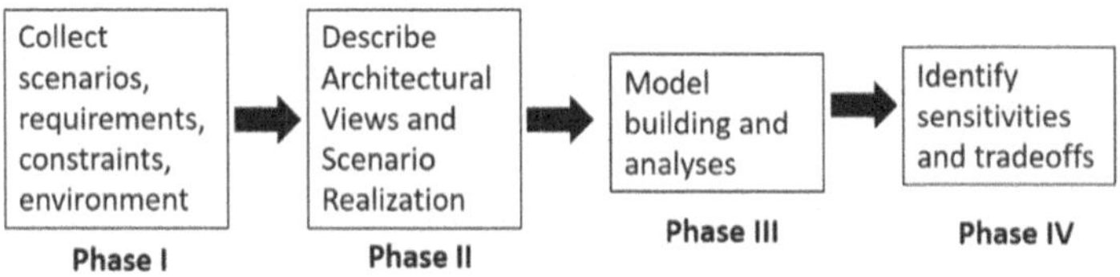

Fig. 5. Phases of ATAM method

After this analysis, the weight factor W_3 for level 3 is calculated for all technologies, according to Eq. 5 and 6, where z_i is the risk and z_j is the parameter obtained from simulations and risk analysis. After this evaluation, the weight factor W_3 for level 3 is found for all technologies, according to Eq. 6.

$$z_{i,j} = \begin{cases} 1 \; if \; z_i \; meets \; z_j \\ 0 \; if \; z_i \; not \; \; meets \; z_j \end{cases} \tag{5}$$

Then,

$$W_3 = \sum z_{i,j} \tag{6}$$

If threshold value μ for level 3 is accomplished, then the technology continues into the selection process; hence, $W_2 \geq \mu$. The final decision is made according to Eq. 7 where W_1, W_2 and W_3 are summed for each of the selected technologies. The high result is the final decision found.

$$T = \sum\nolimits_{i=1}^{3} W_i \tag{7}$$

where Wi is the weight for each level and T is the technology analysed.

3.2 Comparative Example

For the hybrid model, the input data for the selection process is required from the customer and analysed according to the area location. In the case presented, the network sensor application is agriculture for rural zone close to a city. Temperature and humidity are the data collected for the network taking one sample per hour; then, real time transmission data is not needed - which reduces the implementation cost. The system's life is 10 years with fixed installation, flat terrain, and rain conditions. Table 1 summarizes the customer requirements and input data for the selection process.

Table 1. Example scenario requirements

Example Scenario	
Application	Agriculture - monitor humidity and temperature
Type of terrain	Flat
Climate conditions	Rain
Type of installation	Fixed
Type of collected data	Temperature and Humidity
Sample rate	1 sample per hour (low data volume)
Type of area	Rural: 30Km from next city
Infrastructure preinstalled	Cellular
System Life	10 years
Coverage area	50000 m^2
Budget	U$30000
Additional Information	Sensors must be displayed every 150m

Table 2 indicates the possible technologies that can be used for this example and the coverage area parameter (the most important for this application), and Table 3 presents the technologies for the data transportation between the sensed area and the analysis data centre. If terminal technology is needed, the value 1 is assigned and the most convenient technologies are selected. The final selected technology is considered for budgeting purpose.

Table 2. Technologies for the sensor network level

Technology	Range
Zigbee	100m
WIMAX	50km
LORA(long range)	10km
Bluetooth -BLE	50m
WIFI	250m
LTE	2Km

Table 3. Technologies for transporting data from the studied area to the data centre

Terminal Technology		Application requirement
Apply ?	1	0
LTE	0	1
WIFI	0	1
Microwave	0	0
Satellite	0	0
Wired system	0	0

According to customer requirements, the list in Table 2 and $\alpha = 3$, Zigbee and Bluetooth technologies are discarded because of the range for sensor network level. LTE and WIFI are selected for the transportation data from the study area and central node. Then, applying Eq. 1 and 2, W_1 is calculated and results are presented in Table 4.

On the second level (budget study), Eq. 3, 4, and $\beta = 1$ are applied. For the budget study, equipment, installation, transportation, maintenance, accessories and service fees for the terminal are taken into consideration. Comparing the Total and Project Budget columns in Table 5, only WIMAX and LORA technologies are kept in the selection process. Table 5 is summarized the level 2 results.

On the third level, a detailed risk study and network simulation are developed for each technology in the process on level 2 in order to avoid problems in the implementation, costs, and risks. Then, the results provide the technology and the implementation parameters such as topology with the best energy consumption, latency, reliability, redundancy systems, and cybersecurity. Figure 6 shows the parameters analysed on level 3. Neural networks will be used for level 3 results.

Table 4. Results for level 1

Technology	Customer requirements			Results W1
	Range	Data rate	Bandwidth	
Zigbee	0	1	1	2
WIMAX	1	1	1	3
LORA	1	1	1	3
Bluetooth	0	1	1	2
WIFI	1	1	1	3
LTE	1	1	1	3

Table 5. Level 2 results

Technology	Customer requirement					Total (U$)	Project Budget (US)	W2
	Equipment (U$)	Installation and transportation (U$)	Maintain (U$)	Accessories (U$)	Terminal node service fee (U$)			
WIMAX	3000	6500	6000	8200	6000	29700	30000	1
LORA	1700	6500	6000	8200	6000	28400	30000	1
WIFI	1800	6500	6000	8200	12000	34500	30000	0
LTE	3000	6500	6000	8200	10800	34500	30000	0

For the AHP model [8], weighted model classification, range, costs, and the same technologies were chosen for the analysis. According to this model, WIMAX is the best option for the scenario described. However, risks and technical factors were analysed for the implementation part; then, only theoretical parameters and expert opinions were used for getting the best technology. Table 6 presents the results for the AHP model.

For the Hub structure model [6], by comparison model the LORA system was selected for the sensor network area and LTE (Cellular) for the data transportation because of the cost and infrastructure already installed.

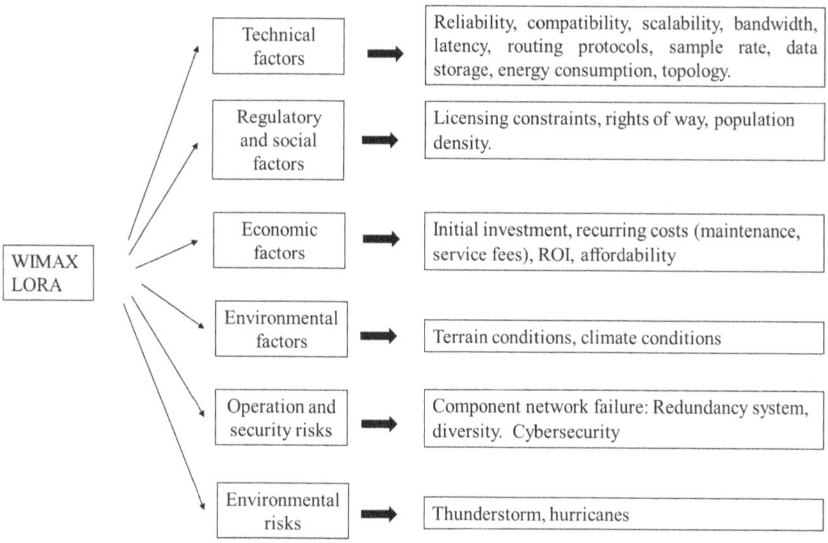

Fig. 6. Parameters analysed on the third level

Table 6. AHP results for the scenario proposed

Technology	Range	Cost	Total
WIMAX	0.623480665	0.33778195	**0.59491079**
LORA	0.219501328	0.56334586	0.25388578
WIFI	0.040447431	0.04943609	0.0413463
LTE	0.116570576	0.04943609	0.10985713
Criteria Average vector	0.9	0.1	

4 Discussion

The Multicriteria Decision Methods (MCDM) are methodologies designed to choose the most appropriate technology for wireless sensor networks based on the quality requirements requested by the customer. Weighted and comparison-based models have been used in technology selection; however these models include quantitative approaches only. The quantitative part involves problem identification, development of questionnaires from experts to provide high-precision answers, creation of surveys to collect customer and project requirements, technology literature review and creation of the decision matrix. However, simulations and detailed analysis of the technologies selected are frequently not carried out, putting the implementation and optimal operation of the Project at risk. Unforeseen environmental, technological, economic, or even regulatory events may occur in the implemented project, incurring on more investment, high operating costs, or problems in the communication systems due to quality factors.

Table 7. Comparison between weighted models, by comparison models and Hybrid model

Model parameters	Weighted models	By comparison models	Hybrid Model
Analyze systems already installed	•	•	•
Involve locality parameters	•	•	•
Parameters based on expert opinions	•	---	---
Use theoretical data	•	•	•
Real data used for the technology selection	---	•	•
Technologies (alternatives) simulation	---	---	•
Risk, trade-offs and sensitive points analysis	---	---	•
Failure analysis (simulation)	---	---	•
Analysis of quality parameters	---	---	•

In the hybrid model, the technologies chosen as viable alternatives are thoroughly analyzed both in theory and in practice (simulation), taking the risks as decision parameters and finding the solutions in order to avoid additional problems and costs in the execution and operating processes. According to the results from each selection model, for the hub structure model and AHP only expert opinions and surveys are applied. Topology, routing protocols, and risks are not studied for these processes, increasing the failure probability and errors in the technology selected. Table 7 shows the comparison between the methods currently used and the hybrid model proposed in this article [10–18].

5 Conclusion

As explained in the previous sections, all decision models have some common characteristics such as identifying the goal of the project, getting the client requirements (input and desired output parameters), selecting technological alternatives, applying mathematical models and taking the final decision [12, 13, 15].

In weighted models, weights are assigned to each parameter based on expert knowledge and literature reviews. Comparison-based models choose the alternatives focusing on the infrastructure installed in big cities or towns close to the survey area. Neither model performs an analysis of the qualitative parameters of the project such as signal delay, packet traffic congestion, reachability, feasibility, life of communications links, throughput, number of packets, system speed, risks, and sensitive points of failure. The analysis of these parameters is important to prevent problems, failures and high costs in the implementation and operation process.

Although the hybrid model can be long and tedious, it analyses the possible risks of the Project, the failures caused by technical or environmental factors, and is able to define solutions for the problems found before choosing a technological alternative and reaching the stage of project implementation.

Acknowledgments. I thank Texas Instruments Foundation for the awards received to support my studies and research.

References

1. Gasiea, Y., Emsley, M., Mikhailov, L.: Rural telecommunications infrastructure selection using the analytic network process. J. Telecommun. Inf. Technol. **2**, 28–42 (2010)
2. Kaptur, V., Mammadov, E.: Methodology of selectin appropriate technologies for constructing telecommunication access networks. In: 2nd International Scientific Practical Conference Problems of Info-communications Science and Technology (PIC S&T), Kharkiv, pp. 90–92. IEEE (2015). https://doi.org/10.1109/INFOCOMMST.2015.7357278
3. Ruder, K.A., Pretorius, M.W., Maharaj, B.T.: A technology selection framework for the telecommunications industry in developing countries. In: IEEE International Conference on Communications, Beijing, pp. 5490–5493. IEEE (2008). https://doi.org/10.1109/ICC.2008.1029
4. Edwards, S., Duncan, W., Plante, J., et al.: Regional telecommunications review 2018 (2018). https://www.yilgarn.wa.gov.au/documents/299/agenda-attachments-june-2018. Accessed 15 July 2020
5. Tam, M., Tummala, V.: An application of the AHP in vendor selection of a telecommunications system. Omega Int. J. Manage. Sci. **29**(2), 171–182 (2001). https://doi.org/10.1016/S0305-0483(00)00039-6
6. Torres, M., González, V.: Decision models for selecting wireless sensor networks technologies: a survey. Paper presented at the 24th world multi-conference on systemics, cybernetics and informatics: WMSCI 2020, Orlando, Florida (2020)
7. Rahim, S.A.: Supplier selection in the Malaysian telecommunication industry. Dissertation, Brunel University, London, UK (2013)
8. Taherdoost, H.: Decision making using the Analytic Hierarchy Process (AHP); A step by step approach. Int. J. Econ. Manage. Syst. **2**, 244–246 (2017)
9. Saaty, T.L.: Decision making with the analytic hierarchy process. Int. J. Serv. Sci. **1**(1), 83–98 (2008)
10. Alam, M., Jebran, J., Hossain, A.: Analytical Hierarchy Process (AHP) approach on consumers' preferences for selecting telecom operators in Bangladesh. Inf. Knowl. Manage. **12**(4), 7–18 (2012)
11. Adebiyi, S., Oyatote, E.: An analytic hierarchy process analysis: application to subscriber retention decisions in the Nigerian mobile telecommunications. Int. J. Manage. Econ. **48**, 63–83 (2016). https://doi.org/10.1515/ijme-2015-0035
12. Becker, J., Becker, A., Sulikowski, P., Zdziebko, T.: ANP-based analysis of ICT usage in Central European enterprises. Procedia Comput. Sci. **126**, 2173–2183 (2018). https://doi.org/10.1016/j.procs.2018.07.231
13. Pramod, V.R., Banwet, D.K.: Analytic network process analysis of an Indian telecommunication service supply chain: a case study. Serv. Sci. **2**(4), 281–293 (2010). https://doi.org/10.1287/serv.2.4.281
14. Butenko, V., Nazarenko, A., Sarian, V., et al.: Telecommunication standardization sector of ITU: Technical Paper, International Telecommunication Union (ITU) Series Y.2000, pp. 1–94 (2014)

15. Suganthi, L.: Multi expert and multi criteria evaluation of sectoral investments for sustainable development: an integrated fuzzy AHP, VIKOR/DEA methodology. J. Sustain. Cities Soc. **43**, 144–156 (2018). https://doi.org/10.1016/j.scs.2018.08.022
16. Kazman, R., Klein, M., Clements, P.: ATAM: method for architecture evaluation. Technical Report CMU/SEI-2000-TR-004. Carnegie Mellon, Software Engineering Department (2020). https://resources.sei.cmu.edu/library/asset-view.cfm?assetid=5177. Accessed 20 June 2020
17. Kaur, R.: Trade-off analysis for evolution paths in software architecture. Dissertation, Lovely Professional University, India (2015)
18. Kim, J., Ahn, B.: Extended VIKOR method using incomplete criteria weights. Expert Syst. Appl. Int. J. **126**, 124–132 (2019). https://doi.org/10.1016/j.eswa.2019.02.019

Optimizing Maintenance Policies of Computed Tomography Scanners with Stochastic Failures

Andrés Felipe Cardona Ortegón and William J. Guerrero(✉)

Facultad de Ingeniería, Universidad de La Sabana,
campus puente del Común, Chía, Colombia
william.guerrero1@unisabana.edu.co

Abstract. The maintenance policies of computed tomography scanners in most hospitals are based on empirical knowledge and often following the manufacturer's advice. In developing countries, the frequency of use of this equipment may be superior than recommended due to scarcity of resources, which could affect the optimal maintenance policy. Taking into account the budget and capacity of the equipment that public and private hospitals have in the administration, it is crucial to find an adequate decision support system that serves as a tool for the design of maintenance policies. The objective of this research is to develop an optimization model that allows making better decisions when preparing maintenance policies. The computed tomography scan is used in several diagnostic procedures, on different specialties, as it is a non-invasive exploration of the body. A continuous-time Markov chain model is proposed to model the different states of the equipment (working, requiring preventive maintenance, broken down). An optimization model is proposed with the objective of maximizing the benefit generated by the operating equipment and requiring maintenance. Budget constraints are considered. Two optimization methods are proposed and compared to solve the optimization model: an exhaustive search algorithm to understand the behaviour of the solution surface generated by the objective function and a meta-heuristic based on gradient-ascent to find near-optimal solutions in a reasonable time.

Keywords: Healthcare · Maintenance · Hospital logistics · Optimization · Metaheuristics · Markov chain

1 Introduction

The medical equipment is highly relevant to the quality of health services provided to people. The equipment downtime due to repair processes, maintenance, and damage becomes a problem for healthcare services and an inconvenience for patients and their families. In emerging economies, due to the small number of available equipment, healthcare systems are often forced to use the medical

D. Trentesaux et al. (Eds.): SOHOMA 2021, SCI 987, pp. 331–342, 2021.
https://doi.org/10.1007/978-3-030-80906-5_24

equipment more often than recommended by the manufacturer, and thus downtime is likely to happen regularly. Therefore, the costs associated with maintenance are significant, not only for repairing the equipment, but also considering the impact on key performance indicators in hospitals, such as the length-of-stay of patients.

In the hospital context, reducing medical equipment downtime time becomes important when we consider the problems that late diagnosis leads to. Including that in the end, this delay will affect both the doctor's practice and the patient. In the worstcase scenario, the patient can die due to a lack of correct treatment on time. In other words, as described by [19], the quality of the health service could be affected by the availability of medical equipment. Additionally, diagnostic errors can be generated due to problems with the equipment, as mentioned by [7].

In the Colombian context, healthcare resources are very limited and must be used efficiently. As a matter of fact, within the law that concerns the medical practice, doctors have 20 min for the direct care of one patient according to article 97 of resolution 5261 (1994) [18]. Also, hospitals have scarce equipment, either because the infrastructure does not allow them to have more, or because of the limited budget; in both cases investments are necessary. Further, the high number of patients requiring specialized equipment to obtain an accurate diagnosis for the treatment of their illnesses, implies that it is of vital importance that the medical equipment works properly. Thus, it is important and necessary that this equipment has long available times. This is an opportunity to apply Operations Research techniques to improve decision making for equipment maintenance in hospitals.

In this paper, the optimization of medical equipment maintenance policies is studied, aiming to support a healthcare system's efficient operation. We consider that the effective use and timely maintenance of medical equipment will increase the available time of use, and this translates into a reduction of the downtime of the equipment and a greater number of patients served by the TC scan.

To deal with this problem properly, mathematical optimization is crucial to improve the decision-making processes about the use of medical resources. Unfortunately, as indicated in the studies by [12], out of 34 research documents, 64% of them are empirical, 19% are prioritization proposals of medical devices, and only 17% propose optimization models. Scarce literature about this problem is available to the best of our knowledge. This situation motivates the development of studies in the field of the optimization of maintenance policies specifically for medical equipment, which have different dynamics than other types of equipment.

The medical equipment studied is the computerized tomography (CT) scanner. This is a tool used in several diagnostic processes, of different specialties, as it is a non-invasive exploration of the human body. The equipment is used to take internal images of the head, thorax, and limbs. Generally, hospitals do not have a large number of CT scans, considering the important purchase value of this equipment and maintenance requirements. Depending on the brand and type, it exceeds US $ 800,000 according to [3]. Due to the high number of patients that require a CT Scan exam, it is necessary that this type of equipment is available to support the diagnostic process as much as possible. The CT Scanner uses a motorized x-ray source that rotates around the patient, shooting beams of x-rays through the body that are captured by digital x-ray detectors and sent to a computer that can interpret them. Thus, high resolution images of the internal organs of the body are provided.

The failure times of the CT scanner and the medical equipment in general are random. Likewise, the repair and maintenance times are random and depend in the first case on the failure that occurs and in the second case on the moment in which the maintenance is scheduled. Therefore, providing a policy for the maintenance of the equipment is a challenge, especially in emerging economies where the equipment may be overused.

Markov Chains are an appropriate tool to model this type of system, considering the assumption on random events that need to be analyzed. This technique is proposed due to the Markov memory-less property, which states that the probability of future events do not depend on the past history, but only on the current state of the system [21]. In our case, we assume that future failures of a CT Scanner do not depend on previous events, but only depend on the current state of the machine.

The contributions of this paper are the following: First, a Markov chain model is constructed to represent the dynamics and the state of a group of identical CT scanners, assuming that a single technician is available to perform the maintenance. Second, the proposed model of Markov chains is structured as a nonlinear optimization model. Finally, the objective function of the optimization model is analyzed, demonstrating that it is concave; for this reason, a meta-heuristic based on gradient-ascent is implemented to compute near-optimal maintenance policies.

The structure of the article is as follows: Sect. 2 presents a literature review. Section 3 describes the proposed Markov chain model. Section 4 presents the optimization model associated with the decision-making process of designing the optimal maintenance policies. Section 5 describes the solution methodologies. Section 6 describes the computational experiments and results performed using the proposed techniques. Finally, conclusions and future research directions are provided in Sect. 7.

2 Literature Review

Recent progress in Industrial Internet-of-Things (IIoT) driven by the Industry 4.0 revolution has opened new research opportunities to optimize the maintenance policies of equipment, given the possibility of tracking many signals from sensors and centralizing a large amount of real-time information in order to make opportune decisions that can avoid expensive unplanned stops in factories [30]. These technologies allow to increase service levels while reducing operational costs [22] caused by urgent repairs of equipment.

This progress impacts every sector of the economy. In fact, there is a relationship between the operational problems presented in the productive sector and the hospital sector, [13] and [27]. Thus, it is possible to adapt the methodologies and technologies used to optimize decisions at commercial companies to tackle similar hospital issues, with the target of improving the use of resources, providing a better service, and saving lives.

The need to provide better service does not exist only in the manufacturing industry. The hospital sector is presenting high pressures to improve its service quality and competitiveness. Therefore, different works to improve decision-making in the hospital sector arise, [1,2,6,9,10,17]. An example is presented by [8] that uses Markov chains, minimizing the cost of inventory management of a medical center without affecting the service levels.

Additionally, both the quantity and the complexity of the pieces of equipment used in the health care sector has increased in recent years, [5], and likewise, these require more complex, and more expensive maintenance. Therefore, optimizing the maintenance policies has attracted the interest of researchers.

In [11] it is concluded that substantial savings in costs can be achieved by considering monitoring conditions (MC) which provide valuable information about the condition of the production plant at the end of each cycle or production run. These results are transferable and applicable to the hospital sector, as mentioned by [13].

It is pertinent to give the definitions of preventive maintenance and corrective maintenance. According to [24], preventive maintenance is a technique especially aimed at supporting activities for all types of equipment, performed periodically to review the equipment to prevent unforeseen stops in the production line. The corrective maintenance is the set of activities required to restart the equipment once it has failed or malfunctions.

According to [26], the preferred maintenance strategy in the surveyed hospitals is the strategy of preventive maintenance. The least preferred strategy is the corrective maintenance policy since costs are higher. Nevertheless, some hospitals are interested in cost savings achieved by reducing the frequencies of preventive maintenance.

According to [31] for each of the cases tested in his research, the optimal policy is one that manages the maintenance calendar, that is scheduled maintenance and a process of statistical control. Further, there are potential cost savings when using information resulting from statistical control processes. The reason for that is the interdependence between the process of statistical control and the maintenance procedures which justifies the possibility of considering both criteria jointly. Also, [20] and [28] provide an assessment of the effectiveness of equipment maintenance practices in public hospitals and the prioritization of medical equipment for maintenance decisions.

In addition, [14], conclude that the age of some medical equipment does not influence whether the machine requires more frequent maintenance or not. This is because the analyzed medical equipment has a life expectancy of 15 years, but often hospitals renew their equipment every 5 years. As a result, more long-term research needs to be performed for old equipment.

For the analysis and optimization of equipment maintenance policies, there are different models and techniques that are applicable depending on the dynamics of the equipments in operation. As pointed out by [12], most of the methodologies used to solve this problem are empirical, followed by prioritization of equipment, and finally optimization models. They also mentions that research in this field has increased in the last decade.

Mainly two modelling approaches are used in the literature to address the issue: mixed-integer programming models and the models based on Markov decision processes. The substantial difference between theme is the assumption on the nature of the information: deterministic respectively stochastic. However, the problem of maintenance policies should not be addressed in a deterministic way because the times of the failures do not follow a given pattern, but on the contrary they are probabilistic. Therefore, as described by [15], a Stochastic process is a probability model that describes the evolution of a system that develops randomly over time. This is the reason for which a stochastic process fits better for the study. An example is the study carried out by [29], where they employ semi-Markovian decision processes for the optimization of maintenance policies. A standard Markov process is used with the generalization of the distribution between transitions.

Choosing Markov chains for this type of problem is appropriate since as [4] mentions Markov chains can model the behaviour of different types of systems by means of combinatorial techniques. They can also model repairs in a natural way (repairs of individual or group components, a variable number of technicians who repair, sequential or partial repair), and the handling of equipment in detail. Likewise, [25] describe a Markov decision process (MDP) in which a decision maker does not know with certainty the exact effect of executing some specific action. Their model is adapted to the context of inspection and maintenance policies to be improved, considering the characteristics of the machines, so the model can be better adjusted.

Simulation is another technique that can be used to understand the dynamics and impact of maintenance decisions. Simulation techniques consist in the development of a logical-mathematical model of a system such that an imitation of the operation of a process or a system is obtained, [16]. Some recent examples of simulation models applied to healthcare are provided by [23]. The use of simulation brings advantages and disadvantages. The main advantage is that, once the model is built, it is possible to make changes quickly to analyze different policies and scenarios. It is usually cheaper to improve the system via simulation, than to do it directly in the real system. Also, it is easier to understand and visualize simulation methods than purely analytical methods. Simulation models can analyze systems of greater complexity or greater detail.

The disadvantages of simulation models include the fact that oftentimes these models require a lot of time to develop and validate. Also, a large number of computational runs are required to find good solutions, and it is difficult to validate simulation models. The simulation models do not produce optimal solutions. Because of the reasons exposed, the simulation does not optimally solve the problem. This becomes the strongest reason why simulation is not chosen as the solution method for this problem.

In conclusion, we have not found in the literature a work associated with the modelling and optimization of equipment maintenance policies with stochastic failures that use Markov chains to model their operating status, or metaheuristics to solve problems in this context. Therefore, there will be a new research task to evaluate the adjustment of this type of model with respect to reality.

3 Markov Chain Model

Next, we describe the continuous-time Markov chain that represents the system of multiple CT scanners and their repair process. The following assumptions are considered: first, the system has a fixed number of machines, that is, it is assumed that the new machines are not bought or sold. There is only one technician to repair the group of machines and he performs one repair at a time. The technician's travel time is zero. That is, he or she arrives instantly to the site where machines are located. Machine failures are random with non-memory property. The group of machines is identical. Priority is given to corrective maintenance over preventive maintenance. No partial repairs are made. Preventive maintenance reduces the possibility of failures. Once the machine is repaired, it is in perfect condition. Finally, update maintenance is not performed.

In our system, the CT scanner can be in one of the three conditions listed in Table 1:

Table 1. Possible conditions for a CT Scanner

State	Description
Perfect	The machine works perfectly and there is no need to make maintenance
Deteriorated	The machine is working but preventive maintenance is due
Damaged	The machine is not working

The following notations are introduced:

- n: number of identical machines
- γ: damage rate of a machine in perfect condition [times per month]
- δ: deterioration rate of a machine in perfect condition [times per month]
- θ: damage rate of a machine that is not in perfect condition (preventive maintenance is due) [times per month]
- β: preventive maintenance rate [times per month]
- α: corrective maintenance rate [times per month]

The state variables of the Markov chain model are as follows: Let (X_t, Y_t, Z_t) be a continuous time Markov chain; the state variables are define as follows:

- X_t = number of machines in perfect condition at instant t.
- Y_t = number of machines in deteriorated condition at instant t.
- Z_t = number of damaged machines at instant t.

The space state for this Markov chain is defined in Eq. 1.

$$S = \{(X_t, Y_t, Z_t) | X_t + Y_t + Z_t = n, X_t, Y_t, Z_t \geq 0, \forall t \geq 0\} \quad (1)$$

The transition rate matrix between two states $i = (X_{1(t)}, Y_{1(t)}, Z_{1(t)})$ to state $j = (X_{2(t)}, Y_{2(t)}, Z_{2(t)})$ is defined by the following Eq. 2:

$$R_{i,j} = \begin{cases} \delta X_{1(t)}, & if\ X_{2(t)} = X_{1(t)} - 1, Y_{2(t)} = Y_{1(t)} + 1, Z_{2(t)} = Z_{1(t)} \\ \gamma X_{1(t)}, & if\ X_{2(t)} = X_{1(t)} - 1, Y_{2(t)} = Y_{1(t)}, Z_{2(t)} = Z_{1(t)} + 1 \\ \theta Y_{1(t)}, & if\ X_{2(t)} = X_{1(t)}, Y_{2(t)} = Y_{1(t)} - 1, Z_{2(t)} = Z_{1(t)} + 1 \\ \beta \cdot W_{1(t)}, & if\ (N - Z_{1(t)}) > 0\ ,\ X_{2(t)} + 1 = X_{1(t)}, \\ & Y_{2(t)} = Y_{1(t)} - 1, Z_{2(t)} = Z_{1(t)} \\ \alpha, & if\ X_2(t) + 1 = X_1(t), Y_2(t) = Y_1(t), \\ & , Z_2(t) = Z_1(t) - 1 \\ 0, & otherwise. \end{cases} \quad (2)$$

In the first line of the Eq. 2, the transition is represented where the number of machines in operation is reduced and the number of deteriorated machines is increased by assigning the rate δ multiplied by the number of machines in operation before the transition. In the second line, the transition in which the number of machines in operation is reduced and the number of machines requiring corrective maintenance is increased is assigned the rate γ multiplied by the number of machines in operation before the transition. In the third line, the transition in which the number of machines in operation requiring preventive maintenance is reduced and the number of machines requiring corrective maintenance is increased is assigned the rate θ multiplied by the number of machines in operation requiring preventive maintenance before the transition. On the fourth line, the transition in which the number of running machines requiring preventive maintenance is reduced and the number of running machines is increased is assigned the rate β multiplied by the number of technicians not assigned to corrective maintenance before the transition. Consider that $W_{1(t)} = \min(Y_{1(t)}, N - Z_{1(t)})$. In the fifth line the transition in which the number of machines requiring corrective maintenance is reduced and the number of machines in operation is increased is assigned the rate α multiplied by the minimum between the number of technicians and machines requiring corrective maintenance before the transition.

4 Optimization Model

We describe the optimization model for the damage of the machines and their repair. The following parameters are considered (all values in \$):

- C_1: benefit per unit of time of having a machine in operation.
- C_2: benefit per unit of time of having a deteriorated machine in operation.
- C_3: cost per unit of time of having a machine not working because it requires corrective maintenance.
- C_4: cost per unit of time to perform corrective maintenance.
- C_5: cost per unit of time to perform preventive maintenance.
- Q: budget per unit of time for maintenance.

The decision variables of the proposed model are α and β. Also the auxiliary decision variables of the proposed model $\bar{X}_t$ to model the expected number of machines in perfect condition, $\bar{Y}_t$ are the expected number of machines in deteriorated condition, and $\bar{Z}_t$ the expected number of damaged machines. The optimization model is defined as follows:

$$MAX(C_1\bar{X} + C_2\bar{Y} - C_3\bar{Z} - C_4\alpha - C_5\beta) \tag{3}$$

subject to:

$$\alpha > 0 \tag{4}$$

$$\beta \geq 0 \tag{5}$$

$$\alpha C_4 + \beta C_5 \leq Q \tag{6}$$

The objective function in Eq. 3 is calculated as the sum of benefits obtained from having machines in operation minus the cost of having damaged machines and the costs of performing the maintenance. Equation 4 and 5 state that the preventive and corrective maintenance rates must be positive. Equation 6 limits the cost for making maintenance operations to be lower than the available budget.

5 Solution Methodologies for Solution Finding

Two methodologies are evaluated and compared. The first one finds the optimal solution with a Brute-Force approach, that requires significant computing resources and important computation time. The second methodology proposed finds a good solution in a reasonable time. It is a Hill-Climbing algorithm that finds near-optimal solutions.

The first methodology evaluates the objective function value for every value of α and β within the search space defined by the constraints of the problem. The second solution method starts with a random solution that is built with random values of α and β. Then, neighbour solutions are explored to search for a better solution. That is, four solutions are evaluated: $(\alpha + \Delta, \beta)$, $(\alpha, \beta + \Delta)$, $(\alpha - \Delta, \beta)$, $(\alpha, \beta - \Delta)$, where Δ is the step size to evaluate neighbour solutions. The best solution found among these four, satisfying the constraints, is kept. The procedure is repeated for a fixed number of iterations.

6 Computational Results

The R language was used to implement the algorithms (version i386 3.4.1). The workstation is a computer with Intel®Core ™ 2 DUO E7500 @ 2.93 GHz processor with 2.00 GB DDR2 RAM.

Figure 1 presents the number of n states in the space S of the Markov chain as a function of the number of machines in the system. As presented, the number of states increases polynomially. In the case with 30 machines, 496 states are modelled.

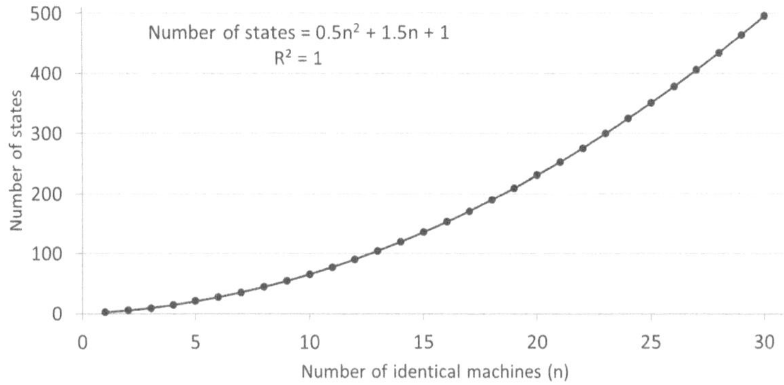

Fig. 1. Number of states in the Markov chain model as a function of the number of machines

The test instance set is randomly generated to evaluate the performance of the meta-heuristic method. In these instances $\delta = \gamma = \theta = 10$, $C_1 = C_2 = C_3 = 1$ and $C_4 = C_5 = 0.01$. Different instance sizes are used from 1 to 20.

Table 2 shows the objective function value of the proposed metaheuristic in column 2, and the Brute-Force method in column 3. Column 4 presents the percentage gap between both methods. It can be seen that the percentage gap for all instances is below 0.06%, which signifies a consistently good performance of the proposed meta-heuristics. A Friedman test confirms that there is no statistical difference between the solution quality of the metaheuristic and the Brute Force method (p-value >0.1). Thus, near-optimal solutions are computed by the proposed metaheuristic.

Further, the proposed Hill-Climbing algorithm presents a significant reduction in computational time when compared to the Brute-Force algorithm. Figure 2 depicts the CPU time in seconds for both methods in logarithmic scale with an average decrease of 99%. For the largest instances ($n = 20$ machines) the computational time of the Hill Climbing method is 289 s (4.8 min), while the Brute-Force algorithm takes more than 32,000 s (8.8 h).

Table 2. Comparative Results for Solution Methodologies

n	Hill-Climbing	Brute Force	GAP
1	0.2005	0.2006	0.05348%
2	0.8300	0.8301	0.00957%
3	1.4802	1.4803	0.00349%
4	2.1432	2.1433	0.00169%
5	2.8251	2.8251	0.00067%
6	3.5269	3.5269	0.00092%
7	4.2468	4.2468	0.00041%
8	4.9818	4.9819	0.00035%
9	5.7293	5.7293	0.00008%
10	6.4870	6.4870	0.00009%
11	7.2529	7.2529	0.00007%
20	14.3669	14.3669	0.00009%

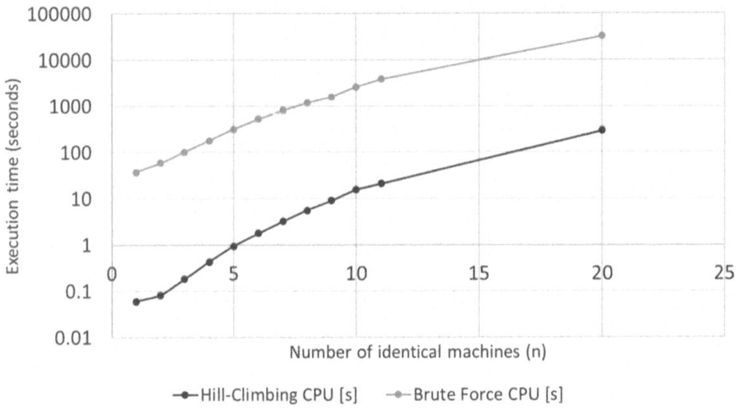

Fig. 2. Comparison of computation times between the Hill-Climbing method and the brute-force algorithm in seconds for instances with n machines. Logarithm scale.

7 Conclusions and Future Research

In this work, we have proposed a continuous time Markov chain model that represents the dynamics of the computed tomography scanners, an optimization model that aims to maximize the benefit generated by maintenance policies of the machines, and two solution methods: a meta-heuristic based on Hill-Climbing and an exhaustive search algorithm (Brute-Force).

This problem is relevant for hospitals in emerging economies that are forced to use CT scanners more frequently than the manufacturer's recommended levels and have a limited budget to perform the maintenance activities of medical equipment.

The results prove that the proposed algorithm is competitive by computing near-optimal solutions (with an optimality gap below 0.06%) with computational times that respond to decision-makers' requirements (below 5 min for 20 machines).

Future research will be directed towards the design of a bi-objective optimization model that maximizes the benefit generated by the machines and minimizes the variance of the auxiliary decision variables. The objective is to have a more robust system in which the system of machines is more reliable. Also, we propose to design a Markov chain model that considers heterogeneous machines and to study the deterioration of the machines over time. This model may also be applied in other sectors than healthcare. Further, implementing simheuristics as a solution method is also a future research direction.

References

1. Aguilar-Escobar, V.G., Garrido-Vega, P.: Gestión lean en logística de hospitales: estudio de un caso. Revista de Calidad Asistencial **28**(1), 42–49 (2013)
2. Aguilar-Escobar, V.G., Garrido-Vega, P., Godino-Gallego, N.: Mejorando la cadena de suministro en un hospital mediante la gestión lean. Revista de Calidad Asistencial **28**(6), 337–344 (2013)
3. ASC Communications. 12 Statistics on CT Scanner Costs (2012). Accessed 15 May 15 (2020)
4. Boyd, M.A., Lau, S.: An introduction to Markov modeling: concepts and uses. Technical report, NASA Ames Research Center Moffett Field, CA (1998)
5. María Carmen Carnero: Multicriteria model for maintenance benchmarking. J. Manuf. Syst. **33**(2), 303–321 (2014)
6. Lasprilla, N.G.: Diseño de un método meta-heurístico para resolver el problema de asignación de turnos de enfermería (nsp) con soft-constraints. Technical report, Escuela Colombiana de Ingeniería Julio Garavito (2017)
7. Graber, M.L., Franklin, N., Gordon, R.: Diagnostic error in internal medicine. Arch. Intern. Med. **165**(13), 1493–1499 (2005)
8. Guerrero, W.J., Yeung, T.G., Guéret, C.: Joint-optimization of inventory policies on a multi-product multi-echelon pharmaceutical system with batching and ordering constraints. Eur. J. Oper. Res. **231**(1), 98–108 (2013)
9. Guerrero, W.J., Velasco, N., Amaya, C.A.: Multi-objective optimization for interfacility patient transfer. In: Production Systems and Supply Chain Management in Emerging Countries: Best Practices, pp. 81–95. Springer (2012)
10. Hernández-Nariño, A., Medina-León, A., Nogueira-Rivera, D., Negrín-Sosa, E., Marqués-León, M.: La caracterización y clasificación de sistemas, un paso necesario en la gestión y mejora de procesos. particularidades en organizaciones hospitalarias. Dyna **81**(184), 193–200 (2014)
11. Jafari, L., Makis, V.: Optimal lot-sizing and maintenance policy for a partially observable production system. Comput. Ind. Eng. **93**, 88–98 (2016)
12. Jamshidi, A., Rahimi, S.A., Ait-kadi, D., Bartolome, A.R.: Medical devices inspection and maintenance; a literature review. In: IIE Annual Conference. Proceedings, p. 3895. Institute of Industrial and Systems Engineers (IISE) (2014)
13. Jiménez, A.M., Guerrero, J.G., Amaya, C.A., Velasco, N.: Optimización de los recursos en los hospitales: revisión de la literatura sobre logística hospitalaria. Technical report. Universidad de Los Andes, Colombia (2008)

14. Khalaf, A.B., Hamam, Y., Alayli, Y., Djouani, K.: The effect of maintenance on the survival of medical equipment. J. Eng. Des. Technol. **11**, 142–157 (2013)
15. Kulkarni, V.G.: Introduction to Modeling and Analysis of Stochastic Systems. Springer, New York (2011)
16. Lowery, J.C.: Getting started in simulation in healthcare. In: 1998 Winter Simulation Conference. Proceedings (Cat. No. 98CH36274), vol. 1, pp. 31–35. IEEE (1998)
17. Mardani, A., et al.: Application of decision making and fuzzy sets theory to evaluate the healthcare and medical problems: a review of three decades of research with recent developments. Expert Syst. Appl. **137**, 202–231 (2019)
18. Ministerio de Salud de la Republica de Colombia. Resolución número 5261 de 1994 (1994). https://www.minsalud.gov.co/
19. Mosadeghrad, A.M.: Factors affecting medical service quality. Iran. J. Public Health **43**(2), 210 (2014)
20. Mwanza, B.G., Mbohwa, C.: An assessment of the effectiveness of equipment maintenance practices in public hospitals. Procedia Manuf. **4**, 307–314 (2015)
21. Norris, J.R., Norris, J.R.: Markov Chains, vol. 2. Cambridge University Press, Cambridge (1998)
22. Noureddine, R., Solvang, W.D., Johannessen, E., Yu, H.: Proactive learning for intelligent maintenance in industry 4.0. In: International Workshop of Advanced Manufacturing and Automation, pp. 250–257. Springer (2019)
23. Oakley, D., Onggo, B.S., Worthington, D.: Symbiotic simulation for the operational management of inpatient beds: model development and validation using δ-method. Health Care Manag. Sci. **23**(1), 153–169 (2020)
24. Alzate, N.O.: Conceptos básicos sobre mantenimiento preventivo y mantenimiento correctivo. Technical report, Facultad de Minas, Universidad Nacional de Colombia, sede Medellin (2013)
25. Papakonstantinou, K.G., Shinozuka, M.: Optimum inspection and maintenance policies for corroded structures using partially observable Markov decision processes and stochastic, physically based models. Probab. Eng. Mech. **37**, 93–108 (2014)
26. Rani, N.A.A., Baharum, M.R., Akbar, A.R.N., Nawawi, A.H.: Perception of maintenance management strategy on healthcare facilities. Procedia-Soc. Behav. Sci. **170**, 272–281 (2015)
27. Rodríguez, D.R., Altamiranda, Á.S., Daza-Escorcia, J.M., Ferrer-Vázquez, M.: Revisión de los desafíos en las operaciones logísticas de los hospitales. In: Congreso internacional de Ingeniería Corporación Politécnico Costa Atlántica CIIPCA 2013 (2013)
28. Taghipour, S., Banjevic, D., Jardine, A.K.S.: Prioritization of medical equipment for maintenance decisions. J. Oper. Res. Soc. **62**(9), 1666–1687 (2011)
29. Tomasevicz, C.L., Asgarpoor, S.: Optimum maintenance policy using semi-Markov decision processes. In: 2006 38th North American Power Symposium, pp. 23–28. IEEE (2006)
30. Wang, K.: Intelligent predictive maintenance (IPdM) system–industry 4.0 scenario. WIT Trans. Eng. Sci. **113**, 259–268 (2016)
31. Xiang, Y.: Joint optimization of x control chart and preventive maintenance policies: a discrete-time Markov chain approach. Eur. J. Oper. Res. **229**(2), 382–390 (2013)

A Multi-agent Optimization Approach to Determine Farmers' Market Locations in Bogotá City, Colombia

Daniela Granados-Rivera[1](✉) and Gonzalo Mejía[2]

[1] Maestría en Diseño y Gestión de Procesos, Universidad de La Sabana, Chía, Colombia
danielagrri@unisabana.edu.co

[2] Faculty of Engineering, Universidad de La Sabana, Km 7, Autopista Norte de Bogotá D.C., Chía, Cundinamarca, Colombia
gonzalo.mejia@unisabana.edu.co

Abstract. This paper describes the implementation of a mixed optimization method in multi-agent system approach, that determines farmers' market (FM) locations. Occasionally, the optimization models that locate facilities do not consider the interactions on medium term between agents in the market. For this reason, the mixed method considers the consumer patterns of buying fruits and vegetables (F&V) in a certain probability for each household from each seller. This probability was calibrated using a multi-agent simulation to evaluate the initial solution given by the mathematical model. FM's location is a relevant topic due to the fact that this type of market is a good strategy to address food insecurity in areas with a low density of food supply. The model was implemented in an area of study in the city of Bogotá, Colombia. The results showed a satisfactory identification of sites with a limited number of supply channels. In these areas, the model located FM and thus increased food access.

Keywords: Farmers' market · Facility location problem · Short supply chain · Food insecurity · Food logistics

1 Introduction

In recent years, food security has taken significant importance due to increasing worrying indices [1]. The problem is directly related to low F&V consumption caused by low budget, extensive distance, assortment, a.o. Authors of [2] summarize the reasons for this situation by the lack of food availability (amount of F&V present in an area) and food access (physical, economic, and social access to food) [3]. Many governments have implemented strategies to solve this problem, e.g., social establishments, food banks, and support from foundations. However, they do not answer the main challenge of creating more efficient F&V supply circuits in the supply chain. These circuits are often complex structures involving several stakeholders [4]. That is why there is a trend to develop alternative channels to simplify the operations in the F&V supply. One of the most popular of them is FM.

D. Trentesaux et al. (Eds.): SOHOMA 2021, SCI 987, pp. 343–355, 2021.
https://doi.org/10.1007/978-3-030-80906-5_25

FM is a strategy to improve food access in food deserts (areas with low food availability and accessibility) [5, 6]. These markets are based on a short supply chain in which farmers commercialize directly for clients, without intermediaries [7]. The markets are located in pre-established sites with a weekly schedule [8], and have been implemented since long time in many countries such as the United States [9], New Zealand [10], Brazil [11], and Chile [12].

Several works have studied FM as a strategy dealing with food deserts; the scope of these research works is to evaluate the impact of the location of FM on food accessibility in an area around this location [8–13]. These works concluded that FM is a potential strategy to improve food access. However, Schupp *et al.* [14] found that FMs are seldom located in food deserts. Much of the research focuses on evaluating the current location, but does not determine where these markets should be located. Only four works addressed the issue from the perspective of two methods: multi-agent models [15, 16] and optimization approach [17, 18].

Locating FMs is an important strategic decision from a logistics perspective, due to the need for flexible planning and the complexity of their operation constraints [8, 19]. The optimization approach mentioned above provides a feasible solution to these issues. However, this approach does not consider the consumers' patterns, as the agent-based model do. The patterns change in general the FM's optimal locations since the addition of market's behaviour may generate a dynamic model that takes into account real needs [16]. A model type in the facility location domain involves the consumer choice, being formulated as Competitive Facility Location Problem (CFLP); however only short term needs are evaluated.

Considering these aspects, the described research work started by identifying a gap in the literature about FMs' optimal locations aimed at improving food access and availability. That is why we proposed the implementation of a mixed model based on an optimization approach and an agent-based framework that calculates the locations of FMs in five UPZ (territorial units of the town) of Bogotá: Los Cedros, Toberín, Country Club, Britalia, and El Prado, in order to improve F&V access and availability measured with the coverage of F&V demand.

2 Background

This section offers a brief analysis from the literature review performed; two significant approaches were found and will be further described.

2.1 Facility Location Problem (FLP)

FLP consists of selecting one or more facilities to serve a set of consumers [20]. In the specific case of FM location studied by Tong *et al.* [17], two variations of FLP *p*-median based on geographical positions were considered. The variations feature spatial and temporal constraints to minimize consumers' travel distances. A similar study was done by Widener *et al.* [18], who employed a configurable *p*-median to reduce the travel costs of a residential cluster to select optimal FM locations. Both studies minimize costs or distances. In this paper, we use a model to maximize the coverage of the F&V demand inspired by the research work of Balcik *et al.* [21]. Different from this research, we evaluated the probability through simulation, ensuring the middle term feasibility of selected FM locations.

2.2 Multi-agent Systems

Multi-agent systems (MAS) have been used to implement several social and environmental strategies [22]. They implement the interactions among agents involving autonomy, social ability, reactivity, and proactivity. Agents follow three steps: 1) analysis of the initial status of the environment, 2) decision-making and executing the decision, and 3) evaluation of the decision according to some rules [23]. For the specific case of FM, multi-agent systems have been implemented in two studies. The first was developed by Widener *et al.* [16] who evaluated the introduction of FMs in Buffalo-New York as a potential public policy. The authors found that few FM were necessary to enhance food security in this area. The second work was initiated by Na *et al.* [15], who also employed an agent-based model finding the best FMs locations according to consumers' patterns in Los Angeles. For now, there a few studies that implement multi-agent systems to deal with FLP in general. The primary objective of the studies is to evaluate the consumers' behaviour, knowing existing market locations [24, 25]. Our proposal represents an innovation as it integrates a method that provides feedback to the simulation solution in order to improve the FLP results.

2.3 Multi-agent Optimization

In the literature, there are reported combinations of optimization methods with multi-agent systems that improve the computational efficiency of the solution search [26–28]. Yet, these methods are not focused on the multi-agent system to simulate the interaction between stakeholders. A few studies, such as Barbati *et al.* [28], implemented a multi-agent system to solve FLP using the market locations as active agents and the demand points as passive agents.

Nevertheless, these studies did not consider preferences on the active agents' decisions, so that the interactions only responded to changes in the locations. For the investigation scope of our study, we found only one work that implements the stakeholders' decisions in a mixed model of multi-agent optimization. This work was developed by Fikar *et al.* [29], who implemented a mixed-integer programming model (MILP) to define the number of transfer points to open and the number of vehicles to use in a last-mile distribution of relief goods in a case of humanitarian logistics.

This solution of the model solution was evaluated in parallel by an agent-based system that analyzed different promising settings of transfer points. We implemented a similar system combining an FLP solved by MILP with a multi-agent system. We used a step-by-step mixed process instead of a simulation in parallel of the MILP. There is no study about FM location in the literature review performed that evaluates the possible middle-term effects.

That is why our proposal represents a good approximation to estimate the FM locations feasibility considering the customer's behavioural change on middle-term with weekly organizing FMs.

3 Proposal

This section presents the design methodology divided into three parts: i) MILP used to locate FM, ii) multi-agent simulation employed to evaluate the solution, and iii) the iterative step-by-step mixed process.

3.1 FM Facility Locating Based on MILP Model

We defined the problem statement as a set of households (consumers) $H = \{1 \ldots j\}$ geographically distributed within a food desert area. They must be provided with F&V daily demand d_{jt} in multiple days $T = \{1 \ldots t\}$ by the available FM in this day in some location $FM = \{1 \ldots i\}$ that has a capacity Q_i. The coverage is defined by a_{ji} in which a household $j \in H$ is covered by FM located in $i \in FM$, if this household is within the maximum coverage distance of FM. Each FM has a coverage level l_{ji} according to the closeness of each household $j \in H$ as shown in the study of Balcik *et al.* [29]. Considering that FM is not the only way of supply, each household $j \in H$ has a probability p_{ji} to choose these markets rather than other retailers. Finally, we consider a maximum number N of markets that can open all considered days T and a minimum percentage β to cover the households' demand.

The model decision variables are X_{it} which is a binary variable that defines: if FM is situated in location i in a day t; if the binary variable Z_{jt} specifies that household j is covered in a day t; and if Y_{jit} defines the demand percentage of household j covered by FM located at i in a day t. The objective function (1) maximizes the quality of the household coverage level.

$$Max \sum_{i \in M} \sum_{j \in H} \sum_{t \in T} d_{jt}\, Y_{jit}\, l_{ji}\, p_{ji} \tag{1}$$

$$\sum_{i \in FM} a_{ji} X_{it} \geq Z_{jt} \quad \forall j, t\, ;\; j \in H,\; t \in T \tag{2}$$

$$\sum_{i \in FM} a_{ji} X_{it} \leq M\, Z_{jt} \quad \forall j, t\, ;\; j \in H,\; t \in T \tag{3}$$

$$\sum_{i \in FM} Y_{ijt} \leq Z_{jt} \quad \forall j, t\, ;\; j \in H,\; t \in T \tag{4}$$

$$Y_{ijt} \leq X_{it} c_{ji} \quad \forall i, j, t\, ;\; i \in FM,\; j \in H,\; t \in T \tag{5}$$

$$\sum_{i \in FM} \sum_{j \in H} Y_{ijt} d_{jt} \geq \beta \sum_{j \in H} d_{jt} \quad \forall t\, ;\, t \in T \tag{6}$$

$$\sum_{j \in H} Y_{ijt} d_{jt} \leq Q_i \quad \forall i, t\, ;\; i \in FM,\; t \in T \tag{7}$$

$$X_{it-1} + X_{it} \leq 1 \quad \forall i, t\, ;\; i \in FM,\; t \in T \tag{8}$$

$$\sum_{i \in M} \sum_{t \in T} X_{it} \leq N \quad \forall i, t \,; i \in FM \,, t \in T \tag{9}$$

$$Y_{ijt} \geq 0 \quad \forall i, j, t \,; \; i \in FM \,, \; j \in H \,, \; t \in T \tag{10}$$

$$X_{it} \,, \; Z_{jt} \; \in \; \{0, 1\} \quad \forall i, j, t \,; \; i \in FM \,, \; j \in H \,, \; t \in T \tag{11}$$

The constraint sets (2) and (3) establish if a household can be covered within the range of FM coverage, in which M is a sufficiently large number for logical constraints. Te constraint set (4) ensures that any proportion of the demand will not be covered if a household cannot be covered. Constraint set (5) specifies that the supply of F&V can only be done from an opening FM for a household within the coverage range. Constraint set (6) guarantees that the covered proportion of F&V's demand is greater or equal to the percentage fixed previously. Constraint set (7) prevents that the FM's sales are greater than their capacities. Constraint set (8) prohibits the opening of an FM on consecutive days. Constraint set (9) limits FM to open according to the pre-established maximum value N for the planning horizon. Finally, constraint sets (10) and (11) refer to the non-negativity and binary restrictions for the decision variables.

3.2 The Multi-agent Simulation

We consider two groups of agents $a \in (H \cup FM \cup NA \cup OS)$: households $h \in H$, and sellers such as FM $s \in FM$, nanostores $s \in NA$, supermarkets $s \in OS$, and central markets $s \in OS$. The active agents are the households $h \in H$. Additionally, the passive agents are the sellers $s \in (FM \cup NA \cup OS)$ who only sell their product to the active agents. We assume that nanostores and FM have a limited capacity, whereas supermarkets and

Table 1. The algorithm employed for the multi-agent system

```
Algorithm 1
For t=1 to T'
    ∀s ∈ (FM ∪ OS) //sellers review stock
    ∀h ∈ H //households review stock and select market
     If t ∈ SD_h  // if t is a shopping day of household h
        Let PS_ht = FM'_t ∪ NA'_h ∪ OS ∪ OS  //be the set of potential sellers of household h on day t
        s* ← h. select (PS_ht)
                 j=t-1
        Q'_ht =   Σ   d_hj          //calculation of order quantity of household h on shopping day t
                j=t-T_a
        Q_hs*t = min{ Q'_ht , I_s*t } // household h purchases a quantity from seller s*
        I_s*t = max (0, I_s*t − Q_hs*t)  //chosen seller s* updates its stock
     Else// t ∉ SD_h
          Q'_ht = 0
          Q_hs*t = 0
     End If
     Generate a random demand d_ht with a normal distribution ND(μ_h, σ_h)
    I_ht = I_ht-1 + Q_hst − d_ht
End
```

central markets have not. The buying choice probability gets a weight for price, distance, and assortment factors based on the discrete choice model [30]. The model calibration was done according to Guarín's survey [31] about the preferences of buying places in Bogotá. A planning horizon was defined to run the simulation. For each period $t \in T'$, all sellers $s \in (FM \cup NA \cup OS)$ update their stocks I_{st}. Next, the households $h \in H$ shop according to their stocks I_{ht-1} and shopping days $t \in SD_h$. They select a place to buy considering the nanostores within a walking distance NA'_h and the open FM FM'_t and determine the buying quantity Q'_{ht} based on an intake rate d_{hj}. At the end of the period, all agents update their stocks I_{at}. This process is shown in Table 1.

3.3 Iterative Step-By-Step Mixed Process

The combination of optimization and the multi-agent system is realized in an iterative step-by-step mixed process. Figure 1 shows this sequential process. The first step consists in running the MILP mathematical model to determine the optimal locations. The second step is the multi-agent simulation that evaluates the environment interactions related to the newly located FM. With the result of the simulation, in the third step the new probabilities p'_{ji} are calculated. Then, the differences Dif_{ji} beten the original probability p_{ji} of the mathematical model and the new calculated p'_{ji} of the simulation are determined. If the maximum absolute difference $Max(Dif_{ji})$ is greater than 5%, the parameter p_{ji} is updated using the calculated probabilities p'_{ji}. With the new probabilities, the process goes back to the first step. If the maximum absolute difference $Max(Dif_{ji})$ is less than 5%, the process stops. Finally, the solution results.

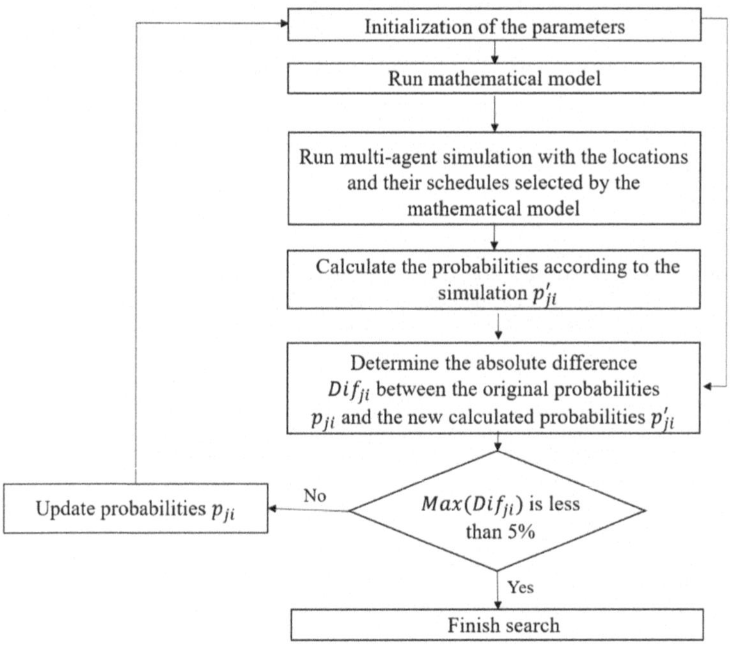

Fig. 1. Proposed flow chart of the iterative step-by-step multi-agent optimization process

4 Case Description

In Bogotá, the capital of Colombia, located in the central region of the country, it is known that 4.9% of the households do not eat any of three meals for one or more days [32]. FM was created as a strategy by the Office of Economic Development of Bogotá District (SDDE, in Spanish [33]) because the basic food basket is the most expensive in the country, and Bogotá is not a great producer of F&V [34, 35]. This situation creates limited food access, placing Bogotá people amongst the lowest F&V consumers, at half of FAO recommendations [1].

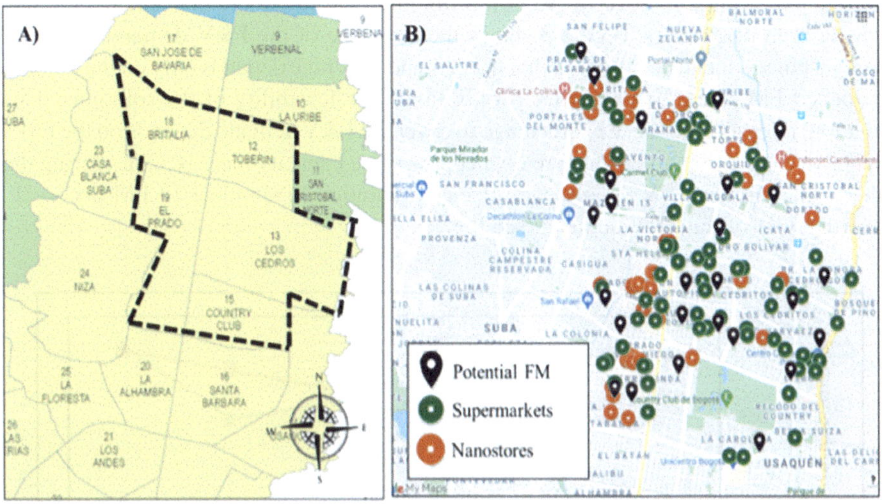

Fig. 2. The area selected for this research. A): Delimitation of UPZ [42]. Norma Urbana UPZ. [Figure]. Retrieved at http://www.sdp.gov.co/gestion-territorial/norma-urbana/normas-urbanisti cas-vigentes/upz. B): Location of nanostores, supermarkets, and potential locations of FMs.

The area selected included five UPZs: Los Cedros, Toberín, Country Club, El Prado, and Britalia, see Fig. 2A in the north of the urban centre of Bogotá around Parque Alcalá FM. This market is the second one selling the most quantity of F&V [36]. The population in this area is 325,803 inhabitants. Households' sites were selected through a mesh that established 568 points, and thus the population was distributed proportionally in these points. We assumed an average F&V daily intake of 200 g per person with a standard deviation of 25 g [34]. The demand was generated with a normal distribution [37]. On the other hand, possible FM's locations were public parks conveniently located in the studied area, which led to 26 points selected (see Fig. 2B). The FMs' capacity is established according to the area of each possible location. Also, we identified 55 nanostores and 72 supermarkets (see Fig. 2B). All locations were established by scanning Google Maps™ and the calculating the distances with Harvesine formula [38], with a multiplicative factor to approximate real distances considering the road structure or any geographic conditions [39]. With these distances, the coverage was determined. The prices of the sellers were determined according to [40, 41].

5 Results

The mathematical model was implemented in the software tool IBM ILOG CPLEX Optimization Studio version 12.8.0.0, and the simulation model was running in Eclipse IDE for Java Developers Version: 2019-06 (4.12.0). Both models used a laptop with 4GB RAM and a 2GHz processor.

5.1 Parameters Settings

To define the maximum number N of FM to open and the minimum percentage of the demand of F&V to cover β, we ran different instances of the mathematical model varying both parameters. Figure 3 shows the coverage of the F&V demand in two of these instances. It can be observed that the demand coverage depends only on the number of opened FM since the parameter β only limits the feasibility of the solutions. This situation happens because the percentage to cover the F&V demand depends on the FM's capacity. The more markets that are opened, the more capacity there is. Considering the other sellers in the F&V supply, we selected a solution to simulate 17 FM per week, covering 30% of the F&V demand (see Fig. 3).

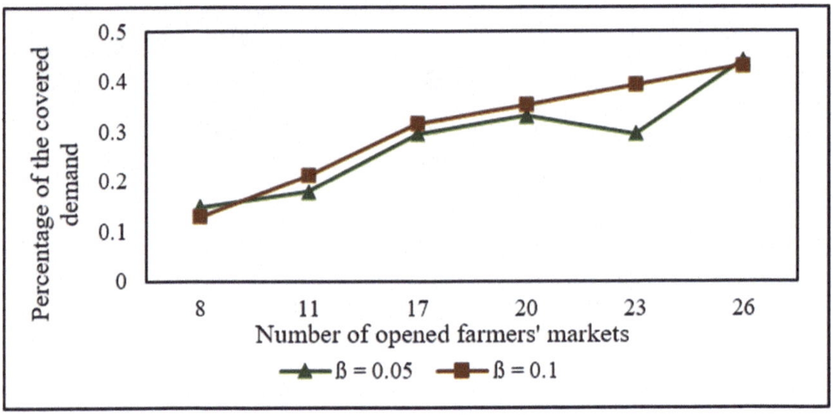

Fig. 3. Percentage of the demand covered according to the number of opened FM and the minimum percentage of the demand to cover β

5.2 Comparison of the Iterations in the Step-by-Step Process

To begin the iterative process, we generated initial random probabilities for the parameter p_{ji}. Table 2 illustrates the variation of each iteration. The maximum absolute differences between the probabilities were decreasing for each iteration. After three iterations, the process converged, indicating that the probabilities used in the mathematical model were equal to the probabilities generated in the multi-agent simulation. It means that the locations given by the FLP considered the maximization of the F&V coverage based on consumer patterns.

Table 2. Results of the iterative step-by-step process for the selected instance

Iteration	Maximum absolute difference	Average of the absolute differences	Percentage of covered demand
1	36.05%	3.26%	29.76%
2	11.56%	2.06%	33.23%
3	0.00%	0.00%	33.08%

According to the simulation, the selected locations are more likely to be chosen. Surprisingly, the percentage of covered demand and the rating of coverage increase each iteration. This situation may happen for changes in the configuration of the opened FM.

5.3 Impact on the Network Configuration of FM

Analyzing the opened locations' configuration, we found that updating the probabilities caused some places of the initial scenario to be closed. Figure 4A illustrates the configuration in the initial situation and the locations closed in the iterative process. Although there are fewer locations after the iterative process (black points), the weekly schedule benefited the increase of the demand coverage because the probability of choosing an FM was more significant in these areas. The reduction of the number of enabling locations for FM is favourable, considering locating markets' costs.

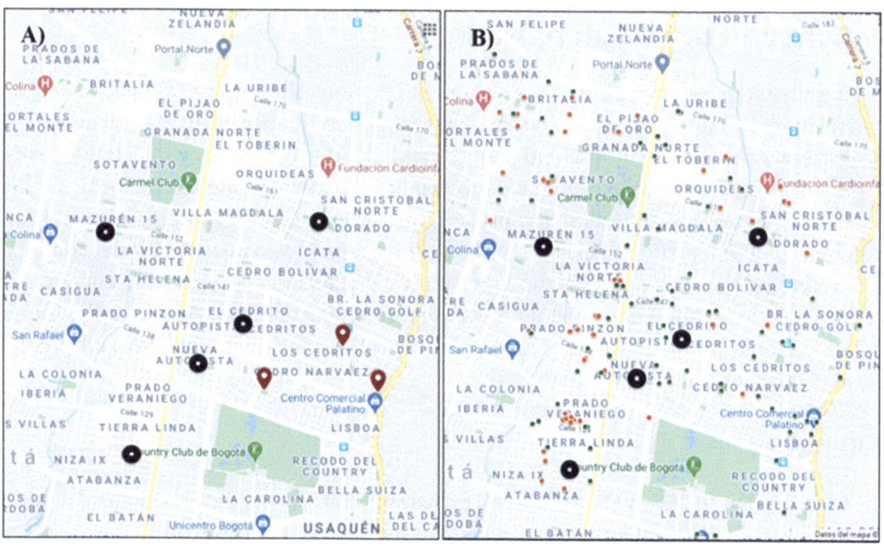

Fig. 4. A) Enabled locations to open FM. All waypoints (black and red) represent the initial configuration. The three red points are the closed locations at the end of the iterative process; B) The density of the F&V supply in the studied area considering the selected FM's locations.

We could also determine that the selected locations were a good solution because the FM was in areas with a low-density of F&V supply. This situation is observed in Fig. 4B, in which around the locations of markets (black points) there is a limited number of sellers (green and orange points). This behaviour proves the impact of using the mixed-model to recognized food deserts and the advantage over studies that only implemented the multi-agent simulation, such as Widener *et al.* [16] and Na *et al.* [15] who only evaluated the FM's locations' impact as a public policy. In contrast, they did not provided feedback on the simulation with the updating of the MILP model to find better FM's sites that enhance food security.

5.4 Managerial Implications

The implementation of farmer markets affects positively household consumption patterns because these can obtain low prices close to their homes; this was also confirmed in the work of Tong *et al.* [17]. According to the simulation, the FM could achieve a market share up to 40% showing the trend to be selected for household purchase. Also, the entry of FMs may help the decentralization of central markets [43]; in Bogotá the most significant part of the F&V market is taken by Corabastos, the leader central market. However, FMs must become a sustainable strategy because until today they are financed by SDDE, a government entity that limited the strategy as outreach. Locating FM in areas with a low density of supply channels ensures profit that allows producers to reinvest. Additionally, in the process of calibrating the simulation, we realized the importance of the price factor on the buying choice. That is why FM implementation must always have low price strategies to encourage F&V consumption.

6 Conclusions and Future Work

The iterative step-by-step mixed process implementations allowed to evaluate the mathematical model solution that calibrates the chosen probability parameter establishing that the solution considers the household patterns and identifies the areas with a low-density of F&V supply. These areas were taken as food deserts, where the model located FMs to extend food access. This situation proved the advantage of considering the market's preference and interaction to establish where FM will be operated. This model could be in the future a tool that analyses the increase of the demand caused by the encouragement of healthy eating habits and by other stimuli for the demand. Additionally, future research should define a learning rate to evaluate the consumers' pattern and predict possible changes.

References

1. FAO, IFAD, UNICEF, WFP, and WHO: The State of Food Security and Nutrition in the World 2020, Rome (2020). http://www.fao.org/3/ca9692en/online/ca9692en.html
2. Mejía-Argueta, C., Benitez-Perez, V., Salinas-Benitez, S., Brives, O., Fransoo, J., Salinas-Navarro, D.: The importance of nanostore logistics in combating undernourishment and obesity. In: 10th International Conference on Production Research - Americas, Communication in Computer and Information Science, vol. 1408 (2019). https://doi.org/10.1007/978-3-030-76310-1

3. Napoli, M., de Muro, P., Mazziota, M.: Towards a food insecurity multidimensional index (FIMI), Master thesis, Roma Tre University, pp. 1–72 (2015)
4. Tsolakis, N.K., Keramydas, C.A., Toka, A.K., Aidonis, D.A., Iakovou, E.T.: Agrifood supply chain management: a comprehensive hierarchical decision-making framework and a critical taxonomy. Biosyst. Eng. **120**, 47–64 (2014)
5. Li, K.Y., Cromley, E.K., Fox, A.M., Horowitz, C.R.: Evaluation of the placement of mobile fruit and vegetable vendors to alleviate food deserts in New York city. Prev. Chronic Dis. **11**(9), 1–9 (2014)
6. USDA: Access to Affordable and Nutritious Food: Measuring and Understanding Food Deserts and their Consequences Report to Congress, Agriculture (2009). https://www.ers.usda.gov/publications/pub-details/?pubid=42729
7. Balaji, M., Arshinder, K.: Modeling the causes of food wastage in Indian perishable food supply chain. Resour. Conserv. Recycl. **114**, 153–167 (2016)
8. Robinson, J.A., Weissman, E., Adair, S., Potteiger, M., Villanueva, J.: An oasis in the desert? The benefits and constraints of mobile markets operating in Syracuse, New York food deserts. Agric. Hum. Values **33**(4), 877–893 (2016). https://doi.org/10.1007/s10460-016-9680-9
9. Dimitri, C., Oberholtzer, L., Zive, M., Sandolo, C.: Enhancing food security of low-income consumers: an investigation of financial incentives for use at farmers markets. Food Policy **52**, 64–70 (2015)
10. Pearson, A.L., Wilson, N.: Optimising locational access of deprived populations to farmers' markets at a national scale: One route to improved fruit and vegetable consumption? PeerJ **2013**(1), 1–13 (2013). https://doi.org/10.7717/peerj.94
11. Nogueira, L.R., et al.: Access to street markets and consumption of fruits and vegetables by adolescents living in São Paulo, Brazil. Int. J. Environ. Res. Public Health **15**(3), 1–12 (2018)
12. Mora, R., Bosch, F., Rothmann, C., Greene, M.: The spatial logic of street markets: an analysis of Santiago, Chile. In: Proceedings of the 9th International Space Syntax Symposium (2013)
13. Strome, S., Johns, T., Scicchitano, M.J., Shelnutt, K.: Elements of access: the effects of food outlet proximity, transportation, and realized access on fresh fruit and vegetable consumption in food deserts. Int. Q. Community Health Educ. **37**(1), 61–70 (2016)
14. Schupp, J.: Wish you were here? The prevalence of farmers markets in food deserts: an examination of the United States, Food. Cult. Soc. **22**(1), 111–130 (2019)
15. Na, H., Kim, S., Langellier, B., Son, Y.J.: Famers markets location-allocation framework for public health enhancement. In: IIE Annual Conference and Expo 2015 (2015)
16. Widener, M.J., Metcalf, S.S., Bar-Yam, Y.: Agent-based modeling of policies to improve urban food access for low-income populations. Appl. Geogr. **40**, 1–10 (2013)
17. Tong, D., Ren, F., Mack, J.: Locating farmers' markets with an incorporation of spatio-temporal variation. Socioecon. Plann. Sci. **46**(2), 149–156 (2012)
18. Widener, M.J., Metcalf, S.S., Bar-Yam, Y.: Developing a mobile produce distribution system for low-income urban residents in food deserts. J. Urban Health **89**(5), 733–745 (2012)
19. Lucan, S.C.: Local food sources to promote community nutrition and health: storefront businesses, farmers' markets, and a case for mobile food vending. J. Acad. Nutr. Diet. **119**(1), 39–44 (2019)
20. Hutchison, D., Mitchell, J.C.: Lecture Notes in Computer Science, vol. 9, no. 3 (1973)
21. Balcik, B., Beamon, B.M.: Facility location in humanitarian relief. Int. J. Logist. Res. Appl. **11**(2), 101–121 (2008)
22. Gimblett, H.R.: Integrating geographic information systems and agent-based modeling techniques for simulating social and ecological processes. In: Integrating Geographic Information Systems and Agent-Based Modeling Techniques for Simulating Social and Ecological Processes (2002)

23. Mejía, G., García-Díaz, C.: Market-level effects of firm-level adaptation and intermediation in networked markets of fresh foods: a case study in Colombia. Agric. Syst. **160**(October 2018), 132–142 (2018)
24. Serafino, P., Ventre, C.: Heterogeneous facility location without money. Theoret. Comput. Sci. **636**(July), 27–46 (2016)
25. Serafino, P., Ventre, C.: Truthful mechanisms for the location of different facilities. In: 13th International Conference on Autonomous Agents and Multiagent Systems. AAMAS 2014, vol. 2, pp. 1613–1614 (2014)
26. Meignan, D., Créput, J.C., Koukam, A.: A cooperative and self-adaptive metaheuristic for the facility location problem. In: Proceedings of the 11th Annual Conference on Genetic and Evolutionary Computation. GECCO-2009, pp. 317–324 (2009)
27. Mouhcine, E., Khalifa, M., Mohamed, Y.: Route optimization for school bus scheduling problem based on a distributed ant colony system algorithm. In: Intelligent Systems and Computer Vision, ISCV (2017)
28. Chatty, A., Gaussier, P., Kallel, I., Laroque, P., Pirard, F., Alimi, A.M.: Evaluation of emergent structures in a 'Cognitive' multi-agent system based on on-line building and learning of a cognitive map, ICAART 2013. In: Proceedings of the 5th International Conference on Agents and Artificial Intelligence, vol. 1, pp. 269–275 (2013)
29. Fikar, C., Gronalt, M., Hirsch, P.: A decision support system for coordinated disaster relief distribution . Expert Syst. Appl. **57**, 104–116 (2016)
30. Train, K.E.: Discrete Choice Methods with Simulation. Cambridge University Press , Cambridge (2009)
31. Guarín, M.: Modelo de negocio para el mercado de frutas en la ciudad de Bogotá, orientado hacia el aprovechamiento de la oferta que brindan los productores de frutas en cundinamarca, Univ. Piloto de Colombia (2016). http://polux.unipiloto.edu.co:8080/00003098.pdf
32. Alcaldía Mayor de Bogotá (2014). Encuesta Multipropósito 2014
33. Dirección de Comercialización de la Agencia de Desarrollo Rural, Metodología y Evaluación de Mercados Campesinos, ISBN 978-958-56571-1-3
34. FAO and Ministerio de Salud y Protección social. Perfil nacional de consumo de frutas y verduras 2013 (2013). ISBN 978-92-5-307534-8
35. Departam. Administrativo Nacional de Estadística, DANE. Pobreza monetaria por departamentos en Colombia, Boletín Técnico Pobr. Monet. Dep., vol. 8, no. 004, pp. 1–28 (2019)
36. Aranda Quimbaya, C.D.: Perfil del comprador y percepción respecto a las frutas y hortalizas que se ofrecen en los mercados campesinos de Plaza de los Artesanos y Parque de Alcalá en ciudad de Bogotá (2019). https://ciencia.lasalle.edu.co/administracion_agronegocios/316
37. S. distrital de planeación de Bogotá: Población UPZ Bogotá - LabUrbano Bogotá (2016). https://bogota-laburbano.opendatasoft.com/explore/dataset/poblacion-upz-bogota/table/?flg=es
38. Koroliuk, M., Connaughton, C.: Analysis of big data set of urban traffic data, Project report, pp. 1–12. Warwick University (2015)
39. Wang, H.: Consumer valuation of retail networks: evidence from the banking industry. SSRN Electron. J. (2012). https://doi.org/10.2139/ssrn.1738084
40. ¿El supermercado o la plaza? Los contrastes de precios de alimentos, El tiempo (2016). https://www.eltiempo.com/archivo/documento/CMS-16506108
41. Acosta Leal, D.A.: Fijación de precios en mercados campesinos de Bogotá. Caso hortalizas frescas de Fómeque y Chipaque (Cundinamarca), Universidad Nacional de Colombia (2014). https://repositorio.unal.edu.co/handle/unal/52259

42. S. distrital de Planeación, Norma Urbana UPZ (2019). http://www.sdp.gov.co/sites/default/files/div_upz.png. Accessed 03 Dec 2019
43. Rajagopal: Street markets influencing urban consumer behavior in Mexico, Lat. Am. Bus. Rev. **11**(2), 77–110 (2010). https://www.tandfonline.com/doi/abs/10.1080/10978526.2010.487028

Author Index

D. Trentesaux et al. (Eds.): SOHOMA 2021, SCI 987, pp. 357–358, 2021.
https://doi.org/10.1007/978-3-030-80906-5

The manufacturer's authorised representative in the EU is Springer Nature Customer Service Centre GmbH, Europaplatz 3, 69115 Heidelberg, Germany. If you have any concerns regarding our products, please contact ProductSafety@springernature.com

Printed and bound by CPI Group (UK) Ltd, Croydon, CR0 4YY
30/09/2026
02225692-0011